T0263518

ADVANCES IN MOLECULAR AND CELL BIOLOGY

Volume 23B • 1998

ION PUMPS

ADVANCES IN MOLECULAR AND CELL BIOLOGY

ION PUMPS

Series Editor: E. EDWARD BITTAR
Department of Physiology
University of Wisconsin
Madison, Wisconsin

Guest Editor: JENS PETER ANDERSEN
Institute of Physiology
University of Aarhus
Aarhus, Denmark

VOLUME 23B • 1998

JAI PRESS INC.

Greenwich, Connecticut   London, England

Copyright © 1998 JAI PRESS INC.
55 Old Post Road No. 2
Greenwich, Connecticut 06836

JAI PRESS LTD.
38 Tavistock Street
Covent Garden
London WC2E 7PB
England

ISBN: 0-7623-0287-9

Printed and bound in Great Britain by
CPI Antony Rowe, Chippenham and Eastbourne

CONTENTS (VOLUME 23B)

v

CONTENTS (Volume 23A)

LIST OF CONTRIBUTORS

Robert Aggeler   Institute of Molecular Biology
                             University of Oregon
                             Eugene, Oregon

Krister Bamberg   Wadsworth Veterans Administration
                             Medical Center
                             Los Angeles, California

Denis Bayle   Wadsworth Veterans Administration
                             Medical Center
                             Los Angeles, California

Roderick A. Capaldi   Institute of Molecular Biology
                             University of Oregon
                             Eugene, Oregon

Leopoldo de Meis   Departamento de Bioquimica Médica
                             Universidade Federal do Rio de Janeiro
                             Rio de Janeiro, Brazil

Peter Dimroth   Mikrobiologisches Institut
                             Eidgenössische Technische Hochschule
                             Zürich, Switzerland

Jan Eggermont   Laboratorium voor Fysiologie
                             Katholieke Universiteit Leuven
                             Leuven, Belgium

Simone Engelender   Departamento de Bioquimica Médica
                             Universidade Federal do Rio de Janeiro
                             Rio de Janeiro, Brazil

Agnes Enyedi
National Institute of Haematology,
Blood Transfusion and Immunology
Budapest, Hungary

Michael Forgac
Department of Cellular and Molecular Physiology
Tufts University
Boston, Massachusetts

Käthi Geering
Institute of Pharmacology and Toxicology
University of Lausanne
Lausanne, Switzerland

Ursula Gerike
School of Biology and Biochemistry
University of Bath
Bath, England

James E. Haber
Rosenstiel Basic Medical Sciences Research
Center and Department of Biology
Brandeis University
Waltham, Massachusetts

Parjit Kaur
Department of Biology
Georgia State Univesity
Atlanta, Georgia

Daniel Khananshvili
Department of Physiology and Pharmacology
Sackler School of Medicine
Tel-Aviv University
Tel-Aviv, Israel

David B. McIntosh
MRC Biomembrane Research Unit and
Department of Chemical Pathology
University of Cape Town Medical School
Cape Town, South Africa

Luc Mertens
Laboratorium voor Fysiologie
Katholieke Universiteit Leuven
Leuven, Belgium

John T. Penniston
Department of Biochemistry and
Molecular Biology
Mayo Clinic/Foundation
Rochester, Minnesota

David S. Perlin
Public Health Research Institute
New York, New York

Luc Raeymaekers
Laboratorium voor Fysiologie
Katholieke Universiteit Leuven
Leuven, Belgium

George Sachs
Wadsworth Veterans Administration
Medical Center
Los Angeles, California

Jai Moo Shin
Wadsworth Veterans Administration
Medical Center
Los Angeles, California

Marc Solioz
Department of Clinical Pharmacology
University of Berne
Berne, Switzerland

Ludo Van Den Bosch
Laboratorium voor Fysiologie
Katholieke Universiteit Leuven
Leuven, Belgium

Hilde Verboomen
Laboratorium voor Fysiologie
Katholieke Universiteit Leuven
Leuven, Belgium

Herman Wolosker
Departamento de Bioquimica Médica
Universidade Federal do Rio de Janeiro
Rio de Janeiro, Brazil

Frank Wuytack
Laboratorium voor Fysiologie
Katholieke Universiteit Leuven
Leuven, Belgium

PREFACE

Both eukaryotic and prokaryotic cells depend strongly on the function of ion pumps present in their membranes. The term ion pump, synonymous with active ion-transport system, refers to a membrane-associated protein that translocates ions uphill against an electrochemical potential gradient. Primary ion pumps utilize energy derived from chemical reactions or from the absorption of light, while secondary ion pumps derive the energy for uphill movement of one ionic species from the downhill movement of another species.

In the present volume, various aspects of ion pump structure, mechanism, and regulation are treated using mostly the ion-transporting ATPases as examples. One chapter has been devoted to a secondary ion pump, the Na^+-Ca^{2+} exchanger, not only because of the vital role played by this transport system in regulation of cardiac contractility, but also because it exemplifies the interesting mechanistic and structural similarities between primary and secondary pumps.

The cation-transporting ATPases fall into two major categories. The P-type ATPases (treated in eight chapters) are characterized by the formation during the catalytic cycle of an aspartyl phosphorylated intermediate, and by their rather simple subunit composition (one catalytic α-chain, supplemented in Na^+,K^+ and H^+,K^+ pumps with a β-chain that has a crucial role in expression of functional pumps at the cell surface). The F/V-type ATPases (treated in three chapters) do not form a recognizable phosphoenzyme intermediate and are composed of multiple different subunits whose individual roles are now beginning to be understood. Similarly, some of the anion-transporting ATPases (treated in a single chapter) seem to be composed of multiple subunits with separate roles in ATP hydrolysis and ion translocation.

Although there are too many structural and functional differences between the various ion transport ATPase families to warrant speculations about a common evolutionary origin of all families, they do seem to resemble each other in some fundamental aspects. Thus, irrespective of whether an ATP-driven pump consists of one or several subunits, it has a large hydrophilic "pump head" protruding from the membrane, which is responsible for the ATP hydrolysis/synthesis reactions. The vectorial transport process is carried out by a more slender membrane-buried part forming a channel-like structure. Energy coupling seems to occur by long-range communication between the membrane-buried and extramembranous domains/subunits, mediated by conformational changes. A major challenge is to elucidate the exact nature of the coupling process and to define the ion-translocation pathway through the membrane-buried part of the protein. As discussed in some of the chapters in this book, the distinction between ion pumps and ion channels now appears less sharp than previously, since several transporters can exhibit both types of function. It appears reasonable to consider an ion pump as a channel equipped with locks and gate controls.

The three-dimensional structure of the extramembranous catalytic F_1 part of F-type ATPases has recently been determined, and this has helped to provide a firm basis for speculations on the energy-coupling mechanism. Similarly, although a high-resolution crystal structure is still a distant goal in the case of P-type ATPases, analysis of the amino acid sequences of these pumps, in conjunction with various site-directed approaches such as mutagenesis and affinity labeling, has paved the way for realistic modeling of the structure of the ATP binding domain and its coupling with the ion transport domain.

The topological features of the membranous part of P-type ATPases have long been a matter of conjecture, but several pieces of evidence now support a 10-transmembrane segment structure in Ca^{2+}-, H^+,K^+-, and Na^+,K^+-ATPases. A core region constituted by six of these transmembrane segments seems to have been retained also during evolution of P-type ATPases involved in heavy-metal-ion transport.

The determinants of the specific ion selectivities of the pumps are poorly understood, but some interesting clues have been provided in studies of hybrids between H^+- and Na^+-transporting F-type ATPases and by comparing heavy-metal-transporting P-type ATPases with other P-type ATPases.

Many ion pumps are known to occur as multiple isoenzymes differing with respect to tissue distribution and functional and regulatory characteristics. Regulation occurs both at the level of gene expression and through changes in kinetic parameters that determine the molecular pump rates. In health as well as in disease, control is exerted by the concentrations of ions to be pumped and by hormonal and developmental influences. Although the molecular details of the regulatory events are now beginnning to be understood, as described in several of the chapters in this volume, there seems to be much more to learn.

Jens Peter Andersen
Guest Editor

COMPARISON OF ATP-POWERED Ca^{2+} PUMPS

John T. Penniston and Agnes Enyedi

I. INTRODUCTION

The plasma membrane Ca^{2+} pump is the only active ion pump responsible for ejecting Ca^{2+} from the cell. It is a P-type ATPase, like the sarcoplasmic reticulum Ca^{2+}

Advances in Molecular and Cell Biology
Volume 23B, pages 249-274.
Copyright © 1998 by JAI Press Inc.
All right of reproduction in any form reserved.
ISBN: 0-7623-0287-9

pump, the sodium-potassium pump and the H^+-K^+-ATPase. All P-type ATPases form and break down an acylphosphate intermediate on the carboxylate side chain of aspartic acid as a part of their ion transport cycle. Both the sarcoplasmic reticulum Ca^{2+} pump and the plasma membrane Ca^{2+} pump remove Ca^{2+} from the cytosol in an ATP-powered, electrogenic, Ca^{2+}/H^+ exchange (Yu et al., 1993; Hao et al., 1994), but they differ substantially from one another: The plasma membrane pump moves only one Ca^{2+} per ATP hydrolyzed (Hao et al., 1994), while the sarcoplasmic reticulum pump moves two Ca^{2+}. When the primary sequences of these pumps are aligned, 68% of the amino acids are different, and the regulatory mechanisms of the two pumps are different, as discussed below. In some kinds of cells (particularly cells of excitable tissue) the plasma membrane Ca^{2+} pump shares with the Na^+-Ca^{2+} exchanger the responsibility of ejecting Ca^{2+} from the cell, but in many other tissues the pump is the sole Ca^{2+} ejector. The pump is embedded in the plasma membrane and moves Ca^{2+} across it but the hydrophilic regions are nearly all on the cytoplasmic face of this membrane. On this face, the substrates of the pump (Mg^{2+}, ATP and Ca^{2+}) are bound and acylphosphate is formed and broken down. These cytoplasmic portions of the molecule consist of four segments. The first is the amino terminal tail which enters the membrane at the first transmembrane domain. After the chain loops back in the second transmembrane domain, the small cytoplasmic domain intervenes between transmembrane regions 2 and 3. This is followed by transmembrane domains 3 and 4, and then by the major cytoplasmic domain, which contains the phosphorylation site. This domain reenters the membrane at transmembrane domain 5, and several (4 or 6) membrane crossings ensue. Finally, the carboxyl terminal tail emerges from the last transmembrane domain. The carboxyl terminal tail of the pump is an inhibitory regulatory region; all indications are that it is this region through which the physiologically important properties of the pump are regulated. The presence of this region inhibits the pump and the degree to which interactions occur between this region and the rest of the pump is a primary determinant of the activity of the pump.

The plasma membrane Ca^{2+} pump is encoded by at least four different genes and additional diversity among the PMCA isoforms is generated by alternate RNA splicing. Alternate splices occur at sites A and C; site A is near the region where the regulation of the pump with acidic lipids is thought to happen and site C is in the middle of the calmodulin binding domain. This diversity of the pump at the mRNA level leads to the notion that as many as 20 isoforms may occur in different tissues and cells and each of them may have unique characteristics. However, the question of which of these isoforms are actually present at the protein level remains unanswered. Attempts are being made to answer this question using isoform-specific antibodies (Caride et al., 1994; Stauffer and Carafoli, 1995).

Until now, these isoforms have been called PMCA1, 2, 3, and 4, and the alternate splice variants have been noted by small letters a, b, c, d ... at site C and w, x, y, z ... at site A. This nomenclature was introduced when the pump was first cloned (Shull and Greeb, 1988, Greeb and Shull, 1989) and has been used by all workers until recently, when a new nomenclature was proposed to replace the old one. This proposal (Carafoli, 1994) affects only the alternate splice variants and replaces the small letters by a capital letter denoting which site (A or C) is spliced and by Roman numerals I, II, III... indicating which exon is inserted. This latter nomenclature gives slightly more information, but its use may cause confusion in understanding the relationship between previous and future studies on the pump. The tissue and cell-specific distribution of mRNAs corresponding to the different isoforms has been demonstrated and extensively studied by PCR in different tissues (Westlin et al., 1991; Magocsi et al., 1992; Howard et al., 1993, 1994; Meszaros and Karin, 1993; Sarkar et al., 1993; Stauffer et al., 1993; Hammes et al., 1994; Zacharias et al., 1995), using the old nomenclature. Changing the nomenclature would make it difficult to compare that PCR data with forthcoming information about the expression of the pump's isoforms at the protein level. For these reasons, we intend to continue to use the old nomenclature, while making clear its relationship to the newly proposed system.

In order to provide a framework for further discussions, let us briefly consider which properties of the pump are physiologically the most important. Of the substrates of the pump, Mg^{2+} and ATP are present in a healthy cell in considerable excess over the normal affinity of this pump for them. On the other hand, the affinity of the activated pump for Ca^{2+} is in the range of cellular free Ca^{2+} concentrations. Since this pump exists mainly to remove Ca^{2+} from the cell, it is apparent that alterations in its Ca^{2+} affinity are most important in affecting its function. A second physiologically important property of the pump is the speed of its response to changes in Ca^{2+} concentration. Since the Ca^{2+} signal in cells consists of a rapid increase in Ca^{2+} level followed by a decrease which may be rapid or slow, the kinetics of the response of the pump are obviously important in helping determine the shape of such a Ca^{2+} signal. This is particularly true when the signal is a rapid spike. As is discussed below, the rate at which calmodulin binds to the pump in the presence of Ca^{2+} may be an important factor in controlling the rate at which the pump is turned on, and, therefore, in controlling the shape of the down curve of the Ca^{2+} spike. Its possession of the carboxyl terminal regulatory region sets the plasma membrane Ca^{2+} pump apart from that of the sarcoplasmic reticulum, which has no such region but which ends just downstream of the last transmembrane region. Instead of having a built in autoinhibitory region, whose inhibition can be relieved by various intracellular processes, the sarcoplasmic reticulum Ca^{2+} pump interacts in some cases with a separate inhibitory protein called phospholamban. In heart muscle, this inhibitor is a separate regulatory molecule which inhibits the pump in its dephosphorylated state.

II. REGULATION OF THE PLASMA MEMBRANE Ca^{2+} PUMP

At the molecular level, the regulation of this pump is most straightforwardly discussed in terms of two regions of the polypeptide chain. These regions are the carboxyl terminus (which consists of two sub-regions, the calmodulin-binding or C-domain and the downstream region), and the lipid binding or L-domain, (in an upstream region of the pump) which regulates at low concentrations of free Ca^{2+}.

A.   Calmodulin Regulation and the Carboxyl-Terminal Region

The Calmodulin-Binding Domain

Calmodulin Regulation.   That a protein factor activated the plasma membrane Ca^{2+} pump was first reported by Bond and Clough (1973) who demonstrated the presence of such a factor in human erythrocyte cytosol. It was later shown that this factor was calmodulin (Gopinath and Vincenzi, 1977; Jarrett and Penniston, 1977); this pump was the third enzyme found to be regulated by calmodulin. The interaction of calmodulin with the plasma membrane Ca^{2+} pump was sufficiently specific and tight that it was possible to utilize a calmodulin-affinity column to purify the pump from human erythrocyte membranes (Niggli et al., 1979) even though this pump constituted only about 0.1% of the total membrane protein. In the presence of excess Ca^{2+}, calmodulin bound with a high affinity (5 nM) to the pump in erythrocyte membranes (Penniston et al., 1980). Localization of the calmodulin binding domain was first done by observing the effects of proteolytic fragmentation of the pump. Treatment of erythrocyte membranes with trypsin reduced the size of the pump protein as measured by SDS-gel electrophoresis and caused an activation similar to that caused by calmodulin. It was, therefore, proposed that proteolysis removed the calmodulin-binding domain, thereby activating the enzyme (Enyedi et al., 1980). Proteolysis was subsequently shown to produce three fragments (called 90 kDa, 81 kDa, and 76 kDa) (Carafoli and Zurini, 1982), of which the 90 kDa acted like the intact enzyme while the 81 kDa and 76 kDa were activated and did not bind calmodulin (Zurini et al., 1984). It was later shown that the 81 kDa fragment had a Ca^{2+} curve like that of the calmodulin activated enzyme and the 76 kDa fragment was more highly activated, similar to the activation caused in the presence of acidic phospholipid (Enyedi et al., 1987).

Since the molecular mass of the intact pump was known, even before its sequencing, to be about 138 kDa (Graf et al., 1982), it was surprising that a fragment of about 50 kDa could apparently be removed from the enzyme without affecting its activity. From the proteolysis patterns of the pump, it was inferred that the fragment removed to produce the 90 kDa portion was at the end of the molecule opposite the end which bound calmodulin (Zurini et al., 1984). When cloning of the pump subsequently revealed its sequence (Shull and Greeb, 1988; Verma et al., 1988), it became apparent that the removal of such a large fragment from the end opposite the calmodulin-binding end would require the removal of a region which contained a sequence which was highly conserved with the other P-type ATPases. Since it was evident that the pump could not function in the absence of those sections, it became clear that the only solution to this apparent paradox is that the two fragments produced by the first trypsin cleavage must remain together after the cleavage, a result which was rendered more probable by the imbedding of both fragments in the plasma membrane. The subsequent identification of the exact cleavage site producing the various proteolytic fragments (Zvaritch et al., 1990) confirmed these inferences. Expression of a 105 kDa Ca^{2+} pump from which a substantial part of the amino-terminus was removed resulted in an enzyme which was expressed but inactive; this was interpreted as demonstrating that the "90 kDa" fragment must remain together with the smaller product of trypsin proteolysis (Heim et al., 1992). However, the inactivity of this 105 kDa fragment could be due to misfolding or other causes.

While the question of the removal of the 50 kDa proteolytic fragment has been resolved only to the degree mentioned above, the function of the fragment removed on the transition from 90 kDa to 81 kDa is much clearer, since its removal activated the enzyme and eliminated its calmodulin-binding capacity (Zurini et al., 1984). Because of this, it was inferred that this portion contained the calmodulin-binding domain. Additional, clearer information came from proteolysis of the pump by calpain or chymotrypsin. It was observed that these proteases did not remove a 50 kDa fragment, but removed only a 10-15 kDa fragment from one end of the pump. This proteolysis also caused a calmodulin-like activation of the pump; indicating that this small part of the molecule contained the regions which bound calmodulin and inhibited the pump (Au, 1987; Wang et al., 1988a, 1988b; James et al., 1989c; Papp et al., 1989). The calpain-activated enzyme could be further stimulated by acidic lipids, showing the presence of the lipid-activated regulatory region elsewhere in the pump molecule.

Further definition of the calmodulin-binding domain was possible when partial sequences of the cloned calcium pump became available. Inspection of these clearly showed one 28-residue region which had the mixture of basic and hydrophobic residues required for tight binding of calmodulin. This was confirmed by showing that calmodulin cross-linked to the pump at that region (James et al., 1988). This region is sometimes called the C-domain. Studies with synthetic peptides showed that a 28-residue peptide representing the C domain had a very high affinity for calmodulin (Kd = 0.1 nM). In addition, this peptide also reinhibited the proteolytically activated pump, indicating that the calmodulin-binding domain is an auto-inhibitory domain (Enyedi et al., 1989). Two regions where this peptide interacts with the rest of the molecule have been identified by cross-linking (Falchetto et al., 1991; 1992). One of these cross-links is a short distance downstream from the acylphosphate, in a position nearly corresponding to the site at which phospholamban interacts with the sarcoplasmic reticulum Ca^{2+} pump. The second cross-link is in the smaller cytoplasmic domain between transmembrane domains 2 and 3. Extensive physical studies were also done on the interaction of calmodulin and these peptides (Vorherr et al., 1990).

Calmodulin binds tightly to the C-domain and deinhibits the pump, resulting in an increase in its V_{max} and an increase in its apparent affinity for Ca^{2+}. A similar degree of activation and change in Ca^{2+} affinity is caused by removing the portion of the pump including the C-domain and everything downstream from it. A truncated mutant of isoform 4b of the pump lacking these regions (called ct120) showed characteristics essentially identical to those of the pump when activated with excess calmodulin (Enyedi et al., 1993). (Isoform 4b is called isoform 4CI by the proposed nomenclature mentioned above.) Taking ct120 as being a fully activated pump, restoration of the 28 residues of the C-domain (which made a construct called ct92) caused only about 2/3 of the inhibition which existed in the full-length pump (Verma et al., 1994). It was clear from this study that the region responsible for the remaining 1/3 of the inhibition lay in the remaining 92 residues of the downstream region. This additional inhibition could be due to either an extension of the C-domain, or a separate region downstream of the C-domain. Additional constructs are being made to locate the complete inhibitory region.

The ct92 molecule, containing only the 28 residues of the C-domain, had an apparent affinity for calmodulin just as high as that of the complete isoform 4b. This indicated that calmodulin bound just as tightly to this construct as it did to the complete enzyme and, therefore, that the portion of isoform 4b downstream of the C-domain played no role in calmodulin binding.

The upstream 19 residues of the C-domain are conserved in all of the isoforms whereas an alternate splice in the middle of the C-domain changes the structure of its downstream part and the entire carboxyl terminus. This splice inserts a new segment of code at the region of the mRNA which is coding for the C-domain and that leads to the expression of a protein with a more acidic C-domain. This isoform is called form $\underline{a}$ (or CII according to the proposed nomenclature mentioned above). The product without the insert is called form $\underline{b}$ (or CI) and has a C-domain with a more basic character. At the level of mRNA, the $\underline{b}$ form is the most common form among all four isogenes while the $\underline{a}$ form is expressed primarily in brain, smooth muscle, and heart. The functional consequences of this change in the structure of the C-domain first were analyzed by using synthetic peptides (Enyedi et al., 1991). The peptide representing the more basic $\underline{b}$ form of the C-domain had 10 times higher affinity for calmodulin than the peptide representing the more acidic $\underline{a}$ form. Subsequently, both human plasma membrane Ca^{2+} pump isoforms 4b and 4a were expressed in COS-1 cells (Enyedi et al., 1994). In good agreement with the peptide data, isoform 4a needed about a 5-7 times higher concentration of calmodulin for activation than isoform 4b. This reduced calmodulin-affinity of isoform 4a caused a reduced apparent affinity for Ca^{2+} activation when tested at a high calmodulin concentration similar to that existing in the living cell. This change in the Ca^{2+} activation occurs within a Ca^{2+} concentration range which is crucial for physiological cellular function. Thus, the expression of such a low calmodulin-affinity isoform is expected to alter intracellular Ca^{2+} signalling, as is also discussed below.

Another interesting concept is that the alternatively spliced sequence of isoform 1 duplicates the calmodulin-binding site by introducing a sequence homologous to the nonvariable region of the C-domain (Kessler et al., 1992). In this region of isoform 1 some of the positive charges are supplied by histidine residues. Thus, the calmodulin regulation of the alternatively spliced variants of isoform 1 might be pH-dependent.

Although the non-spliced C-domains (b forms) of all four isogenes are quite homologous in the variable region, isoform 2b (2CI) has been said to have much higher affinity (5-10 times) for calmodulin than isoform 4b (Hilfiker et al., 1994). This is unexpected since human isoform 4b differs from 2b in only 2 residues in the C-domain: a lysine and a histidine in 4b are replaced by two arginines in 2b. (Note that this differs somewhat from the situation shown in Figure 1, which shows the rat isoforms.) This finding may suggest the involvement of regions outside of the C-domain in calmodulin regulation; such regions may be different among the isoforms.

Proteolysis. In addition to the role of the C-domain in binding calmodulin, it is possibly involved in several other modes of regulation of the pump. It is already evident from this discussion that proteolytic removal of the C-domain activates the pump. Trypsin and chymotrypsin are unlikely to act intracellularly, but calpain has been proposed as a possible physiological activator of the pump (Au, 1987; Wang et al., 1988a, 1988b). This Ca^{2+}-activated protease exists in the cytosol of some kinds of cells and activates the Ca^{2+} pump by limited proteolysis. In the absence of calmodulin, calpain digestion removes the calmodulin-binding domain (James et al., 1989c; Papp et al., 1989). In the presence of calmodulin, calpain digestion produced a slightly larger fragment which still bound calmodulin (Wang et al., 1991), indicating that binding of calmodulin may mask one or more of the sites where calpain cuts. When incubated with the peptide representing the calmodulin-binding domain, calpain was capable of cutting this peptide in several locations (James et al., 1989c), suggesting that the calpain cleavage sites may be in this domain. It is believed that the action of calpain is promoted by PEST domains which are rich in proline, glutamate, aspartate, serine, and threonine. The role of these domains has been thoroughly reviewed by Wang et al., (1989). Two such domains exist in the b isoforms of the plasma membrane Ca^{2+} pump, flanking the calmodulin-binding domain and probably making it a target for calpain action. However, a recent study (Molinari et al., 1995) of calpain digestion of a fragment of the pump reached a different conclusion. The fragment was the entire carboxyl terminus, including both PEST domains; it was made by expression in *E. coli.* Mutation of the PEST domains led these workers to the conclusion that the PEST domains did not play a significant role in calpain proteolysis.

Phosphorylation by Protein Kinases. Evidence for activation of the pump by a protein kinase was first introduced by Smallwood et al. (1988), and observation of the incorporation of phosphorus into the carboxyl terminus was made by Wang et al. (1991). Both groups used protein kinase C on the Ca^{2+} pump from erythrocytes. By studying the amount of phosphate incorporated into different proteolytic frag-

ments of the pump the latter group was able to show that at least two phosphorylation sites were involved, one in the downstream region and one in the
calmodulin-binding domain. However, in the experiments by Wang et al. (1991),
the activity changes associated with this phosphorylation were small and their
amount and direction depended very much on the assay conditions, so that it was
not clear whether this phosphorylation caused any change in the activity. They also
studied a peptide representing part of the C-domain and showed that it could be
phosphorylated on its only threonine by protein kinase C. The threonine on which
they saw phosphorylation of the peptide was consistent with the location of phosphorylation they reported in the whole enzyme, but since it was the only serine or
threonine contained in this peptide, protein kinase C had no possibility of phospho-
rylating elsewhere. A subsequent study (Hofmann et al., 1994) compared a synthetic peptide phosphorylated on this threonine with the same peptide not
phosphorylated. This study showed that the phosphopeptide-bound calmodulin
much less well and was also weaker in its inhibition of the enzyme which had been
activated by proteolysis. This study gave interesting information about the properties of the peptide, but did not address the question of whether phosphorylation of
the pump in the membrane occurs on this threonine.

Other Activation of the Pump at the C-Domain.   The C domain was originally
associated only with calmodulin activation of the pump, but recent studies have
also shown that this domain is involved in activation of the pump by lipids (Brodin
et al., 1992; Filoteo et al., 1992). Since lipid binds both to the C-domain and to the
lipid-binding domain (see below), it is difficult to tell what part of the lipid's effect
is due to the binding at each site. However, it is reasonable to suppose that the binding of the lipid to the C-domain has an activatory effect similar to that caused by
calmodulin's binding to this domain. The overall effect of lipid is to increase the apparent affinity of the pump for Ca^{2+} beyond that caused by calmodulin alone. Thus,
the extra effect of lipid on the $K_{1/2}$ for Ca^{2+} is probably due to its binding by the lipid
domain. The highly basic character of the C-domain, combined with its somewhat
hydrophobic character probably explains its preferential interaction with acidic
phospholipids. It is quite clear from the studies mentioned above that part of the
lipid activation is due to the effects of lipid on the C-domain.
    Another possible regulatory mechanism is dimerization of the calcium pump
which involves either the C-domain or the region downstream from it. The effect of
aggregation of the pump on its activity was first observed by Kosk-Koskica and
Bzdega (1988), who found activation at high concentrations of purified pump in the
presence of detergent. That the activation of the pump was indeed due to its self-
association was subsequently confirmed by fluorescence energy transfer experiments (Kosk-Kosicka et al., 1989). Further studies from the same laboratory
showed that calmodulin affected aggregation of the molecule in complicated ways
(Kosk-Kosicka et al., 1990), and that the aggregation of the pump stopped at the
stage of a dimer (Sackett and Kosk-Kosicka, 1993). Work from another lab showed

that removal of the C-domain with calpain digestion prevented dimerization, indicating that either the C-domain or the region downstream of it was essential for this to occur (Vorherr et al., 1991).

It is evident from the above discussion that the C-domain occupies a crucial role in the regulation of the plasma membrane Ca^{2+} pump. Its involvement in activation by proteolysis, by calmodulin and by phospholipid binding has been definitely shown, and it has the possibility of involvement in regulation by phosphorylation at its central threonine and by dimerization as well. These are essentially all the known means of regulating the Ca^{2+} pump and, therefore, the involvement of this domain is very central.

The Downstream Region

Variability. The domain downstream of the C-domain is one of the most variable regions in the plasma membrane Ca^{2+} pump. One source of this variability is the alternate splice in the middle of the C-domain which produces the $\underline{a}$ and $\underline{b}$ variants whose structure differs in everything downstream of the splice. In the C-domain itself, there is some homology between the $\underline{a}$ and $\underline{b}$ variants, but in the downstream region there is little or no homology. In addition to this source of variability, the four different genes coding for the plasma membrane calcium pump produce substantial variability in this downstream region. Figure 1 shows the C-domain and the downstream region for the $\underline{a}$ and $\underline{b}$ variants of the plasma membrane calcium pump produced from all four known rat genes (the human sequences are similar). This figure illustrates a number of important points about the downstream region. In this part of the molecule, up until the position of the alternate splice, all the different forms of the pump are identical. The alternate splice occurs at the point marked by the exclamation point and downstream of this there is considerable variation. The $\underline{b}$ variants, particularly those of forms 1, 2, and 3 show considerable homology. The existence of conserved domains is evident in the $\underline{b}$ variants of all four isoforms. For example, the PEST domains identified by Wang et al. (1989) are present in all four isoforms. Inspection of the diagram shows that other motifs are also conserved in all of the $\underline{b}$ variants, but that the sites which are potential protein kinase substrates vary in their specific sequence from one isozyme to the next. Thus, the sequence identified as the most likely A-kinase phosphorylation site in isozyme 1 (KRNSS) becomes KQNSS in isozyme 2 and has no exact homolog in isozymes 3 and 4. All of the downstream domains, whether in the $\underline{a}$ or $\underline{b}$ variants, are quite high in serine and threonine with the percentage of serine and threonine ranging from 17% to 33% (Penniston and Enyedi, 1994).

The figure shows that the downstream regions of the $\underline{a}$ variants are very different from those of the $\underline{b}$ variants. The PEST sequences are lacking, as are all of the other motifs that can be identified in the $\underline{b}$ variants. Among the $\underline{a}$ variants there is some conservation of motifs, but once again the exact sequence of all the likely protein kinase phosphorylation sites changes from one isoform to the next. Such variations

```
            >.....C domain....!........<
   rPMCA1b  LRRGQILWFRGLNRIQTQIRVVNAFRSSLYRGLRKPRSRSSIHNFMTHPRFRIRDSRPHIPLIDD
   rPMCA2b  LRRGQILWFRGLNRIQTQIRVVKAFRSSLYRGLRKPRSRTSIHNFMAHPRFRIRDSQPHIPLIDD
   rPMCA3b  LRRGQILWFRGLNRIQTQIRVVKAFRSSLYRGLRKPRSKSCIHNFMATPRFLINDYTHNIPLIDD
   rPMCA4b  LRRGQILWVRGLNRIQTQIRVVKVPHS.FRDVIHKSKNQVSIHSFMTQPRYAARIDRMSQSFLNQR
 b Consensus LRRGQILWFRGLNRIQTQIRVVNAFQSGGSIQGALRRQPSIASQH.......HDVTNVSTPTHVV

   rPMCA1a  LRRGQILWFRGLNRIQTQMDVVNAFQSGGSIQGALRRQPSIASQH.......HDVTNVSTPTHVV
   rPMCA2a  LRRGQILWFRGLNRIQTQIRVVNTFKSGASFQGALRRQSSVTSQS.......QDVASLSSPSRVS
   rPMCA3a  LRRGQILWFRGLNRIQTQMDVVSTFKRSGSFQGAVRRRSSVLSQL.......HDVTNLSTPTHVT
   rPMCA4a  LRRGQILWVRGLNRIQTQIDVINKFQTGASFKGVLRRQNLSQQLDVKLVPSSYSRAVASVRTSPS
 a Consensus LRRGQILWFRGLNRIQTQ-DVVN-F---G-SFQGALRRQ-SV-SQ--------DV---S-PT-V-

   rPMCA1b  TDARIDAPTKRNSSPP.....PSPNKNNNAVDSGIHLTIDMNKSATSSS....PGSPLHSLRTSL
   rPMCA2b  TDLRIDAALKQNSSPP.....SSLNKNNSAIDSGINLTTDTSKSATSSS....PGSPIHSLRTSL
   rPMCA3b  TDVDRDRDRLRAPPPP......PPNQNNNAIDSGIYLTTHATKSATSSAFSSRPGSPLKSMDTSL
   rPMCA4b  DSSSLASKSRITKRLSDARTVSQNNTNNNAVDCH...QVQMLAS........HPNSPLQSQRTPV
 b Consensus TD-DR--------PP--------N-NNNA-DSGI-LT----KSATSS-----PGSPLHSLRTSL

   rPMCA1a  FSSSTASTPVGYPSGRCIS
   rPMCA2a  LSNALSSPTSLPPAAAGQG
   rPMCA3a  LSAAKPTSAAGNPSGRSIP
   rPMCA4a  TSSAVTPPPVGNQSGQSIS
 a Consensus LS-A------G-PSG--I-
```

Figure 1. Alignment of the carboxyl terminal sequences of the _a_ and _b_ splices of the rat plasma membrane Ca²⁺ pump isoforms. Threonines and serines are shown in bold type. Aspartates, and glutamates in boxes and arginines and lysines are underlined. The position of the alternate splice is shown with an exclamation point.

make it likely that each of the eight variants shown in Figure 1 is regulated differently by protein kinases. This variation may be one of the main functions of the different gene products, since they resemble one another closely throughout much of the rest of the molecule.

Protein Kinases. The action of protein kinases on the plasma membrane calcium pump's activity has been comprehensively covered in a recent review (Wuytack and Raeymaekers, 1992). We will not duplicate that discussion, but will add our own observations, and summarize newer data.

G kinases: It appears that any effects of G kinase on the plasma membrane Ca²⁺ pump are not exerted by direct phosphorylation. Considerable confusion was caused by the observed phosphorylation of a protein having almost the same molecular weight as the Ca²⁺ pump, but the consensus of several laboratories is now that G kinase activates the pump by acting on some other protein, a conclusion that was reinforced by a recent communication (Yoshida et al., 1992).

A kinase: Phosphorylation sites for A kinase have been identified in the downstream region. The first identification of an A kinase site was at the sequence KRNSS in this region (James et al., 1989b). At the time this study was done, the implications of the existence of multiple isozymes were not fully appreciated. Although this sequence belongs to isozyme 1, it was isolated from purified erythrocyte membrane Ca²⁺ pump which is primarily isozyme 4 but with some isoform 1. Because the purified erythrocyte pump contains two isozymes, it is difficult to draw any conclusions about the relationship between activity and phosphorylation from a study such as this. Two additional phosphorylation sites for A kinase were identified in an expressed protein fragment which included only the downstream region of pump isozyme 1b (also called 1CI) (James et al., 1992). Because these were measured on a small portion of the pump, the phosphorylation sites actually used in the intact pump may be different.

C kinase: Phosphorylation by this kinase has also been located in the downstream region of the pump, (Wang et al., 1991). This study carried out a detailed analysis of the proteolytic fragments which showed approximately equal amounts of phosphorylation within the C-domain and in the downstream domain, at an unidentified location as was already mentioned above. However, the amount of activation (or in some cases inhibition) observed due to C kinase phosphorylation in this study was small and depended very much on the conditions.

Another study (Kuo et al., 1991) examined the phosphorylation of the pump in cultured rat aortic endothelial cells. A phorbol ester caused a very large increase in the amount of pump mRNA, and lesser increases in the amount of pump protein and in the amount of pump phosphorylation. Presumably all of these effects were related to C kinase activity, but the effect on pump activity was not measured.

Other Aspects. As was discussed under the C-domain, the downstream region of the pump may be involved in self-inhibition. Mutants designed to investigate this phenomenon are currently being made. A mass spectrometry study (Hofmann et al., 1993) of the downstream domain of the plasma membrane calcium pump isoform 1b showed that it contained at least two calcium ions. This study was performed on the same downstream fragment of the pump mentioned above, expressed in *E. coli.* Mass spectrometry indicated the presence of Ca^{2+} which was not removable by EGTA in the last 37 amino acids of the expressed protein fragment. Fluorescent experiments on the dansylated downstream fragment indicated that at least one Ca^{2+} bound to the fragment causing an increase in the fluorescence of the dansyl derivative. The authors of this study suggested that the EGTA-sensitive site might somehow control the access of Ca^{2+} to its catalytic site, while the EGTA-insensitive site may be structural.

B. The Lipid Binding Domain

The stimulation of the Ca^{2+} pump by acidic lipids was first reported by Niggli et al. (1981), but it was at first thought that calmodulin and acidic lipids had essentially identical effects. A subsequent study by Choquette et al. (1984) showed that acidic lipid has a stronger effect on the enzyme than calmodulin does. That this extra effect of lipid could be due to an interaction with a separate portion of the enzyme was indicated by the experiments of Enyedi et al. (1987), who showed that proteolytic cleavage of the enzyme down to a fragment of 81 kDa had an activatory effect similar to that of excess calmodulin while further cleavage to 76 kDa had an additional activation effect similar to that caused by acidic lipids. The mapping of the cleavage points of trypsin on the purified erythrocyte Ca^{2+} pump (Zvaritch et al., 1990) showed that cleavage of the 81 kDa fragment to the 76 kDa fragment involved the removal of a 44 amino acid segment just upstream of the putative transmembrane region M3. This finding focused attention on this region as a lipid-binding domain, and inspection of this region indicated that a portion at its downstream end had the plus charge necessary to interact with acidic lipids. Synthetic peptides representing the sequence of this domain were made by two laboratories (Brodin et al., 1992; Filoteo et al., 1992). Both of these studies measured the interaction of the synthetic peptides with acidic phospholipids and both found that the peptide representing the calmodulin-binding domain interacted with lipid more strongly than did the peptide representing the lipid-binding domain, but that both peptides did interact strongly with acidic lipid. That the peptide representing the lipid-binding domain bound the lipid less strongly was further shown by the fact that peptide was able to compete with the portion of the lipid activation caused by that upstream site and not with the portion caused by the C-domain (Brodin et al., 1992). Another result of these studies was the observation that the lipid-binding domain also binds calmodulin, but much more weakly than does the C-domain (Filoteo et al., 1992). Binding of calmodulin at this site did not cause activation.

The overall picture is, therefore, as follows: The C-domain and the lipid domains both bind calmodulin and both bind lipid. The C-domain binds calmodulin and lipid more strongly than does the lipid-binding domain. In the case of calmodulin binding, only the binding at the C-domain appears to be effective. It was not possible to detect a second calmodulin-binding site by any measurements of the effects of calmodulin on enzyme activity (Filoteo et al., 1992), although measurements of calmodulin binding on solubilized enzyme did disclose two calmodulin-binding sites (Zurini et al., 1984). A peptide representing the C-domain bound calmodulin about 8,000 times more tightly than did a peptide representing the lipid-binding domain (Filoteo et al., 1992), suggesting that the affinity of the lipid-binding domain for calmodulin may be too low to bind significant amounts of calmodulin *in vivo*.

With respect to lipid, the C-domain also binds the lipid more tightly, but the lipid-binding domain binds lipid tightly enough to have an effect on *in vitro* measurements of activity, at concentrations that suggest that *in vivo* effects of lipid may also occur at this site. In the presence of saturating amounts of lipid both the C-domain and the lipid-binding domain will be saturated and the apparent affinity of the enzyme for Ca^{2+} will be greater than that in the presence of only calmodulin or only enough lipid to bind to activate at the C-domain.

It is evident from the above discussion that proteolytic activation of the lipid binding domain can, in principle occur. However, the enzyme which activates the pump by removing this domain is trypsin, which would not be expected to have access to the pump *in vivo*. Calpain, which may act intracellularly, does not cleave the positions necessary to remove the lipid-binding domain. Therefore, the proteolytic activation at the lipid-binding domain is probably a laboratory phenomenon which does not occur *in vivo*.

III. REGULATION OF THE SARCOPLASMIC RETICULUM Ca^{2+} PUMP

A. Regulation by Phospholamban

Turning from the plasma membrane Ca^{2+} pump to that of the sarcoplasmic reticulum, we find a very different regulatory mechanism which, nonetheless, seems to bear some relationship to that of the plasma membrane pump. Regulation of the sarcoplasmic reticulum Ca^{2+} pump appears to occur primarily via phosphorylation of a regulatory protein, phospholamban, which is expressed almost exclusively in cardiac muscle (Tada and Kadoma, 1989). The monomer of phospholamban consists of 52 amino acid residues with a calculated molecular weight of 6,080. It contains one transmembrane hydrophobic region and appears to associate into a pentamer in the membrane, with the hydrophilic domains facing the cytoplasm. Phospholamban is phosphorylated by protein kinases on serines and threonines in its cytoplasmic domain. The dephosphorylated form of phospholamban inhibits the sarcoplasmic reticulum Ca^{2+} pump while phosphorylation, which can occur at numerous sites in a pentamer, relieves the inhibition (James et al., 1989a). Phospholamban can inhibit isoforms 1 or 2 of the pump, but not isoform 3 (Toyofuku et al., 1993). It is ordinarily expressed only along with isoform 2 in heart muscle.

Two positions in the sarcoplasmic reticulum pump at which phospholamban interacts have been investigated by site-directed mutagenesis of the sarcoplasmic reticulum calcium pump. The upstream one of these interactions has been well-defined, and involves a hydrophilic portion of the pump about 50 residues downstream of the aspartic acid which is phosphorylated by ATP (Toyofuku et al., 1994). The second region of interaction is a rather ill-defined one about 200 amino acid residues further downstream, but still in the large hydrophilic domain (Toyofuku et al., 1993). Thus, phospholamban and the sarcoplasmic reticulum Ca^{2+} pump are both integral membrane proteins with transmembrane regions; in the inhibited state, their hydrophilic regions interact with each other. When the amino acid sequences of the sarcoplasmic reticulum and plasma membrane Ca^{2+} pumps are aligned, the upstream interaction site of phospholamban with sarcoplasmic reticulum is near the downstream site at which the C-domain interacts with the plasma membrane Ca^{2+} pump. The distance between these two interaction sites in the alignment is about 40 residues, with the plasma membrane pump interaction site being further downstream, as shown in Figure 2.

Figure 2. Alignment of a portion of the rabbit sarcoplasmic reticulum Ca^{2+} pump, isoform 2 (SERCA2) with the human plasma membrane Ca^{2+} pump, isoform 4b (PMCA4b), showing the aspartate which forms the acylphosphate (P), a site of interaction of SERCA with phospholamban (#) (Toyofuku et al., 1994) and a site of interaction of PMCA with the C domain (*) (Falchetto et al., 1991).

B. Relationship between Regulation of the two Ca^{2+} Pumps

It is clear from the discussion so far that there are striking differences in the mode of regulation of the plasma membrane and sarcoplasmic reticulum Ca^{2+} pumps. The plasma membrane pump is regulated directly by the binding of calmodulin to a region in its carboxyl terminal domain and by activation by protein kinase in that domain. The sarcoplasmic reticulum pump in cardiac muscle is inhibited by interaction with a separate molecule (phospholamban) and is activated by phosphorylation of that molecule by a protein kinase. Calmodulin-dependent protein kinase and A kinase both phosphorylate phospholamban. Thus, calmodulin can also regulate the sarcoplasmic reticulum Ca^{2+} pump, but by a more indirect path involving a protein kinase and phospholamban. Despite these differences there is also a relationship between these regulatory mechanisms, which became evident when peptides were utilized as inhibitors of the activated Ca^{2+} pumps. Experiments of this kind showed that peptides representing the inhibitory regions inhibited their respective pumps: a peptide representing the hydrophilic domain of phospholamban inhibited the sarcoplasmic reticulum Ca^{2+} pump (Chiesi et al., 1991, Sasaki et al., 1992) and a peptide representing the C-domain (Enyedi et al., 1989) inhibited the plasma membrane Ca^{2+} pump. A cross relationship between the two pumps was first reported by Chiesi et al. (1991), who found that a peptide representing the C-domain of the plasma membrane pump also inhibited sarcoplasmic reticulum Ca^{2+} uptake. This cross relationship between the two pumps was supported by a subsequent study (Vorherr et al., 1992) in which a peptide representing the phospholamban-binding domain of the sarcoplasmic reticulum Ca^{2+} pump was tested for its ability to relieve inhibition of the sarcoplasmic reticulum pump by both the phospholamban peptide and the plasma membrane peptide. It competed with both of these inhibitions, further suggesting their relatedness. A subsequent study (Enyedi and Penniston, 1993) showed that a slightly different peptide representing the C-domain of the plasma membrane Ca^{2+} pump and a peptide from the Na^+/Ca^{2+} exchanger were very effective as inhibitors of both the sarcoplasmic reticulum and plasma membrane Ca^{2+} pumps whereas calmodulin-binding peptides from non-Ca^{2+} transport proteins were much less effective. All of these studies suggest the presence of a structural relationship between the inhibitors of these Ca^{2+}

transporters and show that, despite the striking differences between these transporters, their regulatory mechanisms are structurally related.

Although most of the studies of regulation of the sarcoplasmic reticulum Ca^{2+} pump focus on the mechanism of phospholamban described here, there is one report that a factor in addition to phospholamban is needed for its effect (Briggs et al., 1992), and a report that regulation of the sarcoplasmic reticulum Ca^{2+} pump can occur by direct phosphorylation of the pump (Xu et al., 1993).

IV. MEANS OF DISTINGUISHING BETWEEN THE TWO Ca^{2+} PUMPS

Another aspect of the relationship between these two Ca^{2+} pumps has been the necessity of distinguishing between them in whole cells and in isolated membrane preparations. Since both pumps carry out ATP-dependent Ca^{2+} transport, and both form Ca^{2+}-dependent phosphorylated intermediates from ATP, it has frequently been hard to distinguish them. We will discuss the use of lanthanum and of thapsigargin for this purpose.

A.   Lanthanum

Although both pumps are inhibited in a comparable way by lanthanum, the amount of acylphosphate intermediate responds differently to lanthanum, and this difference has been used as a means of discriminating between the two pumps. The presence of the phosphoenzyme is normally demonstrated by incubating membranes with a low concentration of highly labeled ATP in the presence of Ca^{2+} for a short time on ice. Because the rate of phosphorylation of the acylphosphate is much greater than that of formation of serine and threonine phosphates, this technique labels only the two Ca^{2+} pumps. The membranes are then dissolved in detergent and run on an SDS gel on which an autoradiogram is performed to visualize the bands of the proteins which were phosphorylated. Although the plasma membrane Ca^{2+} pump has a larger molecular weight than the sarcoplasmic reticulum one, it has sometimes been very difficult to distinguish the two, particularly in circumstances where proteolysis might have taken place. The presence of lanthanum does not significantly change the amount of phosphoenzyme of the sarcoplasmic reticulum Ca^{2+} pump (isoform 1), but it does substantially increase the amount of phosphorylation of isoform 4b of the plasma membrane Ca^{2+} pump. Indeed, it is difficult to observe the phosphorylated intermediate of this latter pump in the absence of lanthanum, primarily because of its very low concentration in the membrane.

This lanthanum-dependent increase in the amount of phosphoenzyme was thought to be a distinguishing characteristic of the plasma membrane Ca^{2+} pump, but recent results have shown that at least certain isoforms of the sarcoplasmic reticulum Ca^{2+} pump can also show this property. In a study on platelets (Enyedi et al.,

1986), the phosphoenzymes of two different Ca^{2+} pumps were observed with molecular weights near 100 kDa. One of these phosphoenzymes showed an increase in the amount of phosphorylated enzyme in the presence of La^{3+}; this enzyme also showed a tryptic fragment of about 80 kDa, about the same size as one of the major tryptic fragments of the plasma membrane Ca^{2+} pump and much larger than the major tryptic fragments of the sarcoplasmic reticulum Ca^{2+} pump. All of these data suggested that this pump was a shortened version of the plasma membrane Ca^{2+} pump, or at least was not a sarcoplasmic reticulum Ca^{2+} pump. However, recent studies have suggested that this band is due to isoform 3 of the sarcoplasmic reticulum Ca^{2+} pump (Bobe et al., 1994). This isoform resembles the plasma membrane Ca^{2+} pump in giving an approximately 80 kDa tryptic fragment and in having a lanthanum-sensitive acylphosphate. Such confusions can now be more easily resolved with antibodies (Adamo et al., 1992) which were not available to earlier workers.

In addition to its role in stabilizing the phosphoenzyme in plasma membrane Ca^{2+} pumps, lanthanum can be used in whole cells to distinguish the roles of the plasma membrane and sarcoplasmic/endoplasmic reticulum Ca^{2+} pumps. When lanthanum is applied at low concentrations to whole cells it inhibits Ca^{2+} entry through channels. At higher concentrations, it also inhibits Ca^{2+} efflux through the pump (Kwan et al., 1990; Toescu and Petersen, 1994). Since it does not penetrate the cell, it has no effect on the sarco/endoplasmic reticulum Ca^{2+} pump. The use of extracellular lanthanum to inhibit the plasma membrane pump was introduced in the early work on erythrocytes (Quist and Roufogalis, 1975; Sarkadi et al., 1976; Szasz et al., 1978), but its use to distinguish the various components of the Ca^{2+} signal is relatively recent. One recent study was able to distinguish the contribution of the plasma membrane Ca^{2+} pump in this way (Toescu and Petersen, 1995); in this study the response of the cells in the absence of Ca^{2+} channel activity was measured by observing their response in a Ca^{2+}-free solution and then the contribution of the pump was observed by blocking it with high concentrations of extracellular lanthanum. The difference in the change in free Ca^{2+} (measured by a fluorescent indicator) between these two conditions can be attributed to the plasma membrane Ca^{2+} pump.

B. Thapsigargin

The tumor promoter thapsigargin is a highly specific inhibitor of the sarcoplasmic/endoplasmic reticulum Ca^{2+} pump (Thastrup et al., 1990). It does not appear to inhibit the plasma membrane Ca^{2+} pump at all (Thastrup et al., 1990; Enyedi et al., 1993), while a study in heart (Kirby et al., 1992) showed no effect of thapsigargin on plasma membrane properties in general. The sensitivity of sarcoplasmic/endoplasmic reticulum Ca^{2+} pumps to thapsigargin seems to depend on the source of the pump (Thastrup et al., 1990) and on the conditions of inhibition (Kijima et al., 1991), with the presence of Ca^{2+} giving some protection against inhibition. However, a concentration of 0.5-1 µm thapsigargin seems sufficient to fully inhibit sarcoplasmic/endoplasmic reticulum Ca^{2+} pumps from various sources under almost all conditions.

Because of these properties, thapsigargin is very useful for observing the activity of the plasma membrane Ca^{2+} pump in both isolated membranes and intact cells. In mixed membrane preparations from mammalian cells, it is usually possible to add an appropriate amount of thapsigargin so that the sarcoplasmic/endoplasmic reticulum Ca^{2+} pumps are almost totally inhibited. This allows the measurement of the activity of the plasma membrane pump in Ca^{2+} uptake without interference from the sarcoplasmic reticulum enzyme (Enyedi et al., 1993). In studies utilizing whole cells, it is also possible to dissect out the contributions of the plasma membrane and sarcoplasmic/endoplasmic reticulum Ca^{2+} pumps with thapsigargin. This has been done with particular effectiveness in pancreatic acinar cells (Zhang et al., 1992; Toescu and Petersen, 1994). These studies have shown a strong contribution of the plasma membrane pump to the removal of Ca^{2+} from the cells, and have also shown a strong stimulation of the pump's activity by carbachol, which caused a 3-fold increase in the Ca^{2+} efflux due to the plasma membrane Ca^{2+} pump at a constant intracellular free Ca^{2+} (Zhang et al., 1992).

V. ROLE OF PLASMA MEMBRANE Ca^{2+} PUMP IN Ca^{2+} SIGNALING

Many models of intracellular Ca^{2+} signalling and calcium spiking have assigned a relatively passive role to the plasma membrane calcium pump or have even ignored it; however, the evidence is growing that it plays an active role in determining the shape of the Ca^{2+} signal, particularly during the decrease of Ca^{2+} from its peak back to a resting level. Aside from human red cells, which contain only the plasma membrane Ca^{2+} pump as a manager of intracellular Ca^{2+}, the clearest studies have been done on pancreatic acinar cells. These cells lack the Na$^+$/Ca^{2+} exchanger (Muallem et al., 1988), so all Ca^{2+} efflux from the cells must be due to the pump. A study of the cholecystokinin-evoked Ca^{2+} spikes was carried out in a nominally Ca^{2+} free solution on single cells (Tepikin et al., 1992). In this study, it was possible to simultaneously measure intracellular and extracellular Ca^{2+} and from this to estimate the Ca^{2+} extrusion rate. Figure 3, taken from that paper, shows the activation of the plasma membrane Ca^{2+} pump during each spike. In order to observe Ca^{2+} extrusion from the cell by the pump, the extracellular Ca^{2+} concentration was kept very low. The authors estimated that under these conditions $39 \pm 12\%$ of the total mobilizable intracellular Ca^{2+} pool was extruded by the pump during the first spike, and the size of the spikes (and the amount extruded) decreased thereafter until the cells were exhausted of Ca^{2+}. They pointed out that if the whole immobilizable pool is actually mobilized during each spike and 39% is extruded, the remaining 61% must be taken up into the stores. However, if less of the immobilizable pool is activated during each spike, as much as 100% of the mobilized Ca^{2+} might be ejected by the pump.

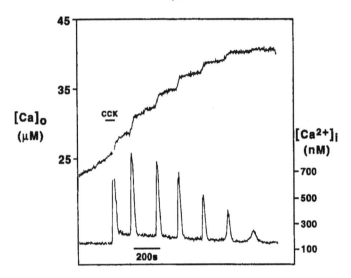

Figure 3. Intracellular Ca^{2+} spikes (lower curve and right axis) and extracellular Ca^{2+} concentration (*upper curve and left axis*) measured simultaneously in a pancreatic acinar cell. The upper curve shows the ejection of Ca^{2+} from the cell by the plasma membrane Ca^{2+} pump during each spike. (From Tepikin et al., 1992).

Under normal conditions (where the extracellular Ca^{2+} concentration is high), Ca^{2+} entry through receptor-operated Ca^{2+} channels should balance Ca^{2+} extrusion and intracellular Ca^{2+} stores could be refilled by the endoplasmic reticulum Ca^{2+} pump. Nonetheless, these experiments indicated that the role of the plasma membrane Ca^{2+} pump is major, and might be dominant in Ca^{2+} spike recovery. Another study from the same laboratory measured the Ca^{2+} dependence of the pump in acinar cells, and showed that it was approximately similar to that observed in isolated membranes (Toescu and Petersen, 1995).

As we mentioned at the beginning of this chapter, the shape of the Ca^{2+}-dependence curve of the pump is a physiologically important property. Clearly, the intensity and duration of a Ca^{2+} signal will be determined in part by the Ca^{2+} dependence of the plasma membrane Ca^{2+} pump. In situations in which the pump requires a higher level of Ca^{2+} for its activation, the Ca^{2+} signal can be expected to be both more intense and longer lasting. As was discussed above, such a situation would occur in cells which express isoform 4a instead of isoform 4b. The action of protein kinases may also affect the Ca^{2+} dependence of the pump, although no accurate data are available.

The nature of the usual assays of the pump is such that the Ca^{2+}-dependence of the enzyme is measured under circumstances in which the enzyme has reached equilibrium with all of its substrates except ATP before the start of the reaction. This gives linear dependence on Ca^{2+} uptake on time, which is essential to accurate measurements. Under physiological situations, however, the rate of calmodulin's binding to the pump is another factor which has substantial importance. The rate constants for binding of calmodulin to the pump in red cell membranes (primarily isoform 4b) were estimated by Scharff and Foder (1982) who followed the ATPase activity continuously by using a reaction which couples ATPase to oxidation of NADH. They started the reaction in such a way that the interaction of the pump with calmodulin did not begin until the start of the measurement of ATPase, and they observed an increase in the ATPase rate with time. At low calmodulin concentrations, the rate was initially very low and continued to increase for a long time before the reaction became linear. At high calmodulin, the period of nonlinearity was very brief. Because of this behavior, they attributed the nonlinearity to a slow binding of calmodulin to the pump. From data of this type, they calculated the on and off rates of calmodulin. They estimated that at 0.7 μM free Ca^{2+} and 4 μM calmodulin the time for 50% occupation of the pump by calmodulin was 33 seconds, a length of time comparable to the lengths of calcium spikes in living cells.

Subsequently, they were able to demonstrate, in resealed ghosts from human erythrocytes, a solitary Ca^{2+} spike (Foder and Scharff, 1992). This spike occurred when the ghosts were loaded with calmodulin; when the calmodulin was inactivated by a binding peptide, the removal of Ca^{2+} was eliminated. This experiment is shown in Figure 4 which is taken from Scharff and Foder (1993). When the erythrocyte membranes were pretreated with trypsin, activating the pump, no spike was observed, as shown in Figure 4. They interpreted this result as meaning that the spike was occurring because of the lag in activation of the pump due to the slow rate

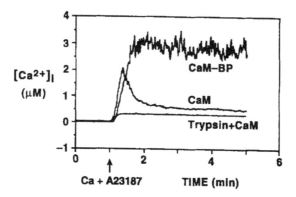

Figure 4. Intracellular Ca^{2+} versus time in resealed erythrocyte ghosts. Since erythrocytes have no Ca^{2+} channels, Ca^{2+} entry into the cell was initiated with the ionophore A23187. In the presence of calmodulin (CaM), a spike ensued, but if the calmodulin was inactivated by a binding peptide (CaM-BP), Ca^{2+} was not removed from the cell. When the pump was constitutively activated by trypsin, only a tiny rise in intracellular Ca^{2+} was seen, and the spike was completely eliminated. (From Scharff and Foder 1993).

of binding of calmodulin to the pump. (The pump is the only means of controlling Ca^{2+} in human erythrocytes.) This explained the elimination of the spike when the pump was continuously activated. The same workers also showed a spike in human neutrophils, in circumstances in which they inhibited the sarcoplasmic reticulum Ca^{2+} pump and took measures to allow Ca^{2+} influx into the cells to occur at a constant rate (Scharff and Foder, 1994). They showed that the descending phase of the spike was abolished by a calmodulin antagonist, but not by a protein kinase C inhibitor nor by sodium free medium. They concluded that activity of protein kinase C or the Na^+/Ca^{2+} exchanger was not necessary for generation of the spike and that it was probably due to delayed activation of the plasma membrane Ca^{2+} pump.

Clearly, the binding rate of calmodulin would be expected to be influenced by the alternate splice in domain C, since the splice affects the affinity of the enzyme for calmodulin. An affinity change probably involves changes in both the on and off rates of calmodulin. The low affinity caused by the A splice would probably be due in part to a slower on rate of calmodulin. Phosphorylation by protein kinases may also affect the rate of binding of calmodulin to the pump. If these interpretations are correct, the plasma membrane Ca^{2+} pump is strikingly different in its time responsiveness from the sarco/endoplasmic reticulum pump, which is normally activated and which must interact with another molecule for inhibition. This would probably result in a very different time dependence of the Ca^{2+} removal by the sarco/endoplasmic reticulum pump, since it would be expected to be activated only by its Ca^{2+}-dependence and not in a time-dependent fashion. Thus, an important feature of the regulation of the Ca^{2+} signal by the calcium pump may be a delay in its activation, which gives the Ca^{2+} signal time to develop and then ends it with relative rapidity. This leads to the speculation that the major role of the endoplasmic reticulum Ca^{2+} pump would be refilling of the intracellular Ca^{2+} stores during the middle of the spike whereas the plasma membrane Ca^{2+} pump may play a crucial role in shaping and ending Ca^{2+} spikes.

VI. SUMMARY

The plasma membrane Ca^{2+} pump is discussed and compared with the sarcoplasmic/endoplasmic reticulum Ca^{2+} pump. Particular attention is paid to the mechanisms by which these pumps are regulated. In the case of the plasma membrane Ca^{2+} pump the carboxyl terminus plays a pivotal role in regulation. This is the part of the molecule where calmodulin binds, protein kinases A and C act and calpain attacks (just to mention the most important regulators). These regulators cause complete or partial activation of the enzyme. Additional modulation of this pump's activity occurs when isoforms are generated from mRNAs with alternate splicing. One alternate splice happens at the carboxyl terminus just in the middle of the calmodulin-binding domain. This splice changes the structure of the whole downstream region producing isoforms with different calmodulin sensitivity and different sites for phosphorylation by protein kinases. In addition to the carboxyl-terminus there is another region of the molecule which is regulated by acidic lipids and may play a role in fine tuning of the plasma membrane Ca^{2+} pump's activity. The regulation of the sarcoplasmic/endo-plasmic reticulum Ca^{2+} pump appears to be much simpler and seems to happen only in heart muscle. Here a separate inhibitory protein is expressed, called phospholamban. The possible roles of these Ca^{2+} pumps in shaping intracellular Ca^{2+} spikes are discussed in the light of the striking differences in their regulation.

This review reflects work through May, 1995.

REFERENCES

Adamo, H. P., Caride, A. J., & Penniston, J. T. (1992). Use of expression mutants and monoclonal antibodies to map the erythrocyte Ca^{2+} pump. J. Biol. Chem. 267, 14244-14249.

Au, K. S. (1987). Activation of erythrocyte membrane Ca^{2+}-ATPase by calpain. Biochim. Biophys. Acta 905, 273-278.

Bobe, R., Bredoux, R., Wuytack, F., Quarck, R., Kovacs, T., Papp, B., Corvazier, E., Magnier, C., & Enouf, J. (1994). The rat platelet 97-kDa Ca^{2+}ATPase isoform is the sarcoendoplasmic reticulum Ca^{2+}-ATPase 3 protein. J. Biol. Chem. 269, 1417-1424.

Bond, G. H., & Clough, D. L. (1973). A soluble protein activator of $(Mg^{2+} + Ca^{2+})$-dependent ATPase in human red cell membranes. Biochim. Biophys. Acta 323, 592-599.

Briggs, F. N., Lee, K. F., Wechsler, A. W., & Jones, L. R. (1992). Phospholamban expressed in slow-twitch and chronically stimulated fast-twitch muscles minimally affects calcium affinity of sarcoplasmic reticulum Ca^{2+}-ATPase. J. Biol. Chem. 267, 26056-26061.

Brodin, P., Falchetto, R., Vorherr, T., & Carafoli, E. (1992). Identification of two domains which mediate the binding of activating phospholipids to the plasma membrane Ca^{2+} pump. Eur. J. Biochem. 204, 939-946.

Carafoli, E., & Zurini, M. (1982). The Ca^{2+}-pumping ATPase of plasma membranes. Purification, reconstitution and properties. Biochim. Biophys. Acta 683, 279-301.

Carafoli, E. (1994). Biogenesis: plasma membrane calcium ATPase: 15 years of work on the purified enzyme. FASEB J. 8, 993-1002.

Caride, A. J., Filoteo, A. G., Enyedi, A., & Penniston, J. T. (1994). Epitope location and isoenzyme specificity of three monoclonal antibodies against the plasma membrane calcium pump. Biophys. J. 66, A120

Chiesi, M., Vorherr, T., Falchetto, R., Waelchli, C., & Carafoli, E. (1991). Phospholamban is related to the autoinhibitory domain of the plasma membrane Ca^{2+}-pumping ATPase. Biochemistry 30, 7978-7983.

Choquette, D., Hakim, G., Filoteo, A. G., Plishker, G. A., Bostwick, J. R., & Penniston, J. T. (1984). Regulation of plasma membrane Ca^{2+}-ATPases by lipids of the phosphatidylinositol cycle. Biochem. Biophys. Res. Commun. 125, 908-915.

Enyedi, A., Sarkadi, B., Szasz, I., Bot, G., & Gardos, G. (1980). Molecular properties of the red cell calcium pump. II. Effects of calmodulin, proteolytic digestion and drugs on the calcium-induced membrane phosphorylation by ATP in inside-out red cell membrane vesicles. Cell Calcium 1, 299-310.

Enyedi, A., Sarkadi, B., Foldes-Papp, Z., Monostory, S., & Gardos, G. (1986). Demonstration of two distinct calcium pumps in human platelet membrane vesicles. J. Biol. Chem. 261, 9558-9563.

Enyedi, A., Flura, M., Sarkadi, B., Gardos, G., & Carafoli, E. (1987). The maximal velocity and the calcium affinity of the red cell calcium pump may be regulated independently. J. Biol. Chem. 262, 6425-6430.

Enyedi, A., Vorherr, T., James, P., McCormick, D. J., Filoteo, A. G., Carafoli, E., & Penniston, J. T. (1989). The calmodulin-binding domain of the plasma membrane Ca^{2+} pump interacts both with calmodulin and with another part of the pump. J. Biol. Chem. 264, 12313-12321.

Enyedi, A., Filoteo, A. G., Gardos, G., & Penniston, J. T. (1991). Calmodulin-binding domains from isozymes of the plasma membrane Ca^{2+} pump have different regulatory properties. J. Biol. Chem. 266, 8952-8956.

Enyedi, A., & Penniston, J. T. (1993). Autoinhibitory domains of various Ca^{2+} transporters cross-react. J. Biol. Chem. 268, 17120-17125.

Enyedi, A., Verma, A. K., Filoteo, A. G., & Penniston, J. T. (1993). A highly active truncated mutant of the plasma membrane Ca^{2+} pump. J. Biol. Chem. 268, 10621-10626.

Enyedi, A., Verma, A. K., Heim, R., Adamo, H. P., Filoteo, A. G., Strehler, E. E., & Penniston, J. T. (1994). The Ca^{2+} affinity of the plasma membrane Ca^{2+} pump is controlled by alternative splicing. J. Biol. Chem. 269, 41-43.

Falchetto, R., Vorherr, T., Brunner, J., & Carafoli, E. (1991). The plasma membrane Ca^{2+} pump contains a site that interacts with its calmodulin binding domain. J. Biol. Chem. 266, 930-936.

Falchetto, R., Vorherr, T., & Carafoli, E. (1992). The calmodulin-binding site of the plasma membrane Ca^{2+} pump interacts with the transduction domain of the enzyme. Protein Sci. 1, 1613-1621.

Filoteo, A. G., Enyedi, A., & Penniston, J. T. (1992). The lipid-binding peptide from the plasma membrane Ca^{2+} pump binds calmodulin, and the primary calmodulin-binding domain interacts with lipid. J. Biol. Chem. 267, 11800-11805

Foder, B., & Scharff, O. (1992). Solitary calcium spike dependent on calmodulin and plasma membrane Ca^{2+} pump. Cell Calcium 13, 581-591.

Gopinath, R. M., & Vincenzi, F. F. (1977). Phosphodiesterase protein activator mimics red blood cell cytoplasmic activator of Ca^{2+} Mg^{2+}-ATPase. Biochem. Biophys. Res. Commun. 77, 1203-1209.

Graf, E., Verma, A. K., Gorski, J. P., Lopaschuk, G., Niggli, V., Zurini, M., Carafoli, E., & Penniston, J. T. (1982). Molecular properties of calcium-pumping ATPase from human erythrocytes. Biochemistry 21, 4511-4516

Greeb, J., & Shull, G. E. (1989). Molecular cloning of a third isoform of the calmodulin-sensitive plasma membrane Ca^{2+}-transporting ATPase that is expressed predominantly in brain and skeletal muscle. J. Biol. Chem. 264, 18569-18576.

Hammes, A., Oberdorf, S., Strehler, E. E., Stauffer, T., Carafoli, E., Vetter, H., & Neyses, L. (1994). Differentiation-specific isoform mRNA expression of the calmodulin-dependent plasma membrane Ca^{2+}- ATPase. FASEB J. 8, 428-435.

Hao, L., Rigaud, J.-L., & Inesi, G. (1994). Ca^{2+}/H^+ countertransport and electrogenicity in proteoliposomes containing erythrocyte plasma membrane Ca-ATPase and exogenous lipids. J. Biol. Chem. 269, 14268-14275.

Heim, R., Iwata, T., Zvaritch, E., Adamo, H., Rutishauser, B., Strehler, E. E., Guerini, D., & Carafoli, E. (1992). Expression, purification, and properties of the plasma membrane Ca^{2+} pump and of its N-terminally truncated 105-kDa fragment. J. Biol. Chem. 267, 24476-24484.

Hilfiker, H., Guerini, D., & Carafoli, E. (1994). Cloning and expression of isoform 2 of the human plasma membrane Ca^{2+} ATPase. Functional properties of the enzyme and its splicing products. J. Biol. Chem. 269, 26178-26183.

Hofmann, F., James, P., Vorherr, T., & Carafoli, E. (1993). The C- terminal domain of the plasma membrane Ca^{2+} pump contains three high-affinity Ca^{2+} binding sites. J. Biol. Chem. 268, 10252-10259.

Hofmann, F., Anagli, J., Carafoli, E., & Vorherr, T. (1994). Phosphorylation of the calmodulin binding domain of the plasma membrane Ca^{2+} pump by protein kinase C reduces its interaction with calmodulin and with its pump receptor site. J. Biol. Chem. 269, 24298-24303.

Howard, A., Legon, S., & Walters, J. R. F. (1993). Human and rat intestinal plasma membrane calcium pump isoforms. Am. J. Physiol. 265, G917-G925.

Howard, A., Barley, N. F., Legon, S., & Walters, J. R. F. (1994). Plasma-membrane calcium-pump isoforms in human and rat liver. Biochem. J. 303, 275-279.

James, P., Maeda, M., Fisher, R., Verma, A. K., Krebs, J., Penniston, J. T., & Carafoli, E. (1988). Identification and primary structure of a calmodulin-binding domain of the Ca^{2+} pump of human erythrocytes. J. Biol. Chem. 263, 2905-2910.

James, P., Inui, M., Tada, M., Chiesi, M., & Carafoli, E. (1989a). Nature and site of phospholamban regulation of the Ca^{2+} pump of sarcoplasmic reticulum. Nature 342, 90-92.

James, P. H., Pruschy, M., Vorherr, T. E., Penniston, J. T., & Carafoli, E. (1989b). Primary structure of the cAMP-dependent phosphorylation site of the plasma membrane calcium pump. Biochemistry 28, 4253-4258.

James, P., Vorherr, T., Krebs, J., Morelli, A., Castello, G., McCormick, D. J., Penniston, J. T., DeFlora, A., & Carafoli, E. (1989c). Modulation of erythrocyte Ca^{2+}-ATPase by selective calpain cleavage of the calmodulin-binding domain. J. Biol. Chem. 264, 8289-8296.

James, P., Hofmann, F., & Carafoli, E. (1992). Localization of a high-affinity Ca-binding site of the plasma membrane Ca^{2+} pump and a second phosphorylation site of the cAMP-dependent kinase by electrospray ionisation mass spectrometry. (8th International Symposium on Calcium-binding Proteins and Calcium Function in Health and Disease, p B44).

Jarrett, H. W., & Penniston, J. T. (1977). Partial purification of the Ca^{2+}-Mg^{2+} ATPase activator from human erythrocytes, its similarity to the activator of 3', 5'-cyclic nucleotide phosphodiesterase. Biochem. Biophys. Res. Commun. 77, 1210-1216.

Kessler, F., Falchetto, R., Heim, R., Meili, R., Vorherr, T., Strehler, E. E., & Carafoli, E. (1992). Study of calmodulin-binding to the alternatively spliced C-terminal domain of the plasma membrane Ca^{2+} pump. Biochemistry 31, 11785-11792.

Kijima, Y., Ogunbunmi, E., & Fleischer, S. (1991). Drug action of thapsigargin on the Ca^{2+} pump protein of sarcoplasmic reticulum. J. Biol. Chem. 266, 22912-22918.

Kirby, M. S., Sagara, Y., Gaa, S., Inesi, G., Lederer, W. J., & Rogers, T. B. (1992). Thapsigargin inhibits contraction and Ca^{2+} transient in cardiac cells by specific inhibition of the sarcoplasmic reticulum Ca^{2+} pump. J. Biol. Chem. 267, 12545-12551.

Kosk-Kosicka, D., & Bzdega, T. (1988). Activation of the erythrocyte Ca^{2+}-ATPase by either self-association or interaction with calmodulin. J. Biol. Chem. 263, 18184-18189.

Kosk-Kosicka, D., Bzdega, T., & Wawrzynow, A. (1989). Fluorescence energy transfer studies of purified erythrocyte Ca^{2+}-ATPase, Ca^{2+} regulated activation by oligomerization. J. Biol. Chem. 264, 19495-19499.

Kosk-Kosicka, D., Bzdega, T., & Johnson, J.D. (1990). Fluorescence studies on calmodulin binding to erythrocyte Ca^{2+}-ATPase in different oligomerization states. Biochemistry 29, 1875-1879.

Kuo, T. H., Wang, K. K. W., Carlock, L., Diglio, C., & Tsang, W. (1991). Phorbol ester induces both gene expression and phosphorylation of the plasma membrane Ca^{2+} pump. J. Biol. Chem. 266, 2520-2525.

Kwan, C. Y., Takemura, H., Obie, J .F., Thastrup, O., & Putney, J. W. Jr. (1990). Effects of MeCh, thapsigargin, and La^{3+} on plasmalemmal and intracellular Ca^{2+} transport in lacrimal acinar cells. Am. J. Physiol. 258, C1006-C1015.

Magocsi, M., Yamaki, M., Penniston, J. T., & Dousa, T. P. (1992). Localization of mRNAs coding for isozymes of plasma membrane Ca^{2+}-ATPase pump in rat kidney. Am. J. Physiol. 263, F7-F14

Meszaros, J. G., & Karin, N. J. (1993). Osteoblasts express the PMCA1b isoform of the plasma membrane Ca^{2+}-ATPase. J. Bone Miner. Res. 8, 1235-1240

Molinari, M., Anagli, J., & Carafoli, E. (1995). PEST sequences do not influence substrate susceptibility to calpain proteolysis. J. Biol. Chem. 270, 2032-2035.

Muallem, S., Beeker, T., & Pandol, S.-J. (1988). Role of Na$^+$/Ca^{2+} exchange and the plasma membrane Ca^{2+} pump in hormone-mediated Ca^{2+} efflux from pancreatic acini. J. Membr. Biol. 102, 153-162.

Niggli, V., Penniston, J. T., & Carafoli, E. (1979). Purification of the Ca^{2+}-Mg^{2+} ATPase from human erythrocyte membranes using a calmodulin affinity column. J. Biol. Chem. 254, 9955-9958.

Niggli, V., Adunyah, E. S., Penniston, J. T., & Carafoli, E. (1981). Purified Ca^{2+}-Mg^{2+} ATPase of the erythrocyte membrane. Reconstitution and effect of calmodulin and phospholipids. J. Biol. Chem. 256, 395-401.

Papp, B., Sarkadi, B., Enyedi, A., Caride, A. J., Penniston, J. T., & Gardos, G. (1989). Functional domains of the *in situ* red cell membrane calcium pump revealed by proteolysis and monoclonal antibodies. Possible sites for regulation by calpain and acidic lipids. J. Biol. Chem. 264, 4577-4582.

Penniston, J. T., Graf, E., & Itano, T. (1980). Calmodulin regulation of the Ca^{2+} pump of erythrocyte membranes. Ann. NY Acad. Sci. 356, 245-257.

Penniston, J. T., & Enyedi, A. (1994). Plasma membrane Ca^{2+} pump, Recent developments. Cell Physiol. Biochem. 4, 148-159.

Quist, E. E., & Roufogalis, B. D. (1975). Determination of the stoichiometry of the calcium pump in human erythrocytes using lanthanum as a selective inhibitor. FEBS Lett. 50, 135-139.

Sackett, D. L., & Kosk-Kosicka, D. (1993). Determining the size of the active species of the red blood cell Ca^{2+}-ATPase. Biophys. J. 64, A334.

Sarkadi, B., Szasz, I., & Gardos, G. (1976). The use of ionophores for rapid loading of human red cells with radioactive cations for cation-pump studies. J. Membr. Biol. 26, 357-370.

Sarkar, F. H., Ball, D. E., Tsang, W., Li, Y-W., & Kuo, T. H. (1993). Use of the polymerase chain reaction for the detection of alternatively spliced mRNAs of plasma membrane calcium pump. DNA Cell. Biol. 12, 435-440

Sasaki, T., Inui, M., Kimura, Y., Kuzuya, T., & Tada, M. (1992). Molecular mechanism of regulation of Ca^{2+} pump ATPase by phospholamban in cardiac sarcoplasmic reticulum. Effects of synthetic phospholamban peptides on Ca^{2+} pump ATPase. J. Biol. Chem. 267, 1674-1679.

Scharff, O., & Foder, B. (1982). Rate constants for calmodulin binding to Ca^{2+} ATPase in erythrocyte membranes. Biochim. Biophys. Acta 691, 133-143.

Scharff, O., & Foder, B. (1993). Regulation of cytosolic calcium in blood cells. Physiol Rev 73, 547-582.

Scharff, O., & Foder, B. (1994). Delayed activation of plasma membrane Ca^{2+} pump in human neutrophils. Cell Calcium 16, 455-466.

Shull, G. E., & Greeb, J. (1988). Molecular cloning of two isoforms of the plasma membrane Ca^{2+}-transporting ATPase from rat brain. Structural and functional domains exhibit similarity to Na^+,K^+- and other cation transport ATPases. J. Biol. Chem. 263, 8646-8657.

Smallwood, J. I., Gugi, B., & Rasmussen, H. (1988). Regulation of erythrocyte Ca^{2+} pump activity by protein kinase C. J. Biol. Chem. 263, 2195-2202.

Stauffer, T. P., Hilfiker, H., Carafoli, E., & Strehler, E. E. (1993). Quantitative analysis of alternative splicing options of human plasma membrane calcium pump genes. J. Biol. Chem. 268, 25993-26003.

Stauffer, T., & Carafoli, E. (1995). J. Biol. Chem. 270, (in press).

Szasz, I., Sarkadi, B., Schubert, A., & Gardos, G. (1978). Effects of lanthanum on calcium-dependent phenomena in human red cells. Biochim. Biophys. Acta 512, 331-340.

Tada, M., & Kadoma, M. (1989). Regulation of the Ca^{2+} pump ATPase by cAMP-dependent phosphorylation of phospholamban. Bioessays 10, 157-163.

Tepkin, A. V., Voronina, S. G., Gallacher, D. V., & Petersen, O. H. (1992). Pulsatile Ca^{2+} extrusion from single pancreatic acinar cells during receptor-activated cytosolic spiking. J. Biol. Chem. 267, 14073-14076.

Thastrup, O., Cullen, P. J., Drobak, B. K., Hanley, M. R., & Dawson, A. P. (1990). Thapsigargin, a tumor promoter, discharges intracellular Ca^{2+} stores by specific inhibition of the endoplasmic reticulum Ca^{2+}- ATPase. Proc. Natl. Acad. Sci. USA 87, 2466-2470.

Toescu, E. C., & Petersen, O. H. (1994). The thapsigargin-evoked increase in $[Ca^{2+}]$ involves an $InsP_3$-dependent Ca^{2+} release process in pancreatic acinar cells. Pflügers Arch. Eur. J. Physiol. 427, 325-331.

Toescu, E. C., & Petersen, O. H. (1995). Region-specific activity of the plasma membrane Ca^{2+} pump and delayed activation of Ca^{2+} entry characterize the polarized, agonist-evoked Ca^{2+} signals in exocrine cells. J. Biol. Chem. 270, 8528-5835.

Toyofuku, T., Kurzydlowski, K., Tada, M., & MacLennan, D. H. (1993). Identification of regions in the Ca^{2+}-ATPase of sarcoplasmic reticulum that affect functional association with phospholamban. J. Biol. Chem. 268, 2809-2815.

Toyofuku, T., Kurzydlowski, K., Tada, M., & MacLennan, D. H. (1994). Amino acids Lys-Asp-Asp-Lys-Pro-Val402 in the Ca^{2+}-ATPase of cardiac sarcoplasmic reticulum are critical for functional association with phospholamban. J. Biol. Chem. 269, 22929-22932.

Verma , A. K., Enyedi, A., Filoteo, A. G., & Penniston, J. T. (1994). Regulatory region of plasma membrane Ca^{2+} pump. 28 residues suffice to bind calmodulin but more are needed for full auto-inhibition of the activity. J. Biol. Chem. 269, 1687-1691.

Verma, A. K., Filoteo, A. G., Stanford, D. R., Wieben, E. D., Penniston, J. T., Strehler, E. E., Fischer, R., Heim, R., Vogel, G., Matthews, S., Strehler-Page, M., James, P., Vorherr, T., Krebs, J., & Carafoli, E. (1988). Complete primary structure of a human plasma membrane Ca^{2+} pump. J. Biol. Chem. 263, 14152-14159.

Vorherr, T., James, P., Krebs, J., Enyedi, A., McCormick, D. J., Penniston, J. T., & Carafoli, E. (1990). Interaction of calmodulin with the calmodulin-binding domain of the plasma membrane Ca^{2+} pump. Biochemistry 29, 355-365.

Vorherr, T., Kessler, T., Hofmann, F., & Carafoli, E. (1991). The calmodulin-binding domain mediates the self-association of the plasma membrane Ca^{2+} pump. J. Biol. Chem. 266, 22-27.

Vorherr, T., Chiesi, M., Schwaller, R., & Carafoli, E. (1992). Regulation of the calcium ion pump of sarcoplasmic reticulum. Reversible inhibition by phospholamban and by the calmodulin-binding domain of the plasma membrane calcium ion pump. Biochemistry 31, 371-376.

Wang, K. K. W., Roufogalis, B. D., & Villalobo, A. (1988a). Further characterization of calpain-mediated proteolysis of the human erythrocyte plasma membrane Ca^{2+}-ATPase. Arch. Biochem. Biophys. 267, 317-327.

Wang, K. K. W., Villalobo, A., & Roufogalis, B. D. (1988b). Activation of the Ca^{2+}-ATPase of human erythrocyte membrane by an endogenous Ca^{2+}-dependent neutral protease. Arch. Biochem. Biophys. 260, 696-704.

Wang, K. K. W., Villalobo, A., & Roufogalis, B. D. (1989). Calmodulin- binding proteins as calpain substrates. Biochem. J. 262, 693-706.

Wang, K. K. W., Wright, L. C., Machan, C. L., Allen, B. G., Conigrave, A. D., & Roufogalis, B. D. (1991). Protein kinase C phosphorylates the carboxyl terminus of the plasma membrane Ca^{2+}-ATPase from human erythrocytes. J. Biol. Chem. 266, 9078-9085.

Westlin, M., Yu, A., & Lytton, J. (1991). A single plasma membrane Ca- ATPase isoform is expressed throughout rat kidney. J. Am. Soc. Nephrol. 2, 755

Wuytack, F., & Raeymaekers, L. (1992). The Ca^{2+}-transport ATPases from the plasma membrane. J. Bioenerg. Biomembr. 24, 285-300.

Xu, A., Hawkins, C., & Narayanan, N. (1993). Phosphorylation and activation of the Ca^{2+}-pumping ATPase of cardiac sarcoplasmic reticulum by Ca^{2+}/calmodulin-dependent protein kinase. J. Biol. Chem. 268, 8394-8397.

Yoshida, Y., Cai, J-Q., & Imai, S. (1992). Plasma membrane Ca^{2+}-ATPase is not a substrate for cGMP-dependent protein kinase. J. Biochem. (Tokyo) 1992, 559-562.

Yu, X., Carroll, S., Rigaud, J.-L., & Inesi, G. (1993). H$^+$ countertransport and electrogenicity of the sarcoplasmic reticulum Ca^{2+} pump in reconstituted proteoliposomes. Biophys. J. 64, 1232-1242.

Zacharias, D., Dalrymple, S. J., & Strehler, E. E. (1995). Transcript distribution of plasma membrane Ca^{2+} isoforms and splice variants in the human brain. Mol. Brain Res. 28, 263-272.

Zhang, B.-X., Zhao, H., Loessberg, P., & Muallem, S. (1992). Activation of the plasma membrane Ca^{2+} pump during agonist stimulation of pancreatic acini. J. Biol. Chem. 267, 15419-15425

Zurini, M., Krebs, J., Penniston, J. T., & Carafoli, E. (1984). Controlled proteolysis of the purified Ca^{2+} ATPase of the erythrocyte membrane. A correlation between the structure and the function of the enzyme. J. Biol. Chem. 259, 618-627.

Zvaritch, E., James, P., Vorherr, T., Falchetto, R., Modyanov, N., & Carafoli, E. (1990). Mapping of functional domains in the plasma membrane Ca^{2+} pump using trypsin proteolysis. Biochemistry 29, 8070-8076.

MOLECULAR MECHANISMS INVOLVED IN THE REGULATION OF Na-K-ATPase EXPRESSION

Käthi Geering

I. STRUCTURAL AND FUNCTIONAL ASPECTS OF Na^+-K^+-ATPase

Na^+-K^+-ATPase belongs to the family of P-type ATPases which are characterized by transient phosphorylation during the reaction cycle. H^+-K^+-ATPases and Ca^{2+}-ATPases of vertebrates, H^+-ATPases of fungi, protozoa, and plants as well as several other cation-transporting ATPases of lower organisms also belong to this

Advances in Molecular and Cell Biology
Volume 23B, pages 275-309.
Copyright © 1998 by JAI Press Inc.
All right of reproduction in any form reserved.
ISBN: 0-7623-0287-9

family. (For recent reviews of the main functional and structural characteristics of these ATPases, see Lingrel et al., 1990; Inesi and Kirtley, 1992; Glynn, 1993; Carafoli and Stauffer, 1994; Horisberger, 1994; Lingrel et al., 1994; Sachs, 1994.) A common functional property of all P-type ATPases is their involvement in the maintenance of the characteristic ionic composition of the intracellular milieu. The Na^+-K^+-ATPase is the enzyme found in the plasma membrane of all animal cells which couples ATP hydrolytic activity and the derived energy to the transport of 2 K^+ into the cell and 3 Na^+ out of the cell. Animal cells characteristically exhibit low internal Na^+ and high K^+ concentrations and the reverse condition outside the cell. Due to its transport mode, the Na^+-K^+-ATPase is responsible for the maintenance of the electrochemical gradients of K^+ and Na^+, which are crucial for cellular homeostasis. The electrochemical gradient of Na^+ provides the energy for a number of secondary active transport systems which couple facilitated Na^+ entry into the cell with the uptake or extrusion of many other substrates (Figure 1A). For instance, Na^+/Ca^{2+} or Na^+/H^+ exchange systems permit the transport of Ca^{2+} or H^+ out of the cell against their electrochemical gradients. The same principle is used to accumulate various ions or nutrients above their equilibrium constant in the cell via Na^+ co-transport systems. Maintenance of the membrane potential, nutrient uptake, regulation of both cell volume and pH are thus assured by the activity of the Na^+-K^+-ATPase. In addition to these general functions which are fundamental for the survival of all animal cells, the Na^+ and K^+ gradients maintained by the Na^+-K^+-ATPase are also important for specialized cellular functions. For instance, in the vasculature and the heart, the Na^+ gradients influence the contractility of the myocytes since they drive the Na^+/Ca^{2+} exchanger, which is an important determinant of intracellular Ca^{2+} concentration. Other examples of specialized functions which depend entirely on Na^+-K^+-ATPase activity are the generation and control of the action potential in excitable tissues such as neurons or muscle cells and the reabsorptive and secretory processes in epithelia of the kidney, colon, and exocrine glands. In the latter case, vectorial transcellular transport becomes possible due to the asymmetric distribution of transport systems in the apical (facing the external milieu) and the basolateral (facing the internal milieu) membrane of the epithelial cell. The Na^+-K^+-ATPase which is located in the basolateral membrane of renal epithelia, drives the entry of Na^+ into the cell via apical Na^+ channels, Na^+/H^+ exchangers or $Na^+/K^+/Cl^-$ cotransporters. It facilitates net reabsorption of Na^+ from the tubular lumen back to the bloodstream and thus assures a constant extracellular fluid volume and composition. Furthermore, the Na^+-K^+-ATPase plays a key role in the regulation of ion transport across the plasma membrane as an integral part of distinct cellular programs involved in developmental processes.

Considering its multiple basic and specialized functions, it seems obvious that Na^+-K^+-ATPase activity must be finely tuned to meet the constantly changing physiological demands in different tissues. It might also be predicted that deregulation of the function of the Na^+-K^+-ATPase could have severe consequences

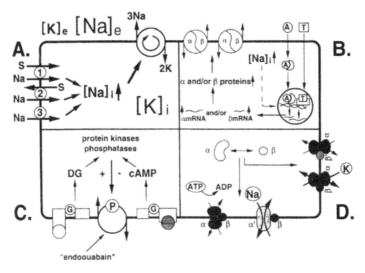

Figure 1. Regulatory mechanisms of Na$^+$-K$^+$-ATPase expression and function. (**A**) Acute control of transport activity by intracellular Na$^+$. Na$^+$-K$^+$-ATPase is responsible for the characteristic Na$^+$ and K$^+$ gradients existing between the intra- and extracellular milieu of animal cells. The Na$^+$ gradient provides the energy for the transport activity of Na$^+$-dependent cotransporters, exchange transport systems or Na channels. Na$^+$-K$^+$-ATPase senses the changes in intracellular Na$^+$ concentrations produced by an increased activity of the Na$^+$-dependent transporters and promptly adapts its transport activity to reestablish low intracellular Na$^+$ concentrations. $[Na]_i$, $[K]_i$ and $[Na]_e$, $[K]_e$ = intracellular and extracellular Na$^+$ and K$^+$ concentrations, respectively. (**B**) Long-term adaptation of the number of Na$^+$-K$^+$-pumps at the cell surface. Corticoid hormones such as aldosterone (A) and thyroid hormones (T) bind to intracellular receptors and the hormone-receptor complexes interact with specific hormone-responsive elements on the α and/or β subunit genes to increase α and/or β mRNA and ultimately α and/or β proteins. Similarly, cells may respond to a prolonged increase in intracellular Na$^+$ by a yet ill-defined transcriptional control of α and/or β subunit genes. A quality control at the level of the ER assures the formation and expression at the cell surface of stoichiometric α-β complexes. (**C**) Short-term adaptation of the transport activity of Na$^+$-K$^+$-pumps. Agonists acting through G-protein coupled receptors modulate the level of intracellular second messengers such as diacylglycerol (DG) or cAMP. Protein kinases and/or phosphatases controlled by second messengers may modulate the transport activity of Na$^+$-K$^+$-pumps through phosphorylation or dephosphorylation of the α subunit. Similarly, an acute control of the transport activity of Na$^+$-K$^+$-pumps might be achieved by endogenous ouabain-like substances ("endoouabain"). (**D**) Control of the functional state of Na$^+$-K$^+$ pumps by subunit assembly and subunit composition. Assembly of α and β subunits at the level of the ER permits the functional maturation of the catalytic α subunit. The type of α or β isoforms contained in the α-β complexes may determine kinetic parameters such as apparent affinities for ATP, Na$^+$ or K$^+$ and thus the transport activity of functional Na$^+$-K$^+$-pumps expressed at the cell surface. For more details and references see text.

277

on body homeostasis and might be involved in several pathological situations (see Weder, 1994; Rose and Valdes, 1994).

The Na^+-K^+-ATPase is regulated by a wide variety of mechanisms which affect its transport activity and/or its expression at the cell surface. Recent experimental evidence suggests that some of the most fundamental control mechanisms are closely related to the heterodimeric structure of the Na^+-K^+-pump molecules and to the structural diversity of the heterodimers. The Na^+-K^+-pump consists of a catalytic α subunit, which is common to all P-type ATPases, and a glycoprotein β subunit which is exclusively found in Na^+K^+- and H^+-K^+-ATPases (Figure 2). The α subunit spans the membrane 8 - 10 times and exposes the N- and C-termini to the cytoplasm. It is responsible for the primary functional properties of these enzymes. For example, it hydrolyzes ATP, binds cations and phosphate, and probably forms the pore for ion transport. In addition, the α subunit is the target of specific inhibitors, such as the cardiac glycosides. The β subunit is a type II glycoprotein which has one transmembrane domain, a short cytoplasmic tail, and a large ectodomain containing several sugar chains and 3 disulphide bridges. The role of the β subunit in Na^+-K^+ and H^+-K^+-ATPases has long been a mystery but now it turns out to be of crucial importance for the regulation of the functional expression of these two enzymes. In addition, three distinct isoforms of the α ($\alpha1$, $\alpha2$, and $\alpha3$) and the β subunit ($\beta1$, $\beta2$, and $\beta3$) have been identified (Shull et al., 1986a, b; Martin-Vasallo et al., 1989; Gloor et al., 1990; Good et al.,1990) which form, in various combinations, tissue-specific α-β complexes (Sweadner, 1992; Zlokovic et al., 1993; Cameron et al., 1994; Hundal et al.,1994), the expression of which is under complex hormonal control. In addition, the idea emerges that the transport activity of α-β complexes can be affected both by the α (Sweadner, 1985; Jewell and Lingrel, 1991; Berrebi-Bertrand and Maixent, 1994; Munzer et al., 1994) and the β subunit (Eakle et al., 1992; Schmalzing et al., 1992; Jaunin et al., 1993; Jaisser et al., 1994) (Figure 1D). Thus, fundamental processes including subunit assembly and the isoform composition of the cell surface expressed Na^+-K^+-pumps interact cooperatively with hormones and neurotransmitters in the control of expression and activity and/or in the long-term adjustment of the cellular Na^+-K^+-pump pool size (Figure 1).

In addition to the isoform composition of the α-β heterodimer, examples of Na^+-K^+-ATPase activity regulation include dependence on the intracellular Na^+ concentration (Akera and Brody, 1985); regulation by covalent but reversible modifications, for example, phosphorylation/dephosphorylation reactions directly or indirectly mediated by protein kinases (Bertorello and Katz, 1993); and finally, the possible regulation by putative endogenous ligands such as ouabain (Hamlyn and Manunta, 1992; Doris, 1994). This chapter mainly focuses on the regulation of Na^+-K^+-ATPase expression. Emphasis will be put on the analysis of the molecular mechanisms by which the number of functional Na^+-K^+-pump molecules can be modulated at the cell surface. The variable physiological needs of Na^+-K^+-ATPase regulation in various tissues are discussed in more detail by Clausen and Everts (1989), Schneider (1992), Zeidel (1993) and O'Donnell and Owen (1994).

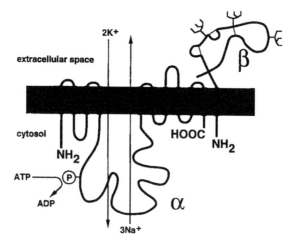

Figure 2. Model of the Na^+-K^+-ATPase structure and topology. The α subunit is proposed to span the membrane 10 times and exposes the NH_2- and the COOH-termini to the cytoplasmic face. The β subunit has one transmembrane domain, a short cytoplasmic NH_2-terminal tail and a large COOH-terminal ectodomain containing 3 disulphide bridges and several sugar chains.

II. REGULATION OF Na^+-K^+-ATPase EXPRESSION

A. Regulation by Selective Subunit Expression

One of the most fundamental processes which controls the formation of functional Na^+-K^+-pumps and thus the level of the cell surface expression, is the process of subunit oligomerization.

α-β Interaction: Consequences on the Maturation of Na^+-K^+-ATPase

There is increasing evidence that the oligomeric structure of the Na^+-K^+-ATPase and in particular the β component is of fundamental importance in the regulation of expression of functional enzymes (see McDonough et al., 1990; Geering, 1991). Earlier studies had shown that the minimal functional enzyme unit is an α-β complex (Brotherus et al., 1983). It was thought that the β subunit was not directly implicated in the catalytic cycle and that it might play some role in the integration and the correct orientation of the α subunit in the membrane. Biosynthesis studies *in vitro* revealed that, as in other oligomeric membrane proteins, the subunits of Na^+-K^+-ATPase are synthesized from distinct mRNAs and inserted into ER membranes during their synthesis independent of each other (see Geering, 1991). During the process of translation, β subunits acquire 3-7 core-sugars depending on the number of consensus asparagine residues found in the extracytoplasmic domain of various β subunits (see Geering, 1990). Other cotranslational modifications of the β subunit include the formation of the 3 disulphide bridges which are essential for the acquisition of a correct tertiary structure of the polypeptide (Kirley, 1989; Miller and Farley, 1990; Noguchi et al. 1994).

The studies of Tamkun and Fambrough (1986) on the biosynthesis of native Na^+-K^+-ATPase in cultured cells suggested that α and β subunits assemble during or soon after synthesis. It was therefore tempting to speculate that this assembly process might be responsible for certain conformational changes of the α subunit which temporarily closely follow this event. First, it was observed that the newly synthesized α subunit in cultured cells is highly trypsin sensitive in a manner similar to the α subunit synthesized in an *in vitro* translation system. The α subunit acquires trypsin resistance and shows an ability to adopt cation dependent conformations, which are characteristic of the mature enzyme, after a short lag following synthesis but still residing at the level of the ER (Geering et al., 1987). In this connection, Caplan et al. (1990) reported that newly synthesized Na^+-K^+-ATPase shows Mg^{2+}- and P_i-dependent ouabain-binding while Na^+, Mg^{2+}, and ATP-dependent binding is only attained 10 minutes after completion of Na^+-K^+-pump synthesis. These data suggested that the α subunit undergoes posttranslational modification before it exhibits its full catalytic activity. Other experimental evidence also favored structural maturation of the α subunit. For example, in tunicamycin-treated cells, in which coreglycosylation of newly synthesized proteins is inhibited, not only the β subunit but also the nonglycosylated α subunit become more trypsin sensitive than in nontreated cells and the amount of newly synthesized α and β subunits significantly decreases (Zamofing et al., 1988, 1989). On the basis of such results, we put forward the hypothesis that an integral glycoprotein, probably the β subunit, might be necessary for the efficient cellular accumulation of the α subunit.

Evidence that the α subunit does indeed require the β subunit in order to become functionally active was obtained when it became possible to express individual or combined subunits from mRNAs, derived from cloned α and β cDNAs in *Xenopus* oocytes or yeast. Noguchi et al. (1987) were the first to show that α subunits can only produce functional Na^+-K^+-pumps in *Xenopus* oocytes when expressed together with β subunits. In the meantime, many efforts were made to elucidate the functional role of the β subunit in the posttranslational processing of the α subunit. The hypothesis that assembly of the β subunit might be responsible for a structural rearrangement of the newly synthesized α subunit as a first step toward the functional maturation of the Na^+-K^+-ATPase could be confirmed. The α subunit synthesized alone proves to be highly trypsin-sensitive but, when expressed together with the β subunit, it becomes trypsin-resistant, and able to undergo cation-dependent conformational changes (Geering et al., 1989; Jaunin et al. 1992) as well as exhibit Na^+-K^+-ATPase activity at the level of the ER (Geening et al., 1996). Furthermore, the unassembled α subunit is retained in the ER (Takeyasu et al., 1988; Jaunin et al.,

1992) where it is rapidly degraded (Ackermann and Geering, 1990). Only α subunits which assemble with β subunits escape degradation and are transported to the cell surface where they become functional Na^+-K^+-pumps (Jaunin et al., 1992) (Figure 3). The α subunit thus follows the fate of many other oligomeric proteins which is that misfolded or unassembled subunits are subjected by the ER to quality control (see Rose and Doms, 1988; Hurtley and Helenius, 1989).

The mechanisms by which the β subunit stabilizes the α subunit are not yet known. To explain the dramatic change in trypsin sensitivity, it is necessary to postulate that important structural rearrangements occur in the α subunit upon assembly. This could possibly include the posttranslational insertion of membrane segments. On the basis of evidence that about 40% of the total α subunit population of α cRNA injected oocytes can be recovered in a high-speed supernatant (presumably in a soluble form), Noguchi et al. (1990) postulated that the β subunit might be needed as a receptor to facilitate the insertion of the α subunit into ER membranes by forming stable α-β complexes with the nascent α subunit. However, other data clearly show that the α subunit can insert into the membrane, at least in an elementary form, without the β subunit. First, integral association with ER membranes, *in vitro*, definitely occurs cotranslationally and independent of the β subunit (Geering et al., 1985b; Homareda et al., 1989; Cayanis et al., 1990). Second, we

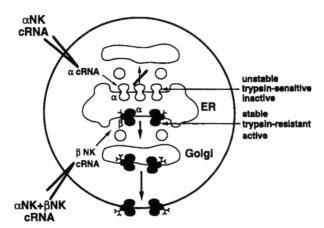

Figure 3. Subunit assembly and maturation of Na^+-K^+-ATPase. The structural and functional maturation of Na^+-K^+-ATPase can be studied in *Xenopus* oocytes after injection of α cRNA alone or together with β cRNA. The fate of the newly synthesized α and β proteins in pulse-chase labeled oocytes is followed by immunoprecipitation with specific antibodies. Unassembled exogenous α subunits synthesized in the absence of β subunits are retained at the level of the ER. They are highly trypsin-sensitive, devoid of Na^+-K^+-ATPase activity and are rapidly degraded. Only α subunits which are coexpressed and assemble with β subunits escape cellular degradation, become trypsin-resistant and functionally active and are transported to the cell surface (for more details see text).

have identified an endogenous α subunit pool in *Xenopus* oocytes which is not associated with β subunits (Geering et al., 1989) and which is retained in the ER in a stable though immature, trypsin-sensitive form (Geering et al., 1989; Jaunin et al., 1992). Even when the synthesis of the oocyte's α subunit is stopped with antisense oligonucleotides, β subunits, expressed by cRNA injection, assemble with the presynthesized immature α subunits, thus rendering them trypsin-resistant permitting expression of more ouabain-binding sites at the cell surface (Ackermann and Geering, 1992). These data clearly show that the α subunit is able to insert in an assembly competent form into ER membranes independent of the β subunit and, at least under certain physiological conditions, maintain its ability to assemble into functional enzyme complexes after synthesis.

Very little is known about the membrane topology of the immature, unassembled α subunits and the conformational changes that follow β assembly. It is of some interest that Homareda et al. (1993) reported that a truncated α subunit (including the transmembrane segments M 1 to 4) interacts in an alkali-resistant form with ER membranes while another truncated α subunit comprising M5 and subsequent, putative transmembrane segments, can be easily extracted. These results suggest that M1 and M3 but not M5, contain signal transfer sequences. One possible interpretation of these results could be that the immature form of the α subunit is anchored in the membrane by its first four transmembrane segments, and assembly with the β subunits facilitates the correct insertion of the C-terminus. A domain which is involved in the interaction with the β subunit has been identified in the C-terminal part of the α subunit, namely a 26 amino acid sequence which is similar across species and isoforms and which is most likely exposed on the extracellular face of the membrane between the putative membrane segments 7 and 8 (see Fambrough et al., 1994). It is not without significance that a similar suggestion was made concerning the location of the interaction site in the α subunit of H^+-K^+-ATPase (Shin and Sachs, 1994).

The corresponding interaction site in the β subunit has not been definitively identified. Though some studies suggest that the ectodomain of the β subunit is necessary and sufficient for assembly with the α subunit (Renaud et al., 1991), other studies demonstrate that the cytoplasmic and/or transmembrane regions might also be involved in a stable interaction with the α subunit (Jaunin et al., 1993; Eakle et al., 1994). With respect to the ectodomain, both a hydrophobic domain in the most C-terminal part of the β subunit (Beggah et al., 1993) and a region of 96 amino acids immediately adjacent to the membrane (Hamrick et al., 1993) have been suggested as possible interaction sites. Most likely there are several regions in the β subunit which participate in the formation of stable α-β complexes.

After this brief account of the importance of subunit assembly for the maturation, intracellular transport and cell surface expression of the Na^+-K^+-ATPase, the examples which will now be discussed illustrate how the equimolar synthesis of α and β subunits is used by the cell to regulate the functional Na^+-K^+-ATPase pool.

Early Developmental Regulation of Na^+-K^+-ATPase

As already explained, selective modulation of the transcription and/or the translation of one of the subunits of the heterodimeric Na^+-K^+-ATPase could influence the expression of functional Na^+-K^+-pumps. The early developmental regulation in *Xenopus* oocytes and embryos provides a particular striking example of how Na^+-K^+-pumps are up- or down-regulated, in this case through translational control of the β subunit. Besides having functional Na^+-K^+-pumps in the plasma membrane, *Xenopus* oocytes possess an α subunit pool in the ER which is inactive for lack of association with β subunits (Geering et al., 1989; Jaunin et al., 1992). Indeed, *Xenopus* oocytes do not synthesize stoichiometric amounts of α and β subunits. Compared to α subunits, only little $\beta3$ and nearly undetectable amounts of $\beta1$ subunits are translated. The disparate rates of α and β subunit synthesis are, however, not due to different mRNA levels since similar amounts of α and $\beta3$ and small amounts of $\beta1$ poly(A)$^+$ mRNAs are found (Burgener-Kairuz et al., 1994) (Figure 4). It is likely that in *Xenopus* oocytes, a functional, stable Na^+-K^+-ATPase pool does accumulate during the growth phase, whereas in fully grown oocytes, the expression of new Na^+-K^+-pumps is stopped, mainly because $\beta1$ and $\beta3$ mRNAs become unavailable for translation. The phenomenon of mRNA "masking," though not well understood, has been described for other mRNAs in oocytes of various species. It possibly involves binding of mRNA to certain maternal proteins (see Sonenberg, 1994).

It is difficult to explain the biological significance of the unusual mode of Na^+-K^+-ATPase synthesis in *Xenopus* oocytes. However, there is evidence that the inactive α subunit pool might be recruited by functional Na^+-K^+-pumps following fertilization to meet the increased needs of Na^+ transport during early development. Indeed, in amphibian (see Morrill et al. 1975) as well as in mammalian embryos (see Biggers et al., 1988; Watson et al., 1992), vectorial Na^+ transport is first established which drives osmotic water flow and is involved in the formation of the blastocoele, a fluid-filled cavity with an ionic composition similar to that of the extracellular milieu in the adult. The Na^+K^+-ATPase, exclusively located in the basolateral membranes lining the cavity, is a principal mediator of this first transepithelial transport in the developing organism (see Wiley et al., 1990). In *Xenopus* embryos a 2- 3-fold increase in Na^+-K^+-ATPase activity is observed during the period between egg fertilization and morula stage 6 but this increase in Na^+-K^+-pump activity is not accompanied by a rise in the biosynthesis of α subunits (Han et al., 1991). Similarly, the total α, $\beta1$, and $\beta3$ mRNA levels remain relatively constant until midblastula stage 9 (Burgener-Kairuz et al., 1994). However, the poly(A)$^+$ fraction of the $\beta1$, but not of the α or the $\beta3$ mRNA, progressively increases up to the morula stage and the rise in the polyadenylated $\beta1$ transcripts is accompanied by a significant increase in the synthesis of $\beta1$ subunits (Burgener-Kairuz et al., 1994) (Figure 4).

Translation represents the main mechanism by which protein expression is regulated during early development since maternal mRNAs are responsible for all protein synthesis before the onset of transcription. In *Xenopus* oocytes, unlike

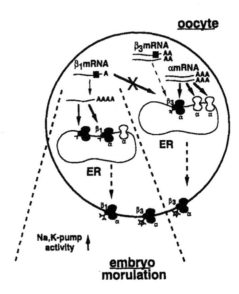

Figure 4. Post-transcriptional regulation of Na^+-K^+-pump expression during early *Xenopus* development. *Xenopus* oocytes express similar amounts of α and $\beta 3$ mRNA and lower amounts of $\beta 1$ mRNA. Translation of $\beta 1$ and $\beta 3$ mRNA is impeded probably through "masking" of the mRNAs. Insignificant amounts of $\beta 1$ and low amounts of $\beta 3$ proteins are synthesized which leads to the formation of functional α-$\beta 3$ complexes at the cell surface and of an unassembled, stable α subunit pool in the ER. After fertilization, during morulation, $\beta 1$ mRNAs are unmasked and become polyadenylated which favors their translatability. Possibly, the newly synthesized $\beta 1$ subunits assemble with the pre-existing unassembled α subunits and contribute to the formation of new functional α-$\beta 1$ complexes at the cell surface (for more details see text).

mammalian embryos, this occurs at the midblastula stage (Newport and Kirschner, 1982). It is now recognized that the translational activity of maternal mRNAs is in general determined by changes in polyadenylation (see Sachs and Wahle, 1993; Wormington, 1993). The developmentally regulated polyadenylation occurs in the cytoplasm and requires the uridine-rich cytoplasmic polyadenylation element (CPE), as well as the nuclear polyadenylation signal AAUAAA in the 3' untranslated region of the mRNA. By using an RNase protection assay, it could be shown that the length of the poly(A)$^+$ tail in the $\beta 1$ mRNA increases significantly in between the period of the fertilized egg and the morula stage (Burgener-Kairuz et al., 1994). In addition, an analysis of the sequence of the 3' untranslated region of the $\beta 1$ mRNA reveals the existence of both elements necessary for developmental polyadenylation. The increase in the Na^+-K^+-ATPase activity observed during early *Xenopus* development, despite the existence of a constant α subunit pool, is likely to be due to:.(1) an increased translation of the $\beta 1$ mRNA mediated by polyadenylation; and (2) the formation of new functional α-$\beta 1$ complexes through association of the newly synthesized $\beta 1$ subunits with the pre-existing immature α subunits (Figure 4). It might be speculated that the change in the isoform composition of Na^+-K^+-ATPase molecules from α-$\beta 3$ to the kidney type α-$\beta 1$ complexes is important for the generation of vectorial Na^+ transport during blastocoele formation.

It is worthwhile to compare the mechanisms found in *Xenopus* embryos with those governing the developmental regulation of the Na^+-K^+-ATPase in mammalian embryos. The latter have similar requirements for an increased Na^+-K^+-pump activity during blastocoele formation. In contrast to *Xenopus* embryos, mouse embryos show an $\alpha 1$ mRNA content which increases significantly from the two cell stage up to completion of the blastocoele (Gardiner et al., 1990; Watson et al., 1990), thus reflecting an early activation of the transcriptional activity. However, the cellular α subunit pool either remains constant when monitored with Western blot analysis (Gardiner et al., 1990) or is first detected at the late morula stage by immunocytochemistry (MacPhee et al., 1994) concomitant with a 2-fold increase in Na^+-K^+-pump activity (Van Winkle and Campione, 1991). This delayed appearance of new functional Na^+-K^+-pumps cannot be explained by impeded mRNA translation since the $\alpha 1$ transcripts appear to be translated throughout the pre-implantation period (MacPhee et al., 1994). In view of the importance of the β subunit in the functional expression of the newly synthesized subunit, MacPhee et al. (1994) have suggested that prior to the late morula stage, α subunits might be unstable and thus undetectable due to an insufficient supply of newly synthesized β subunits, and that the expression of the newly detected Na^+-K^+-ATPases at the onset of cavitation could be triggered by the up-regulation of the β subunit gene. In favor of this hypothesis is the observation that $\beta 1$ transcripts occur in low levels during late morula and are abundant in the blastocyst (Watson et al., 1990), and that the β subunit proteins increase about 2-fold in embryos undergoing morula-blastocyst transition (Gardiner et al., 1990). At present, it is an open question whether the rise in β subunit synthesis is due to new transcription of the $\beta 1$ or another β isoform gene or, as in *Xenopus* embryos, is due to polyadenylation of maternal $\beta 1$ mRNA. In any case, it is interesting to note that in amphibians and mammals, despite some phenomenological differences in mRNA expression, the increase in functional Na^+-K^+-pumps during this developmental period is governed by certain common factors, namely temporal separation of α subunit expression from its transcriptional activation and the limited presence of the β subunit.

There are several parallels to these early regulatory processes of Na^+-K^+-ATPase expression that have been documented during later development and in adult tissues. In the following pages, some of the best examples will be discussed to illustrate the similarities and the differences of the regulatory mechanisms.

Regulation of Na⁺-K⁺-ATPase Gene Expression by Monovalent Cations and Hormones

Intracellular Na^+, corticosteroids, and thyroid hormones are the best-studied modulators of Na^+-K^+-ATPase expression during development and in adult tissues.

A characteristic feature of corticosteroids and thyroid hormones is that they exert their effects mostly through interaction with specific cytosolic or nuclear receptors and subsequent control of gene transcription (see Tsai and O'Malley, 1994) (Figure 1B). In the case of the Na^+-K^+-ATPase, the onset, the relative levels, and the timing in the regulation of the α and β mRNA levels vary considerably in different tissues; they also vary in response to different stimuli. In some cells, these modulators seem to regulate Na^+-K^+-ATPase expression by a linear process including a coordinate increase in α and β subunit gene transcription and translation, and a corresponding rise in functional Na^+-K^+-pumps. In many other situations, the cellular response to the different modulators of the induction of α and β mRNAs or proteins is very distinct and highlights the importance of differential subunit expression as a means of regulating Na^+-K^+-pump activity. No study is yet available in which all the parameters such as α and β gene transcription, α and β mRNA accumulation, stability and translatability, α and β protein expression, and Na^+-K^+-ATPase activity have been determined in parallel for one particular modulator. We shall now focus on some complete studies which provide evidence that up-regulation of the synthesis of either one of the two subunits could be responsible for an increase in Na^+-K^+-ATPase activity. They also provide insight into the problem of variability of the molecular mechanisms which might be involved in Na^+-K^+-ATPase regulation from transcription to translation and expression of functional Na^+-K^+-pumps in different tissues.

Intracellular Na^+ Concentrations. Intracellular Na^+ plays a key role in regulation of the Na^+-K^+-pump activity. Na^+-K^+-pumps have a reserve capacity which is due to submaximal stimulation by the normally low intracellular Na^+ concentration (Jørgensen. 1986). Any alterations in the Na^+ influx immediately changes the activity of the pump. Besides this effect of Na^+, it has been demonstrated that in various tissues, ionic or pharmacological interventions which change the intracellular Na^+ concentration lead to up-regulation of the Na^+-K^+-ATPase which is governed by pretranslational control (Pressley, 1988). In dog kidney cells, incubated in low K^+ medium, the regulatory adjustment of cellular Na^+-K^+-ATPase content involves an increase in α and β subunit synthesis which is mediated by a rapid and coordinate increase in the concentration of α and β mRNAs (Bowen and McDonough, 1987). In other cell types, however, the increase in cellular Na^+-K^+-ATPase observed after an increase in the intracellular Na^+ concentration, is often preceded by a discoordinate, transient, up-regulation of either α or β mRNA. In cultured chicken skeletal muscle (Taormino and Fambrough, 1990) as well as in a pig kidney cell line (Lescale-Matys et al., 1990), there is a more pronounced increase in $\beta 1$ than in $\alpha 1$ mRNA. However, the opposite is true in a liver cell line (Pressley, 1988). In skeletal muscle there is an increase in $\alpha 2$ (Hsu and Guidotti, 1991) and in cultured neonatal rat cardiocytes there is a significantly larger accumulation of $\alpha 1, 2$, and 3 mRNAs than of $\beta 1$ mRNAs (Yamamoto et al., 1993). The up-regulation of β mRNA is associated with a selective increase in the accumulation of β over α proteins (Lescale-Mathys et al., 1990; Taormino and Fambrough, 1990) which probably leads to a more efficient assembly of α-β complexes and is responsible for the increase in Na^+-K^+-ATPase expression observed in these cells after altering the demand for Na^+ transport. Though a similar analysis of α and β subunit expression has not been performed in cells which show a preferential increase in α mRNA, it is conceivable (in the light of the inter-dependence of the α and β subunits in the formation of functional Na^+-K^+-ATPases) that in those cells, the up-regulation of α mRNA, which in some cells is known to lead to a concomitant accumulation of α proteins (Yamamoto et al., 1993), is the factor determining the induction of Na^+-K^+-pump sites following an increase in intracellular Na^+ concentration.

Nuclear runoff (Taormino and Fambrough, 1990) and transfection experiments with chimeric plasmids containing the 5' flanking region of α1, 2, or 3 (Yamamoto et al., 1993, 1994) suggest that in cultured chicken muscle cells and in cardiocytes or vascular smooth muscle cells respectively, the increase in particular mRNA species is accomplished by an increase in the transcription rate of the corresponding gene. Both approaches suggest that Na^+ responsive elements might be located within the 5' flanking sequences of the α and the β genes. Although the mechanism by which Na^+ levels can modulate the transcription of the Na^+-K^+-ATPase genes is not known, changes in Na^+ levels could alter the affinity of regulatory proteins for the binding sites on the genes. It is also not known whether the nature and/or the location of these regulatory elements are different in the α and the β genes and/or whether complimentary regulatory factors act in different tissues to explain the tissue-specific, differential up-regulation of α and β gene transcription following changes in intracellular Na^+.

The complexity of the phenomenon is further underlined by a study which examined whether the early influx of Na^+, observed in serum-treated cells, is involved in the induction of α and β mRNAs. While serum stimulates both α and β mRNA synthesis 2- to 3-fold, only the increase in β mRNA level is dependent on the presence of extracellular Na^+ (Kirtane et al., 1994). However, Na^+-dependent induction of the β mRNA is not transcriptionally regulated. Though a decrease in the degradation rate of the cytoplasmic β mRNA pool was excluded as a possible cause of the accumulation of β mRNA, the conclusion drawn is that differential accumulation of mRNAs in response to changes in intracellular Na^+ is also accomplished by an ill-defined mechanism(s) at the post-transcriptional level. Lastly, an increase in intracellular Na^+ might not necessarily act as the trigger for Na^+-K^+-ATPase gene induction in cells with increased Na^+ influx. An interesting example is the coordinate up-regulation of Na^+-K^+-ATPase and the A system which transports certain amino acids along with Na^+ into cells and which depends, as other Na^+-dependent transporters, on the driving force provided by the Na^+ electrochemical gradient across the cell membrane (see McGivan and Pastor-Anglada, 1994). In particular, it was shown that mutant Chinese hamster ovary cells (CHO cells) which fail to respond to an adaptive control mechanism and have a high content of the A system, undergo a significant increase in ouabain-binding sites and α1 mRNA levels. The A system activity and the amount of α1 mRNA is also increased after amino acid starvation of CHO cells despite a decrease in intracellular Na^+ concentration. Finally, amino acids which act as corepressors of the A system prevent induction of the A system and accumulation of Na^+K^+-ATPase $\alpha1$ mRNA to a similar extent. On the basis of these results, it was postulated that the A system and the Na^+-K^+-ATPase are co-regulated and under the coordinate control of a same regulatory R1 gene (Qian et al., 1991).

Corticosteroids. Corticosteroids, such as glucocorticoids and in particular the mineralocorticoid aldosterone are important regulators of Na^+-K^+-ATPase gene expression in polarized epithelia such as renal tubular and colonic cells which are involved in Na^+-reabsorptive processes (see Katz, 1990; Bastl and Hayslett, 1992; Johnson, 1992) (Figure 1B). In addition, aldosterone and glucocorticoids interact with specific receptors showing significant homology in structure and function (Arriza et al., 1987). Aldosterone binds with a higher affinity to the mineralocorticoid receptor (MR) (Kd about lnM) than glucocorticoids bind to the glucocorticoid receptor (GR) (Kd about 25nM). However, aldosterone can also bind to GR (Kd 25-65nM); cells contain more GR than MR. And, MR, when studied *in vitro*, has an affinity for glucocorticoids which is the same as that for aldosterone. These factors complicate the interpretation of any studies aimed at the identification of hormone-specific molecular mechanisms involved in Na^+-K^+-ATPase expression (for a comprehensive review of the transcriptional control of Na^+ transport by adrenal steroids see Verrey, 1995).

Earlier studies using toad bladder as a model have shown that aldosterone induces a coordinate 2- to 3-fold induction of α and β subunit synthesis as part of the late Na^+ transport response (Geering et al., 1982) which is more closely related to the occupancy of MR than of GR (Geering et al., 1985 a). This is not the case with *Xenopus* kidney A6 cells, since the increase in Na^+ reabsorption observed after aldosterone treatment is mainly mediated by glucocorticoid receptors (Schmidt et al., 1993).

There are indications that at least some of the hormonal specificity *in vivo* is provided by the aldosterone target tissue rather than by the receptors themselves (see Funder, 1990; Stewart and Edwards, 1990). Enzymatic activities such as those of 11-β-hydroxysteroid dehydrogenases, which convert glucocorticoids but not aldosterone to inactive metabolites, have been identified in MR containing tissues (Gaeggeler et al., 1989; Bonvalet et al., 1990). The presence of these degradative enzymes might protect the tissues against an action of glucocorticoids. The MR-mediated aldosterone effects *in vivo* are tissue-specific. Recent experiments with heart, brain, and vascular smooth muscle cells, under conditions favoring the binding of the hormone to MR, reveal a distinct and differential regulation of α and β mRNAs. A similar 3-fold induction in $\alpha1$ and $\beta1$ mRNA accumulation is observed in cultured, neonatal, and adult cardiocytes (Ikeda et al., 1991). Preferential increase in the $\beta1$ over $\alpha1$ mRNA levels has been reported in vascular smooth muscle cells in culture (Ogachi et al., 1993). In addition, there is selective down-regulation of the $\alpha1$ but not of the $\beta1$ mRNA in aldosterone-sensitive cells of the distal nephron in the absence of corticosteroids (Farman et al., 1992). The importance of cell-specific factors that regulate MR-mediated transcriptional and/or posttranscriptional activities is further highlighted by the observation that aldosterone preferentially affects $\alpha 3$ isoform mRNA expression in distinct neuronal populations (Farman et al., 1994).

It is in very few studies that changes in α and β mRNAs have been compared to the changes seen in the respective protein levels. While data on the expression of β proteins are unavailable, there are data indicating that the 3-fold increase in cardiocytes (Ikeda et al., 1991) and the 2.3-fold increase in vascular smooth muscle cells (Oguchi et al., 1993) of the $\alpha 1$ mRNA is associated with a 36% and a 2-fold increase, respectively, in the α protein. Aldosterone also increases the chemical pool of $\alpha 1$ proteins in the renal cortical collecting duct (Welling et al., 1993). The most extensively studied model for aldosterone action on Na^+-K^+-ATPase expression is the *Xenopus* kidney derived A6 cell line. The late phase of aldosterone action which is characterized by a marked increase in vectorial Na^+ transport (Johnson et al., 1986; Verrey et al., 1987; Paccolat et al., 1987; Leal and Crabbé, 1989; Schmidt et al., 1993), is thought to be at least in part mediated by changes in the number of Na^+-K^+-pumps in the basolateral membrane. Aldosterone treatment (300nM) of A6 cells leads to a more pronounced effect on β mRNA than on α mRNA accumulation which is followed by a nearly coordinate increase in the biosynthesis rate of the two proteins (Verrey et al., 1987). There is also a significant increment in the total cellular pool of α and β subunits (Beron and Verrey, 1994) or a corresponding increase in ouabain-binding and Na^+-K^+-ATPase activity (Johnson et al., 1986; Leal and Crabbé, 1989).

We know very little about the molecular mechanism underlying the regulation of Na^+-K^+-ATPase mRNA levels by corticosteroids and the causes for cell-type specific responses. Several lines of evidence speak for a transcriptional control of the mRNAs. In cardiocytes, for example, the 3-fold aldosterone-dependent and MR-mediated increase in $\alpha 1$ mRNA is not due to an increased stability of the mRNA (Ikeda et al., 1991) indicating that $\alpha 1$ mRNA synthesis is induced. This hypothesis is also supported by the fact that the $\alpha 1$ mRNA has a short half-life of 3 h which favors rapid up- or down-regulation by appropriate stimuli. Nuclear run-on experiments with A6 cells have provided evidence that the increases in $\alpha 1$ and $\beta 1$ mRNAs obtained with 300nM aldosterone are preceded by stimulation of the rate of their transcription (Verrey et al., 1989). Similarly, in vascular smooth muscle cells, transfected with a chimeric construct containing the 5' flanking region of the $\alpha 1$ subunit and the luciferase reporter gene, 100nM aldosterone was shown to stimulate $\alpha 1$ gene transcription (Oguchi et al., 1993).

Corticoid-receptor complexes are generally thought to exert their effects by binding to hormone-responsive elements (HRE) upstream to the gene promoter. Glucocorticoid responsive elements (GRE) which interact both with GR and MR have been identified (see Tsai and O'Malley, 1994). $\alpha 1$ (Shull et al., 1990; Yagawa et al., 1990), $\alpha 2$ (Shull et al., 1989; Kawakami et al., 1990b), $\alpha 3$ (Pathak et al.,

1990), $\beta 1$ (Liu and Gick, 1992), and $\beta 2$ (Kawakami et al., 1990a) genes all contain regions in the 5' flanking sequence which resemble quite closely the glucocorticoid responsive elements. The physiological significance of these potential regulatory elements is, however, largely unknown. Wang et al. (1994) produced evidence that the response of Na,K,-ATPase genes in infant kidney to glucocorticoids might be mediated by a specific interaction of glucocorticoid receptors with certain GRE in the promoter region of the α subunit gene. As for the tissue-specific effects of corticosteroids and the interplay between MR and GR with the various other transcription factors for which responsive elements have been identified on α (Kawakami et al., 1990b; Pathak et al., 1990; Shull et al., 1990; Yagawa et al., 1990; Suzuki-Yagawa et al., 1992) and β (Kawakami et al., 1992; Liu and Gick, 1992; Kawakami et al., 1993a) genes, these are problems awaiting elucidation.

For a number of different proteins which are regulated by corticosteroids, it has been shown that protein inhibitors such as cycloheximide or puromycin completely block the induction of mRNA. It has been suggested that in these cases, an intermediate, regulatory protein, induced by corticosteroids is needed for the hormonal activation of the protein in question by acting at the transcriptional and/or post-transcriptional level. For the aldosterone-induced expression of the Na^+-K^+-ATPase mRNAs, the importance of the induction of early, regulatory proteins remains a controversial issue. In A6 kidney cells, treatment with 20ug/ml of cycloheximide, which blocks protein synthesis by more than 97%, clearly abolishes the aldosterone-induced transcriptional activity of the $\beta 1$ subunit gene (Verrey et al., 1989). The interpretation of the results obtained for the α subunit are more conflicting. In the presence of cycloheximide, aldosterone fails to change the basal rate of the $\alpha 1$ transcription in A6 cells (Verrey et al., 1989) or the $\alpha 1$ mRNA accumulation in vascular smooth muscle cells (Oguchi et al., 1993). However, cycloheximide alone considerably induces $\alpha 1$ gene activity or mRNA accumulation, respectively, in the two cell types which may mask the aldosterone-mediated induction. In cultured cardiocytes where cycloheximide has no effect *per se,* the aldosterone-dependent $\alpha 1$ mRNA accumulation is indeed maintained even after protein synthesis inhibition (Ikeda et al., 1991) indicating that in this case the hormone has a direct effect on gene transcription. More experimental data are needed to decide whether the contradictory results reflect variability in the mechanisms of $\alpha 1$ mRNA expression in different tissues or whether the requirements for intermediate regulatory proteins is different for α and β transcription.

Another controversial issue is whether the transient increase in intracellular Na^+ content observed in several cells treated with aldosterone plays a permissive role in Na^+-K^+-ATPase gene expression. In the toad bladder model, amiloride which inhibits apical Na^+ channels and prevents the early aldosterone-mediated Na^+ influx, has no effect on the increased biosynthesis rate of the $\alpha 1$ and the $\beta 1$ subunit upon aldosterone treatment (Geering et al., 1982). These data speak for a direct hormonal action in kidney-type epithelial cells. On the other hand, it has been reported that in cardiocytes, amiloride inhibits both an aldosterone-induced increase in the intracellular Na^+ concentration and $\alpha 1$ mRNA accumulation (Ikeda et al., 1991). Keeping in mind the fact that in the absence of amiloride, the $\alpha 1$ mRNA levels remain elevated despite a decrease in intracellular Na^+, Ikeda et al. (1991) suggest that a prolonged $\alpha 1$ mRNA induction is probably the result of a direct hormonal activation of transcription but that an aldosterone-mediated increase in Na^+ influx via an amiloride-sensitive Na^+/H^+ exchanger may provide an early signal for the mediation of the hormonal action. It remains to be determined whether the existence of interacting, tissue-specific steroid hormone responsive and Na^+-sensitive elements in the $\alpha 1$ promoter are involved in the integration of these signals.

Thyroid Hormone. It is well documented that triiodothyronine (T3) stimulates active Na^+-K^+-transport in thyroid hormone-responsive tissues such as cardiac and skeletal muscle, liver, renal cortex, and adipose tissue which is part of the thermogenic action of thyroid hormone (see Clausen and Everts, 1989; Ismail-Beigi, 1993).

The T3-induced increase in Na^+-K^+-ATPase activity results from a receptor-mediated increase in the number of Na^+-K^+-pumps and not from a change in the affinity of the enzyme for substrates or in its catalytic properties. However, T3 has been implicated in the cellular switch in isoform expression during development and in the regulation of tissue-specific isoform expression in adults which leads to the cell surface expression of α-β complexes with distinct transport characteristics.

The prevalence of pre-translational regulation of Na^+-K^+-ATPase expression by T3 is well documented. In rat kidney (McDonough et al., 1988), rat liver (Gick and Ismail-Beigi, 1990), rat ventricular myocardium (Chaudhury et al., 1987), and rat mesangial cells (Ohara et al., 1993), the increase in Na^+-K^+-ATPase activity is preceded by an increase in the abundance of $\alpha 1$ and $\beta 1$ mRNAs. From recent studies, it seems evident that T3 induces preferentially certain α or β isoforms in a tissue-specific manner. In the kidney, for instance, T3 coordinately increases $\beta 1$ and $\alpha 1$ mRNAs which predicts the observed changes at the protein level (Azuma et al., 1993). On the other hand, in skeletal muscle, T3 has no effect on $\beta 1$ and $\alpha 1$ mRNA or protein levels but instead increases coordinately $\beta 2$ and $\alpha 2$ mRNAs and proteins (Azuma et al., 1993). Similarly, in neonatal rat brain in organ culture, $\alpha 2$-$\beta 2$ complexes appear to be the T3-responsive enzymes (Corthésy-Theulaz et al., 1991). The regulation of Na^+-K^+-ATPase in heart muscle provides a further example for the complex interplay between pre-translational regulatory mechanisms and post-translational regulatory mechanisms, such as subunit assembly for the formation of functional Na^+-K^+-pumps. In adult cardiac myocytes, $\alpha 1$, $\alpha 2$, and $\beta 1$ isoforms are expressed and it can be assumed that the $\beta 1$ assembles with either the $\alpha 1$ or the $\alpha 2$ protein to form functional heterodimers. Euthyroid levels of T3 are necessary for the basic expression of all 3 subunit proteins (Horowitz et al., 1990). T3 induces large changes in $\alpha 2$ and β mRNA levels and only a small increase in $\alpha 1$ mRNA (Gick et al., 1990; Horowitz et al., 1990). In cardiac myocytes from hypothyroid animals, $\alpha 1$, $\alpha 2$, and $\beta 1$ proteins decrease to about half of the levels found in myocytes from euthyroid animals but there is no corresponding decrease in the mRNA levels (Hensley et al., 1992). In euthyroids, the ratio of $\alpha 1$ to $\alpha 2$ mRNA is 1:1.2 (Hensley et al., 1992), while the $\alpha 1$ protein expression significantly dominates the $\alpha 2$ protein expression. Moreover, in the hyperthyroid state, discoordinate changes in the $\alpha 2$ and the $\beta 1$ mRNA versus protein changes are observed. Whereas the $\alpha 2$ mRNA and the $\beta 1$ mRNA levels increase 5- and 12-fold with respect to the hypothyroid state, respectively, the $\alpha 2$ protein increases by 14-fold and the $\beta 1$ protein only by 2.5-fold (Hensley et al., 1992). To reconcile these results, the authors hypothesize that the $\beta 1$ subunit might be rate-limiting in the euthyroid state and would preferentially form complexes with $\alpha 1$ subunits while the excess unassembled α subunits would be degraded. On the other hand, in the hyperthyroid heart, an excess of β subunits over α subunits is synthesized which may render the α subunits rate-limiting for assembly. More $\alpha 2$ proteins are able to associate with β subunits, become stabilized, and accumulate in the cell above the level of the increase of mRNA. The excess unassembled β subunits synthesized would be expected to be degraded for lack of assembly with α subunits in the light of the observations made in a kidney cell study (Lescale-Matys et al., 1993).

In many tissues, the Na^+-K^+-ATPase α isoform and the β subunit genes exhibit a complex pattern of expression during development (Orlowski and Lingrel, 1988; Herrera et al., 1994). In the heart, $\alpha 1$ and $\beta 1$ mRNAs are constitutively expressed at high levels while $\alpha 2$ and $\alpha 3$ mRNAs are present at lower levels and undergo a developmentally regulated transition in expression. The $\alpha 3$ mRNA is initially present in fetal and neonatal heart and is subsequently replaced by the $\alpha 2$ mRNA in juvenile and adult tissue (Orlowski and Lingrel, 1988). The switch in the $\alpha 2$ and $\alpha 3$ mRNA expression is synchronous with the T3-mediated transition in the abundance of cardiac muscle-specific genes, for example, from , β to α myosin heavy (MHC) chain mRNA expression (see Orlowski and Lingrel, 1990). The possibility whether T3 might be responsible for the change in Na^+-K^+-ATPase isoform expression during development in a manner similar to the MHC isoform switch has been investigated. As in the case of administration of T3 to adult hypothyroid rats (Hensley et al., 1992; Gick et al., 1990), treatment of neonatal rats with T3 leads to an increase in $\alpha 1$ and $\beta 1$ mRNA expression in the myocardium (Melikian and Ismail-Beigi, 1991). T3 treatment of euthyroid neonatal rats also increases $\alpha 3$ mRNA abundance in neonatal heart but does not appear to induce a premature expression of $\alpha 2$ mRNA (Melikian and Ismail-Beigi, 1991). Since T3 regulates only the abundance of mRNAs which are already present in this tissue, these workers conclude that the hormone is not involved in the molecular switch of myocardial Na^+-K^+-ATPase α isoform expression. This is strengthened by evidence coming from studies with cultured neonatal rat cardiocytes that T3 does not elicit a "switching" of isoforms. By contrast, *in vitro* studies involving incubation of cultures with T3 markedly increase $\alpha 2$ mRNA in addition to $\alpha 3$ and β mRNA abundance, without influencing $\alpha 1$ mRNA expression (Orlowski and Lingrel, 1990).

Most likely, the change in isoform expression is mediated by a complex interaction of T3 with other hormonal stimuli. Synergistic or additive interactions between different hormones are important factors in Na^+-K^+-ATPase regulation and considerably increase the complexity of this biological phenomenon. In cultured neonatal cardiocytes, for instance, dexamethasone is able to selectively induce $\alpha 2$ mRNA levels and to repress the T3-mediated accumulation of $\alpha 3$ mRNA (Orlowski and Lingrel, 1990). Since in cultured nonmyocardiocytes which contain $\alpha 1$, $\alpha 3$, and β mRNA, T3 appears to regulate only α mRNA levels (Orlowski and Lingrel, 1990), these data suggest that T3 and glucocorticoids might act in concert to bring about the distinct pattern of gene expression during rat cardiac muscle differentiation. Though the molecular pathways by which dexamethasone exerts its inhibitory effect on the T3-mediated $\alpha 3$ mRNA abundance are not understood, two general mechanisms have been proposed which account for similar inhibitory actions of glucocorticoids on other genes. One is that glucocorticoids could interact with GRE in the 5' flanking region of the gene and provoke a negative regulation by interfering with the binding or activity of other transactivating proteins at other important regulatory sites. The other is that glucocorticoids may not affect gene transcription but stimulate the degradation of mRNA. The fact that dexamethasone (Orlowski and Lingrel, 1990), unlike aldosterone (see Ikeda et al., 1991), has no significant effect on $\alpha 1$ mRNA accumulation in neonatal cardiomyocytes suggests that the effect on T3 action is mediated, directly or indirectly, by GR rather than MR.

Several other examples suggest an interplay between the effects of glucocorticoids and T3 on Na^+-K^+-ATPase expression. T3 has permissive effects in mammals on the aldosterone-induced Na^+-K^+-ATPase activity (Barlet and Doucet, 1987) or antagonistic effects in amphibian on aldosterone-induced transepithelial Na^+ transport (Geering et al., 1981). In rabbits, a low Na^+ diet leads to an elevation in serum aldosterone concentrations and a coordinate MR-mediated increase in $\alpha 1$ and $\beta 1$ mRNA and protein levels as well as in ouabain-binding sites in the distal colon (Wiener et al., 1993). From the third day of the diet, T3 levels spontaneously decrease, followed by a down-regulation of $\beta 1$ mRNA and ouabain-binding. Substitution with T3 restores the functional Na^+-K^+-pump pool by preferentially increasing $\beta 1$ mRNA levels (Wiener et al., 1993). In this case, it is suggested that decreased β subunit synthesis resulting from lowered T3 levels, provides a mechanism for escape from the effect of hyperaldosteronism on the number of Na^+-K^+-pumps.

Both transcriptional and post-transcriptional mechanisms have been proposed for the regulation of Na^+-K^+-ATPase subunit mRNA levels by T3, but little conclusive data are available. Since in rat mesangial cells, the turnover of $\alpha 1$ mRNA (half-life: 3 h) does not change in response to T3 (Ohara et al., 1993), the observed accumulation of $\alpha 1$ mRNA probably results from another RNA-processing event or from an increased transcriptional activity of the $\alpha 1$ gene. Gene expression by T3 in rat fetal brain leads to an increase in the relative abundance of different isoforms and is mediated by post-transcriptional mechanisms (Corthésy-Theulaz et al., 1991). Similarly, in transfected. neonatal rat cardiocytes, T3 has no effect on the activity of chimeric genes containing a portion of the 5' end of the rat $\beta 1$ (Liu et al., 1993) or the $\alpha 2$ gene (Huang et al., 1994) linked to the luciferase reporter gene. In addition, the transcription rate of the endogenous $\beta 1$ gene, assessed by nuclear run-on assays is unaffected by T3 (Liu et al., 1993). On the other hand, T3 has been shown to regulate the transcriptional activity of both the $\alpha 1$ and the $\beta 1$ gene in the kidney though post-transcriptional processes may also contribute to the overall regulation (Gick et al., 1988). And, in the human intestinal Caco-2 cell line, T3 increases $\alpha 1$ and $\beta 1$ mRNA levels by 3- and 14-fold, respectively (Giannella et al., 1993). DNA transfection experiments with 5' flanking regions of the human $\beta 1$ gene linked to the chloramphenicol acetyltransferase (CAT) reporter gene show that transfected cells respond to T3 treatment with a 3- to 5-fold increase in CAT activity (Giannella et al., 1993) suggesting that regulation of the $\beta 1$ mRNA occurs mainly at the transcriptional level.

The latter data imply that Na^+-K^+-ATPase genes may contain T3-responsive regulatory elements (TRE) which, under certain circumstances, can directly interact with nuclear T3-receptor complexes and enhance the transcription of Na^+-K^+-ATPase genes. The best-documented example is the human $\beta 1$ gene. Using 5' deletion and site-directed mutagenesis of regions in the 5' flanking region of the $\beta 1$ gene linked to the luciferase reporter gene, one out of three putative TREs was identified as a functional TRE (Feng et al., 1993). The sequence spans 22 bp from -459 and -438 and shows significant homology to the proposed consensus binding site for the T3 receptor (Brent et al., 1989). An analysis of the rat $\beta 1$ gene 5' flanking region reveals a highly homologous though not identical region to the human TRE (Liu and Gick, 1992). So far, it is not clear why only the human (Feng et al., 1993) but not the rat $\beta 1$ TRE (Liu et al., 1993) is functional. Liu et al. (1993) propose that the discrepancy between the results might reflect tissue-specific differences in T3 regulation of the $\beta 1$ gene or more likely, some fundamental differences in the mechanisms governing the T3 regulation of the human and the rat $\beta 1$ gene. Both the $\alpha 1$ (Shull et al., 1990) and the $\alpha 2$ (Shull et al., 1989) gene contain numerous sequences which exhibit limited similarities to putative TREs but nothing is known about their functionality *in vivo*. In addition, in the $\alpha 2$ gene, a negative TRE element mediates a T3-dependent repression of $\alpha 2$/luciferase expression in cardiac myocytes co-transfected with T3 receptors (Huang et al., 1994). Lastly, the observation that no TRE is found within 5.5 kb of the 5' flanking region of the $\beta 2$ gene (Kawakami et al., 1990a) is interesting in view of the selective increase in the accumulation of $\beta 2$ mRNA in skeletal muscle upon T3 treatment (Azuma et al., 1993).

In conclusion, the discussed examples illustrate the variability of mechanisms which govern the pre-translational regulation of Na^+-K^+-ATPase. It becomes increasingly clear that a tight link exists between the up-regulation of Na^+-K^+-ATPase subunit mRNA by various stimuli and the fundamental process of equimolar subunit assembly which ultimately determines whether more or functionally different Na^+-K^+-pumps are expressed at the cell surface. It might be argued that it is economical for a cell not to overproduce one or the other subunit since in their unassembled state subunits are unstable and/or functionally inactive. Often, this requirement is fulfilled by a coordinate up-regulation of α

and β mRNAs and proteins. The question which now arises is, why do many cell types reveal a preferential accumulation in one of the subunit mRNAs upon regulatory stimulation? In fact, even in basal, unstimulated conditions, α and β mRNAs are often present in unequal amounts. The reasons for differential expression of α and β mRNA levels are poorly understood. Analysis of tissue-specific (Kawakami et al., 1992; Suzuki-Yagawa et al., 1992) or developmental-specific (Kawakami et al., 1993b) regulatory elements on the α and β genes is in progress. In many tissues, α mRNA levels exceed β mRNA levels (Young and Lingrel, 1987; Verrey et al., 1989; Gick et al., 1988; McDonough et al., 1988; Gick and Ismail-Beigi, 1990; Horowitz et al., 1990; Taormino and Fambrough, 1990). In the few examples in which the biosynthesis rates of the two subunits have been studied, coordinate synthesis has, however, been reported (Geering et al., 1982; Tamkun and Fambrough, 1986). One exception is the situation occurring in *Xenopus* oocytes, as discussed above (Geering et al., 1989).

Recent experimental evidence suggests that translational control of the α mRNA might be an important factor governing equimolar subunit accumulation in the presence of dissimilar mRNA levels (Devarajan et al., 1992). This study shows that the α mRNA is less-efficiently translated than the β mRNA in a reticulocyte lysate. In addition, the study predicts that a G/C rich region in the 5'-untranslated region could impose a complex secondary structure on the α mRNA which might be responsible for the inhibition of translation (Devarajan et al., 1992). Translational control of the α mRNA, most likely acting in concert with post-translational regulation mechanisms, might be involved in these situations in which an important increase in subunit mRNA levels upon hormonal stimulation is paralleled by a relatively modest increase in Na^+-K^+-ATPase activity. Additionally, the constraint in α mRNA translation might explain, at least to some extent, why in certain cells an increase in functional α-β complexes is preceded by a more substantial accumulation of α mRNA than of β mRNA levels. It remains, however, to be shown whether this regulatory process is of general importance or only exploited in particular physiological contexts. Indeed, it is unlikely that a limited translatability of the α mRNA would be favorable in the upregulation of Na^+-K^+-ATPase complexes in the many examples in which an increase in Na^+-K^+-ATPase is mediated by a more pronounced accumulation of β than of α mRNAs. In these cases, more experimental data are needed to decide whether overproduction of the β mRNA upon regulatory stimulation is merely a compensatory mechanism to overcome under-representation of the β mRNA in the basal condition. Furthermore, it is important to define more precisely the nature and importance of the early post-translational regulatory mechanisms which lead to stoichiometric α-β complex formation.

B. Short-Term Regulation of Na^+-K^+-ATPase Expression

Besides the long-term adaptation mechanisms described above, several other regulatory mechanisms exist which permit a rapid change in the expression of functional Na^+-K^+-pumps at the cell surface. We will now consider, in particular, examples of those mechanisms which document the importance of endocytosis, exocytosis, and activation of "latent" Na^+-K^+-pumps.

Endo- and Exocytosis of Na,K-Pumps

During the normal process of protein turnover, the Na^+-K^+-ATPase is constantly delivered to the plasma membrane followed by removal and degradation of the enzyme in lysosomes (Figure 5). The transit time of the enzyme from the site of synthesis, that is, the ER, to the plasma membrane varies between 2-4 h in somatic cells (Pollack et al., 1981; Tamkun and Fambrough, 1986) and 1-2 days, for example, in *Xenopus* oocytes. On the other hand, the half-life of the α subunit in functional enzyme complexes ranges between 10-18 h in somatic cells (Karin and Cook, 1986; Lescale-Mathys et al., 1993) and several days in *Xenopus* oocytes (unpublished observation). Little attention has been given to changes in the constitutive transit time to the plasma membrane or the degradation rate of the enzyme as a means to regulate the number of Na^+-K^+-pumps at the cell surface. It has been suggested that K^+-starved HeLa cells respond to the stress with a doubling of the number of transport sites by reducing the removal of the pumps from the cell surface and their degradation in lysosomes (see Pollack et al., 1981). Nothing is known, however, of how the cell translates the stress signal into an altered turnover rate of the Na^+-K^+-pumps.

Ample evidence exists that certain cell types establish a pool of spare Na^+-K^+-pumps in the cell interior which might be recruited to the cell surface in response to physiological needs (Figure 5). For instance, more than 50% of the newly synthesized Na^+-K^+-ATPase in chick sensory neurons remains in an intracellular compartment (Tamkun and Fambrough, 1986). In rat exorbital lacrimal glands, Na^+-K^+-ATPase catalytic activity and the α and β subunit immunoreactivity are associated not only with plasma membrane fractions resolved on sorbitol gradients, but also to a large extent with Golgi and endosomal membrane fractions (Bradley et al., 1993). Similarly, an important intracellular pool of functional Na^+-K^+-pumps exists in *Xenopus* oocytes (Schmalzing et al., 1989), which can be recruited to the plasma membrane by low concentrations of Ca^{2+} (Schmalzing and Kroner, 1990). In these two examples, experimental evidence substantiates the physiological relevance of the intracellular N^+-K^+-pump pool in the cellular response to demands for altered Na^+ pumping activity.

In the rat exorbital lacrimal gland, cholinergic stimulation leads to activation of Na^+/H^+ antiporters which in turn lead to a raised cellular Na^+ influx (Lambert et al., 1991). In parallel, additional Na^+-K^+-pumps are recruited to the plasma membrane (Yiu et al., 1991) and an ongoing recycling traffic of the Na^+-K^+-pumps between the plasma membrane and the intracellular compartment is accelerated (Lambert et al. 1993). This provides the means of rapidly restoring the dissipating Na^+ gradient. So far, nothing is known about the molecular mechanisms which mediate the selective exocytosis of Na^+-K^+-pumps from their intracellular storage sites.

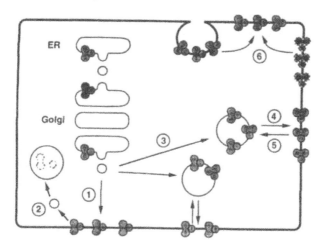

Figure 5. Turnover and short-term regulation of the functional expression of Na^+-K^+-pumps. α and β subunits are synthesized and assembled in the ER and the α-β complexes are constitutively transported via vesicular transport through the Golgi compartment to the plasma membrane (1) α-β complexes are constantly retrieved from the plasma membrane and degraded in lysosomes (2). In certain cell types, α-β complexes enter a regulated, secretory pathway and build up an intracellular pool (3) which will be expressed at the cell surface by exocytosis in response to appropriate stimuli (4). Under certain conditions, α-β complexes can be retrieved from the cell surface by endocytosis and form an intracellular store pool (5). Some cells build up "latent," inactive pools of Na^+-K^+-pumps which can be recruited as functional pumps in response to aldosterone (6). For more details see text.

In the *Xenopus* oocyte and embryo, regulation of endo- and exocytosis is likely to be part of a developmental program to control the number of Na^+- K^+-pumps at the cell surface. During progesterone-induced maturation of *Xenopus* oocytes into fertilizable eggs, Na^+-K^+-pump activity at the plasma membrane is completely down-regulated (Weinstein et al., 1982; Richter et al., 1984; Schmalzing et al., 1990; Pralong-Zamofing et al., 1992). The down-regulation is the result of an endocytotic retrieval of the oocyte's α-β3 complexes from the plasma membrane which are recovered in an active form in an intracellular compartment (Pralong-Zamofing et al., 1992).

The molecular trigger(s) for the specific endocytotic process of the Na^+-K^+-pumps during oocyte maturation is not known. Though activation of protein kinase C by phorbol esters can partially mimic the effect of progesterone (Vasilets et al., 1990; Schmalzing et al., 1991), the down-regulation of the surface Na^+-K^+-pumps observed with phorbol esters is incomplete and is probably mainly due to a general reduction in the cell surface area (Vasilets et al., 1990). Though a specific role of Na^+-K^+-pump down-regulation during meiotic maturation is not yet clear, exposure of oocytes to ouabain seems to facilitate progesterone-induced maturation (Penna and Wasserman, 1987). It is possible that internalization of functional pumps in *Xenopus* oocytes represents an economical regulatory process which preserves oocytes for later stages in development when efficient Na^+ and K^+ transport becomes once more a functional necessity. Indeed, partial recovery of the Na^+-K^+-ATPase activity as well as α subunit immunoreactivity in plasma membrane-enriched fractions are found to occur soon after fertilization (Pralong-Zamofing et al., 1992). Possibly, in *Xenopus* embryos, the re-expression of functional internal pumps at the plasma membrane occurs in concert with functional recruitment of maternal α subunits by newly synthesized β subunits to satisfy the increased demands for vectorial Na^+ transport during blastocoele formation.

Insulin stimulates Na^+-K^+-pump activity in a variety of tissues to cope with the important increase in the hormone-dependent Na^+ influx. Based on differential ouabain-sensitivity of $\alpha1$ and $\alpha2$ subunits, it was suggested that the $\alpha2$ subunit is involved in insulin stimulation of the Na^+-K^+-pump in preparations of rat synaptosomes (Brodsky, 1990), and in adipocytes (Lytton, 1990; McGill and Guidotti, 1991). Experiments with skeletal muscle provide evidence that insulin mediates the mobilization of isoform-specific Na^+-K^+-pump subunits from intracellular compartments to the plasma membrane. Biochemical (Hundal et al., 1992) and morphological (Marette et al., 1993) studies show that skeletal muscle cells contain $\alpha1$, $\alpha2$, $\beta1$, and $\beta2$ isoforms. $\alpha1$ isoforms are mainly located in plasma membranes while $\alpha2$ isoforms constitute an important intracellular pool. Acute insulin stimulation increases the content of $\alpha2$ isoforms in the plasma membrane with a parallel decrease in the intracellular compartment. The observation that insulin treatment also increases $\beta1$, but not $\beta2$ isoforms at the cell surface, suggests that the functional insulin-sensitive Na^+-K^+-pump in skeletal muscle is an $\alpha2/\beta1$ complex (Hundal et al., 1992). It is noteworthy that the $\alpha2$ and the $\beta1$ subunits are not recruited from the same cellular fractions resolved on sucrose gradients. This observation raises the possibility that functional α/β complexes might be formed not only in the endoplasmic reticulum, as described above, but also during their trafficking or in the plasma membrane. Some evidence indicating that subunit assembly might not always be restricted to the ER comes from studies with Na^+-K^+-ATPase subunits expressed in baculovirus-infected insect cells (DeTomaso et al., 1994). In these cells unassociated α and β subunits are targeted to the plasma membrane and are able to assemble as functional α/β complexes after plasma membrane fusion of individual cells expressing either the α or the β subunit (DeTomaso et al., 1994). The physiological relevance of these findings is unclear, but they suggest the interesting possibility that post-ER assembly might represent a novel mechanism for the regulation of Na^+-K^+-ATPase expression.

Activation of "latent" Na-K-Pumps

The intracellular pools of functional Na^+-K^+-ATPase referred to so far are not the only known reservoirs of pumps which can be rapidly mobilized to compensate for the ionic imbalance arising from an increase in Na^+ entry into cells. Indeed, increasing evidence suggests that in certain cell types a functionally inactive Na^+-K^+-ATPase pool exists which can be rapidly activated in response to certain stimuli (Barlet-Bas et al., 1990; Blot-Chabaud et al. 1990; Fujii et al. 1990; Beron and Verrey, 1994) (Figure 5). Silent Na^+-K^+-pumps have mainly been described in aldosterone-responsive cells and are thought to be involved in the mediation of the early Na^+ transport response to aldosterone. In cortical collecting tubules (CCT) of rat and rabbit kidney, an increase in the intracellular Na^+ concentration triggers a rapid increase in Na^+-K^+-ATPase activity and/or ouabain-binding sites which does not depend on RNA or protein synthesis (Barlet-Bas et al., 1990; Blot-Chabaud et al., 1990). The constitution of the "latent" Na^+-K^+-pump pool appears to be under the control of corticoid hormones. An activable but corticoid hormone-independent Na^+-K^+-pump pool could also be demonstrated in *Xenopus* kidney derived A6 cells (Beron and Verrey, 1994). In these cells, a short exposure to aldosterone or an increase in intracellular Na^+ leads to an increase in the initial rate of ouabain binding.

Questions which remain to be answered include the location and the mechanisms of activation of these latent Na^+-K^+-ATPase pools. Barlet-Bas et al. (1990) suggested that in rat CCT an increased intracellular Na^+ concentration elicits rapid mobilization of pre-existing, inactive enzyme molecules to the plasma membrane. However, the results obtained with rabbit CCT (Coutry et al., 1992) and A6 cells (Beron and Verrey, 1994) are more compatible with the view of an *in situ* activation of a pre-existing cell surface pool of silent Na^+-K^+-pumps. Indeed, none of the known mediators of exocytotic processes such as an increase in intracellular Ca^{2+}, interaction with cytoskeletal elements, protein kinase C activation or pH changes were found to influence the recruitment of functional Na-K-pumps (Coutry et al., 1992). At present, information on the molecular nature of the "latent" pool is unavailable. It thus remains an open question whether activation corresponds to the mobilization of previously silent Na^+-K^+-pumps or to a change in the kinetic parameters of already active pumps. In addition, it is not known whether regulatory processes such as phosphorylation/dephosphorylation reactions or interactions with cytoskeletal or other cellular components play a role in mediating the activation or derepression of the "latent" Na^+-K^+-ATPase pool.

In conclusion, this chapter on the regulation of Na^+-K^+-ATPase reveals the existence of a variety of molecular mechanisms which are involved in mediating changes in the number of functional Na^+-K^+-pumps at the plasma membrane. While different modulators may use similar pathways in various tissues, the same modulator may be able to produce the cellular response by various molecular processes. This multitude of regulatory mechanisms of Na^+-K^+-ATPase expression in addition to the control of enzyme activity, ensures the maintenance of ionic homeostasis in each cell and permits N^+ and K^+ transport to fully adapt to continuously changing, physiological conditions.

III. SUMMARY

The maintenance of the Na^+ and K^+ gradients across the plasma membrane is ensured by the Na^+-K^+-ATPase, which couples the energy of ATP hydrolysis to Na^+ and K^+ transport. Since the Na^+ gradients provide the potential energy necessary for many other Na^+-dependent transporters which are essential for basic and specialized cellular functions, it follows that rigorous control of Na^+ and K^+ transport via the Na^+-K^+-ATPase is of crucial importance for cell survival. Both acute and long-term adjustments in the number of functional Na^+-K^+-pumps expressed at the cell surface are important regulatory processes induced by various stimuli. A striking feature of the regulation of Na^+-K^+-ATPase expression is the variety of molecular mechanisms which can mediate the cellular response in various tissues. Moreover, some stimuli are able to regulate Na^+-K^+-pump expression through multiple pathways. Increasing evidence suggests that despite their diversity, the regulatory processes are often closely linked to the requirement of α and β subunit assembly for the formation of functional Na^+-K^+-pumps. Thus, selective subunit regulation at a pre- or post-translational level is a common theme exploited in response to different stimuli. This chapter summarizes most of our present knowledge of mechanisms at the transcriptional, post-transcriptional, translational, and post-translational level which are involved in the regulation of Na^+-K^+-ATPase expression and which ultimately permit adaptational control of the number of functional Na^+-K^+-ATPase at the cell surface.

ACKNOWLEDGMENTS

We would like to thank N. Skarda for secretarial assistance in preparing the manuscript. The personal work presented in this study was supported by grants from the Swiss National Fund for Scientific Research.

REFERENCES

Ackermann, U., & Geering, K. (1990). Mutual dependence of Na^+-K^+-ATPase α-subunits and β-subunits for correct posttranslational processing and intracellular transport. FEBS Lett. 269, 105-108.

Ackermann, U., & Geering, K. (1992). $\beta 1$- and $\beta 3$-subunits can associate with presynthesized α-subunits of *Xenopus* oocyte Na^+-K^+-ATPase. J. Biol. Chem. 267, 12911-12915.

Akera, T., & Brody, T.M. (1985). Estimating sodium pump activity in beating heart muscle. TIPS, 156-159.

Arriza, J.L., Weinberger, C., Cerelli, G., Glaser, T.M., Handelin, B.L., Housman, D.E. Evans, R.M. (1987). Cloning of human mineralocorticoid receptor complementary DNA: structural and functional kinship with the glucocorticoid receptor. Science 237, 268-275.

Azuma, K.K., Hensley, C.B., Tang, M.J., & McDonough, A.A. (1993). Thyroid hormone specifically regulates skeletal muscle Na^+-K^+-ATPase α-2-isoforms and β-2-isoforms. Am. J. Physiol. 265, C680-C687.

Barlet, C., & Doucet, A. (1987). Triiodothyronine enhances renal response to aldosterone in the rabbit collecting tubule. J. Clin. Invest. 79, 629-631.

Barlet-Bas, C., Khadouri, C., Marsy, S., & Doucet, A. (1990). Enhanced intracellular sodium concentration in kidney cells recruits a latent pool of Na^+-K^+-ATPase whose size is modulated by corticosteroids. J. Biol. Chem. 265, 7799-7803.

Bastl, C.P., & Hayslett, J.P. (1992). The cellular action of aldosterone in target epithelia. Kidney Intl. 42, 250-264.

Beggah, A.T., Beguin, P., Jaunin, P., Peitsch, M.C., & Geering, K. (1993). Hydrophobic C-terminal amino acids in the β-subunit are involved in assembly with the α-subunit of Na^+-K^+-ATPase. Biochemistry 32, 14117-14124.

Beron, J., & Verrey, F. (1994). Aldosterone induces early activation and late accumulation of Na^+-K^+-ATPase at surface of A6 cells. Am. J. Physiol. 266, C1278-C1290.

Berrebi-Bertrand, I., & Maixent, J.M. (1994). Immunodetection and enzymatic characterization of the $\alpha3$-isoform of Na^+-K^+-ATPase in dog heart. FEBS Lett. 348, 55-60.

Bertorello, A.M., & Katz, A.I. (1993). Short-term regulation of renal Na^+-K^+-ATPase activity: physiological relevance and cellular mechanisms. Am. J. Physiol. 265, F743-F755.

Biggers, J.D., Bell, J.E., & Benos, D.J. (1988). Mammalian blastocyst: transport functions in a developing epithelium. Am. J. Physiol. 255, C419-C432.

Blot-Chabaud, M., Wanstok, F., Bonvalet, J.P., & Farman, N. (1990). Cell sodium-induced recruitment of Na^+-K^+-ATPase pumps in rabbit cortical collecting tubules is aldosterone-dependent. J. Biol. Chem. 265, 11676-11681.

Bonvalet, J.P., Doignon, I., Blot-Chabaud, M., Pradelles, P., & Farman, N. (1990). Distribution of 11β-hydroxysteroid dehydrogenase along the rabbit nephron. J. Clin. Invest. 86, 832-837.

Bowen, J.W., & McDonough, A. (1987). Pretranslational regulation of Na^+-K^+-ATPase in cultured canine kidney cells by low K^+. Am. J. Physiol. 252, C179-C189.

Bradley, M.E., Azuma, K.K., McDonough, A.A., Mircheff, A.K., & Wood, R.L. (1993). Surface and intracellular pools of Na^+-K^+-ATPase catalytic and immuno-activities in rat exorbital lacrimal gland. Exptl. Eye Res. 57, 403-413.

Brent, G.A., Harney, J.W., Chen, Y., Warne, R.L., Moore, D.D., & Larsen, P.R. (1989). Mutations of the rat growth hormone promoter which increase and decrease response to thyroid hormone define a consensus thyroid hormone response element. Mol. Endocrin. 3, 1996-2004.

Brodsky, J.L. (1990). Insulin activation of brain Na^+-K^+-ATPase is mediated by $\alpha2$-form of enzyme. Am. J. Physiol. 258, C812-C817.

Brotherus, J.R., Jacobsen, L., & Jørgensen, P.L. (1983). Soluble and enzymatically stable (Na^+ + K^+)-ATPase from mammalian kidney consisting predominantly of protomer $\alpha\beta$- units. Preparation, assay and reconstitution of active Na^+, K^+ transport. Biochim. Biophys. Acta 731, 290-303.

Burgener-Kairuz, P., Corthesy-Theulaz, I., Merillat, A.M., Good, P., Geering, K., & Rossier, B.C. (1994). Polyedenylation of Na^+-K^+-ATPase $\beta1$-subunit during early development of *Xenopus laevis*. Am. J. Physiol. 266, C157-C164.

Cameron, R., Klein, L., Shyjan, A.W., Rakic, P., & Levenson, R. (1994). Neurons and astroglia express distinct subsets of Na^+-K^+-ATPase α and β subunits. Mol. Brain Res. 21, 333-343.

Caplan, M.J., Forbush, B., Palade, G.E., & Jamieson, J.D. (1990). Biosynthesis of the Na^+-K^+-ATPase in Madin-Darby canine kidney cells—Activation and cell surface delivery. J. Biol. Chem. 265, 3528-3534.

Carafoli, E., & Stauffer, T. (1994). The plasma membrane calcium pump: functional domains, regulation of the activity, and tissue specificity of isoform expression. J. Neurobiol. 25, 312-324.

Cayanis, E., Bayley, H., & Edelman, I.S. (1990). Cell-free transcription and translation of Na^+-K^+-ATPase α and β subunit cDNAs. J. Biol. Chem. 265, 10829-10835.

Chaudhury, S., Ismail-Beigi, F., Gick, G.G., Levenson, R., & Edelman, I.S. (1987). Effect of thyroid hormone on the abundance of Na, K-adenosine triphosphatase α-subunit messenger ribonucleic acid. Mol. Endocrin. 1, 83-89.

Clausen, T., & Everts, M.E. (1989). Regulation of the Na, K-pump in skeletal muscle. Kidney Intl. 35, 1-13.

Corthésy-Theulaz, I., Mérillat, A.M., Honegger, P., & Rossier, B.C. (1991). Differential regulation of Na^+-K^+-ATPase isoform gene expression by T3 during rat brain development. Am. J. Physiol. 261, C124-C131.

Coutry, N., Blot-Chabaud, M., Mateo, P., Bonvalet, J.P., & Farman, N. (1992). Time course of sodium-induced Na^+-K^+-ATPase recruitment in rabbit cortical collecting tubule. Am. J. Physiol. 263, C61-C68.

DeTomaso, A.W., Blanco, G., & Mercer, R.W. (1994). The α and β subunits of the Na^+-K^+-ATPase can assemble at the plasma membrane into functional enzyme. J. Cell Biol. 127, 55-69.

Devarajan, P., Gilmore-Hebert, M., & Benz, E.J. (1992). Differential translation of the Na^+-K^+-ATPase subunit mRNAs. J. Biol. Chem. 267, 22435-22439.

Doris, P.A. (1994). Regulation of Na^+-K^+-ATPase by endogenous ouabain-like materials. Proc. Soc. Exptl. Biology Med. 205, 202-212.

Eakle, K.A., Kim, K.S., Kabalin, M.A., & Farley, R.A. (1992). High-affinity ouabain-binding by yeast cells expressing Na^+,K^+-ATPase α-subunits and the gastric H^+,K^+-ATPase β-Subunit. Proc. Natl. Acad. Sci. USA 89, 2834-2838.

Eakle, K.A., Kabalin, M.A., Wang, S.G., & Farley, R.A. (1994). The influence of β subunit structure on the stability of Na^+/K^+-ATPase complexes and interaction with K^+. J. Biol. Chem. 269, 6550-6557.

Fambrough, D.M., Lemas, M.V., Hamrick, M., Emerick, M., Renaud, K.J., Inman, E.M., Hwang, B., & Takeyasu, K. (1994). Analysis of subunit assembly of the Na^+-K^+-ATPase. Am. J. Physiol. 266, C579-C589.

Farman, N., Coutry, N., Logvinenko, N., Blot-Chabaud, M., Bourbouze, R., & Bonvalet, J.P. (1992). Adrenalectomy reduces α-1 and not β-1 Na^+-K^+-ATPase mRNA expression in rat distal nephron. Am. J. Physiol. 263, C810-C817.

Farman, N., Bonvalet, J.P., & Seckl, J.R. (1994). Aldosterone selectively increases Na^+-K^+-ATPase α3-subunit mRNA expression in rat hippocampus. Am. J. Physiol. 266, C423-C428.

Feng, J., Orlowski, J., & Lingrel, J.B. (1993). Identification of a functional thyroid hormone response element in the upstream flanking region of the human Na^+-K^+-ATPase β-1 gene. Nucleic Acids Res. 21, 2619-2626.

Fujii, Y., Takemoto, F., & Katz, A.I. (1990). Early effects of aldosterone on Na-K pump in rat cortical collecting tubules. Am. J. Physiol. 259, F40-F45.

Funder, J. W. (1990). Corticosteroid receptors and renal 11β-hydroxysteroid dehydrogenase activity. Semin. Nephrol. 10, 311-319.

Gaeggeler, H.P., Edwards, C.R.W., & Rossier, B.C. (1989). Steroid metabolism determines mineralocorticoid specificity in the toad bladder. Am. J. Physiol. 257, F690-F695.

Gardiner, C.S., Williams, J.S., & Menino, A.R. (1990). Sodium, potassium adenosine triphosphatase α- and β-subunit and α-subunit mRNA levels during mouse embryo development in vitro. Biol. Reprod. 43, 788-794.

Geering, K. (1990). Subunit assembly and functional maturation of Na^+-K^+-ATPase. J. Membrane Biol. 115, 109-121.

Geering, K. (1991). The functional role of the β-subunit in the maturation and intracellular transport of Na^+K^+-ATPase. FEBS Lett. 285, 189-193.

Geering, K., & Rossier, B.C. (1981). Thyroid hormone-aldosterone antagonism on Na^+ transport in toad bladder. J. Biol. Chem. 256, 5504-5510.

Geering, K., Girardet, M., Bron, C., Kraehenbühl, J.P., & Rossier, B.C. (1982). Hormonal regulation of (Na^+,K^+)-ATPase biosynthesis in the toad bladder. J. Biol. Chem. 257, 10338-10343.

Geering, K., Claire, M., Gaeggeler, H.P., & Rossier, B.C. (1985a). Receptor occupancy vs. induction of Na^+-K^+-ATPase and Na^+ transport by aldosterone. Am. J. Physiol. 248, C102-C108.

Geering, K., Meyers, D.I., Paccolat, M.P., Kraehenbuhl, J.P., & Rossier, B.C. (1985b). Membrane insertion of α- and β-subunits of Na^+, K^+-ATPase. J. Biol. Chem. 260, 5154-5160.

Geering, K., Kraehenbuhl, J.P., & Rossier, B.C. (1987). Maturation of the catalytic α-subunit of Na^+-K^+-ATPase during intracellular transport. J. Cell Biol. 105, 2613-2619.

Geering, K., Theulaz, I., Verrey, F., Häuptle, M.T., & Rossier, B.C. (1989). A role for the β-subunit in the expression of functional Na^+-K^+-ATPase in *Xenopus* oocytes. Am. J. Physiol. 257, C851-C858.

Giannella, R.A., Orlowski, J., Jump, M.L., & Lingrel, J.B. (1993). Na^+-K^+-ATPase gene expression in rat intestine and Caco-2 cells: response to thyroid hormone. Am. J. Physiol. 265, G775-G782.

Gick, G.G., Ismail-Beigi, F., & Edelman, I.S. (1988). Thyroidal regulation of rat renal and hepatic Na, K-ATPase gene expression. J. Biol. Chem. 263, 16610-16618.

Gick, G.G., & Ismail-Beigi, F. (1990). Thyroid hormone induction of Na^+-K^+-ATPase and its mRNAs in a rat liver cell line. Am. J. Physiol. 258, C544-C551.

Gick, G.G., Melikian, J., & Ismail-Beigi, F. (1990). Thyroidal enhancement of rat myocardial Na^+-K^+-ATPase: preferential expression of α-2 activity and mRNA abundance. J. Membrane Biol. 115, 273-282.

Gloor, S., Antonicek, H., Sweadner, K.J., Pagliusi, S., Frank, R. Moos, M., & Schachner, M. (1990). The adhesion molecule on glia (AMOG) is a homologue of the β-subunit of the Na^+-K^+-ATPase. J. Cell Biol. 110, 165-174.

Glynn, I.M. (1993). All hands to the sodium pump. J. Physiol. 462, 1-30.

Good, P.J., Richter, K., & Dawid, I.B. (1990). A nervous system-specific isotype of the β-subunit of Na^+,K^+-ATPase expressed during early development of *Xenopus-laevis*. Proc. Natl. Acad. Sci. USA 87, 9088-9092.

Hamlyn, J.M., & Manunta, P. (1992). Ouabain, digitalis-like factors and hypertension. J. Hypertens. 10, S99-SIII.

Hamrick, M., Renaud, K.J., & Fambrough, D.M. (1993). Assembly of the extracellular domain of the Na^+-K^+-ATPase β-subunit with the α-subunit. Analysis of β-subunit chimeras and carboxyl-terminal deletions. J. Biol. Chem. 268, 24367-24373.

Han, Y., Pralong-Zamofing, D., Ackermann, U., & Geering, K. (1991). Modulation of Na^+-K^+-ATPase expression during early development of *Xenopus-laevis*. Dev. Biol. 145, 174-181.

Hensley, C.B., Azuma, K.K., Tang, M.J., & McDonough, A.A. (1992). Thyroid hormone induction of rat myocardial Na^+-K^+-ATPase: αl-, $\alpha 2$-, and βl-mRNA and -protein levels at steady state. Am. J. Physiol. 262, C484-C492.

Herrera, V.L.M., Cova, T., Sassoon, D., & Ruiz-Opazo, N. (1994). Developmental cell-specific regulation of Na^+-K^+-ATPase αl-, $\alpha 2552$-, and $\alpha 3$-isoform gene expression. Am. J. Physiol. 266, C 1301-C1312.

Homareda, H., Kawakami, K., Nagano, K., & Matsui, H. (1989). Location of signal sequences for membrane insertion of the Na^+,K^+-ATPase α subunit. Mol. Cell. Biol. 9, 5742-5745.

Homareda, H., Kawakami, K., Nagano, K., & Matsui, H. (1993). Stabilization in microsomal membranes of the fifth transmembrane segment of the Na^+,K^+-ATPase α subunit with proline to leucine mutation. Biochem. Cell Biol. 71, 410-415.

Horisberger, J.D. (1994). The Na^+-K^+-ATPase: Structure-Function Relationship (CRC Press, eds). In: Molecular Biology Intelligence Unit. R.G. Landes Company, Austin.

Horowitz, B., Hensley, C.B., Quintero, M., Azuma, K.K., Putnam, D., & McDonough, A.A. (1990). Differential regulation of Na^+-K^+-ATPase α- 1, α-2, and β-subunit mRNA and protein levels by thyroid hormone. J. Biol. Chem. 265, 14308-14314.

Hsu, Y.M., & Guidotti, G. (1991). Effects of hypokalemia on the properties and expression of the (Na^+,K^+)-ATPase of rat skeletal muscle. J. Biol. Chem. 266, 427-433.

Huang, F., He, H., & Gick, G. (1994). Thyroid hormone regulation of Na^+K^+-ATPase $\alpha 2$ gene expression in cardiac myocytes. Cell. Mol. Biol. Res. 40, 41-52.

Hundal, H.S., Marette, A., Mitsumoto, Y., Ramlal, T., Blostein, R., & Klip, A. (1992). Insulin induces translocation of the $\alpha 2$- and βl subunits of the Na^+/K^+-ATPase from intracellular compartments to the plasma membrane in mammalian skeletal muscle. J. Biol. Chem. 267, 5040-5043.

Hundal, H.S., Marette, A., Ramlal, T., Liu, Z., & Klip, A. (1994). Expression of β subunit isoforms of the Na$^+$, K$^+$-ATPase is muscle type-specific. FEBS Lett. 328, 253-258.

Hurtley, S.M., & Helenius, A. (1989). Protein oligomerization in the endoplasmic reticulum. Ann. Rev. Cell Biol. 5, 277-307.

Ikeda, U., Hyman, R., Smith, T.W., & Medford, R.M. (1991). Aldosterone-mediated regulation of Na$^+$, K$^+$-ATPase gene expression in adult and neonatal rat cardiocytes. J. Biol. Chem. 266, 12058-12066.

Inesi, G., & Kirtley, M.R. (1992). Structural features of cation transport ATPases. J. Bioenerg. Biomemb. 24, 271-283.

Ismail-Beigi, F. (1993). Thyroid hormone regulation of Na$^+$-K$^+$-ATPase expression. Trends Endocrinol. Metab. 4, 152-155.

Jaisser, F., Jaunin, P., Geering, K., Rossier, B.C., & Horisberger, J.D. (1994). Modulation of the Na,K-pump function by β subunit isoforms. J. Gen. Physiol. 103, 605-623.

Jaunin, P., Horisberger, J.D., Richter, K., Good, P.J., Rossier, B.C., & Geering, K. (1992). Processing, intracellular transport, and functional expression of endogenous and exogenous α-β3 Na$^+$-K$^+$-ATPase complexes in Xenopus oocytes. J. Biol. Chem. 267, 577-585.

Jaunin, P., Jaisser, F., Beggah, A.T., Takeyasu, K., Mangeat, P., Rossier, B.C., Horisberger, J.D., & Geering, K. (1993). Role of the transmembrane and extracytoplasmic domain of β-subunits in subunit assembly, intracellular transport, and functional expression of Na,K-pumps. J. Cell Biol. 123, 1751-1759.

Jewell, E.A., & Lingrel, J.B. (1991). Comparison of the substrate dependence properties of the rat Na$^+$-K$^+$-ATPase α1, α2, and α3 isoforms expressed in HeLa cells. J. Biol. Chem. 266, 16925-16930.

Johnson, J.P. (1992) . Cellular mechanisms of action of mineralocorticoid hormones. Pharmacol. Therapeut. 53, 1-29.

Johnson, J.P., Jones, D., & Wiesmann, W.P. (1986). Hormonal regulation of Na$^+$-K$^+$-ATPase in cultured epithelial cells. Am. J. Physiol. 251, C186-C190.

Jørgensen, P.L. (1986). Structure, function, and regulation of Na$^+$,K$^+$-ATPase in the kidney. Kidney Intl. 29. 10-20.

Karin, N.J., & Cook, J.S. (1986). Turnover of the catalytic subunit of Na$^+$-K$^+$-ATPase in HTC cells. J. Biol. Chem. 261, 10422-10428.

Katz, A.I. (1990). Corticosteroid regulation of Na$^+$-K$^+$-ATPase along the mammalian nephron. Sem. Nephrol. 10, 388-399.

Kawakami, K., Okamoto, H., Yagawa, Y., & Nagano, K. (1990a). Regulation of Na$^+$,K$^+$-ATPases. II. Cloning and analysis of the 5' flanking region of the rat NKA β2 gene encoding the β-2 subunit. Gene 91, 271-274.

Kawakami, K., Yagawa, Y., & Nagano, K. (1990b). Regulation of Na$^+$,K$^+$-ATPases .I. Cloning and analysis of the 5'-flanking region of the rat NKA α2 gene encoding the α-2 subunit. Gene 91, 267-270.

Kawakami, K., Suzuki-Yagawa, Y., Watanabe, Y., & Nagano, K. (1992). Identification and characterization of the cis-elements regulating the rat AMOG (adhesion molecule on glia)/Na$^+$-K$^+$-ATPase β2 subunit gene. J. Biochem. 111, 515-522.

Kawakami, K., Watanabe, Y., Araki, M., & Nagano, K. (1993a). Spl binds to the adhesion molecule on glia regulatory element that functions as a positive transcription regulatory element in astrocytes. J. Neurosci. Res. 35, 138-146.

Kawakami, K., Yanagisawa, K., Watanabe, Y., Tominaga, S., & Nagano, K. (1993b). Different factors bind to the regulatory region of the Na$^+$,K$^+$-ATPase α1-subunit gene during the cell cycle. FEBS Lett. 335, 251-254.

Kirley, T.L. (1989). Determination of three disulfide bonds and one free sulfhydryl in the β subunit of (Na,K)-ATPase. J. Biol. Chem. 264, 7185-7192.

Kirtane, A., Ismail-Beigi, N., & Ismail-Beigi, F. (1994). Role of enhanced Na$^+$ entry in the control of Na$^+$-K$^+$-ATPase gene expression by serum. J. Membr. Biol. 137, 9-15.

Lambert, R.W., Bradley, M.E., & Mircheff, A.K. (1991). pH-sensitive anion exchanger in rat lacrimal acinar cells. Am. J. Physiol. 260, G517-G523.

Lambert, R.W., Maves, C.A., Gierow, J.P., Wood, R.L., & Mircheff, A.K. (1993). Plasma membrane internalization and recycling in rabbit lacrimal acinar cells. Invest. Ophthalmol. Visual Sci. 34, 305-316.

Leal, T., & Crabbé, J. (1989). Effects of aldosterone on $(Na^+ + K^+)$-ATPase of amphibian sodium-transporting epithelial cells (A6) in culture. J. Steroid Biochem. 34, 581-584.

Lescale-Matys, L., Hensley, C.B., Crnkovic-Markovic, R., Putnam, D.S., & McDonough, A.A. (1990). Low K^+ increases Na^+-K^+-ATPase abundance in LLC-PK1/C14 cells by differentially increasing-β, and not-α, subunit messenger RNA. J. Biol. Chem. 265, 17935-17940.

Lescale-Matys, L., Putnam, D.S., & McDonough, A.A. (1993). Na^+-K^+-ATPase α-1-subunit and β-1-subunit degradation—evidence for multiple subunit specific rates. Am. J. Physiol. 264, C583-C590.

Lingrel, J.B., Orlowski, J., Shull, M.M., & Price, E.M. (1990). Molecular genetics of Na^+-K^+-ATPase. Prog. Nucleic Acid Res. Mol. Biol. 38, 37-89.

Lingrel, J.B., Van Huysse, J., Obrien, W., Jewell-Motz, E., Askew, R., & Schultheis, P. (1994). Structure-function studies of the Na^+-K^+-ATPase. Kidney Intl 45, S32-S39.

Liu, B., & Gick, G. (1992). Characterization of the 5' flanking region of the rat Na^+/K^+-ATPase β1 subunit gene. Biochim. Biophys. Acta 1130, 336-338.

Liu, B., Huang, F.L., & Gick, G. (1993). Regulation of Na^+-K^+-ATPase β1 mRNA content by thyroid hormone in neonatal rat cardiac myocytes. Cellul. Mol. Biol. Res. 39, 221-229.

Lytton, J. (1985). Insulin affects the sodium affinity of the rat adipocyte (Na^+,K^+)-ATPase. J. Biol. Chem. 260, 10075-10080.

MacPhee, D.J., Barr, K.J., DeSousa, P.A., Todd, S.D.L., & Kidder. G.M. (1994). Regulation of Na^+,K^+-ATPase a subunit gene expression during mouse preimplantation development. Dev. Biol. 162, 259-266.

Marette, A., Krischer, J., Lavoie, L., Ackerley, C., Carpentier, J.L., & Klip, A. (1993). Insulin increases the Na^+-K^+-ATPase α 2-subunit in the surface of rat skeletal muscle: morphological evidence. Am. J. Physiol. 265, C1716-C1722.

Martin-Vasallo, P., Dackowski, W., Emanuel, J.R., & Levenson, R. (1989). Identification of a putative isoform of the Na^+-K^+-ATPase β subunit. J. Biol. Chem. 264, 4613-4618.

McDonough, A.A., Brown, T.A., Horowitz B., Chiu R., Schlotterbeck J., Bowen J., & Schmitt C.A. (1988). Thyroid hormone coordinately regulates Na^+-K^+-ATPase α- and β-subunit mRNA levels in kidney. Am. J. Physiol. 254, C323-C329.

McDonough, A.A., Geering, K., & Farley, R.A. (1990). The sodium pump needs its β-subunit. FASEB J. 4, 1598-1605.

McDonough, A.A., & Farley, R.A. (1993). Regulation of Na, K-ATPase activity. Curr. Opin. Nephrol. Hypertens. 2, 725-734.

McGill, D.L., & Guidotti, G. (1991). Insulin stimulates both the α1 and the α2 isoforms of the rat adipocyte (Na^+,K^+) ATPase. Two mechanisms of stimulation. J. Biol. Chem. 266, 15824-15831.

McGivan, J.D., & Pastor-Anglada, M. (1994). Regulatory and molecular aspects of mammalian amino acid transport. Biochem. J. 299, 321-334.

Melikian, J., & Ismail-Beigi, F. (1991). Thyroid hormone regulation of Na^+-K^+-ATPase subunit-mRNA expression in neonatal rat myocardium. J. Membr. Biol. 119, 171-177.

Miller, R.P., & Farley, R.A. (1990). β-subunit of $(Na^+ + K^+)$-ATPase contains 3 disulfide bonds. Biochemistry 29, 1524-1532.

Morrill, G.A., Kostellow, A.B., & Murphy, J.B. (1975). Role of Na^+,K^+-ATPase in early embryonic development. Ann. NY Acad. Sci. 242, 543-559. Munzer, J.S., Daly, S.E., Jewell-Motz, E.A., Lingrel, J.B., & Blostein, R. (1994). Tissue- and isoform-specific kinetic behavior of the Na^+-K^+-ATPase. J. Biol. Chem. 269, 16668-16676.

Newport, J., & Kirschner, M. (1982). A major developmental transition in early *Xenopus* embryos: II. Control of the onset of transcription. Cell 30, 687-696.

Noguchi, S., Mishina, M., Kawamura, M., & Numa, S. (1987). Expression of functional (Na$^+$K$^+$-ATPase from cloned cDNAs. FEBS Lett. 225, 27-32.

Noguchi, S., Higashi, K., & Kawamura, M. (1990). A possible role of the β-subunit of (Na,K)-ATPase in facilitating correct assembly of the α-subunit into the membrane. J. Biol. Chem. 265, 15991-15995.

Noguchi, S., Mutoh, Y., & Kawamura, M. (1994). The functional roles of disulfide bonds in the β-subunit of (Na,K)-ATPase as studied by site-directed mutagenesis. FEBS Lett. 341, 233-238.

O'Donnell, M.E., & Owen, N.E. (1994). Regulation of ion pumps and carriers in vascular smooth muscle. Phys. Rev. 74, 683-721.

Oguchi, A., Ikeda, U., Kanbe, T., Tsuruya, Y., Yamamoto, K., Kawakami, K., Medford, R.M., & Shimada, K. (1993). Regulation of Na$^+$-K$^+$-ATPase gene expression by aldosterone in vascular smooth muscle cells. Am. J. Physiol. 265, H1167-H1172.

Ohara, T., Ikeda, U., Muto, S., Oguchi, A., Tsuruya, Y., Yamamoto, K., Kawakami, K., Shimada, K., & Asano, Y. (1993). Thyroid hormone stimulates Na$^+$-K$^+$-ATPase gene expression in cultured rat mesangial cells. Am. J. Physiol. 265, F370-F376.

Orlowski, J., & Lingrel, J.B. (1988). Tissue-specific and developmental regulation of rat Na$^+$-K$^+$-ATPase catalytic α isoform and β subunit mRNAs. J. Biol. Chem. 263. 10436-10442.

Orlowski, J., & Lingrel, J.B. (1990). Thyroid and glucocorticoid hormones regulate the expression of multiple Na$^+$-K$^+$-ATPase genes in cultured neonatal rat cardiac myocytes. J. Biol. Chem. 265, 3462-3470.

Paccolat, M.P., Geering, K., Gaeggeler, H.P., & Rossier, B.C. (1987). Aldosterone regulation of Na$^+$. transport and Na$^+$-K$^+$-ATPase in A6 cells: role of growth conditions . Am. J. Physiol. 252, C468-C476.

Pathak, B.G., Pugh, D.G., & Lingrel, J.B. (1990). Characterization of the 5'-flanking region of the human and rat Na$^+$-K$^+$-ATPase α-3 gene. Genomics 8, 641-647.

Penna, M.J., & Wasserman, W.J. (1987). Effect of ouabain on the meiotic maturation of stage IV-V *Xenopus laevis* oocytes. J. Exp. Zool. 241, 61-69.

Pollack, L.R., Tate, E.H., & Cook, J.S. (1981). Turnover and regulation of Na$^+$-K$^+$-ATPase in HeLa cells. Am. J. Physiol. 241, 173-183.

Pralong-Zamofing, D., Qi-Han, Y., Schmalzing, G., Good, P., & Geering, K. (1992). Regulation of α1-β3 Na$^+$-K$^+$-ATPase isozyme during meiotic maturation of *Xenopus-laevis* oocytes. Am. J. Physiol. 262, C1520-C1530.

Pressley, T.A. (1988). Ion concentration-dependent regulation of Na,K-pump abundance. J. Membr. Biol. 105, 187-195.

Qian, N.X., Pastor-Anglada, M., & Englesberg, E. (1991). Evidence for coordinate regulation of the A-system for amino acid transport and the mRNA for the α1-subunit of the Na$^+$,K$^+$-ATPase gene in Chinese hamster ovary cells. Proc. Natl. Acad. Sci. USA 88, 3416-3420.

Renaud, K.J., Inman, E.M., & Fambrough, D.M. (1991). Cytoplasmic and transmembrane domain deletions of Na$^+$-K$^+$-ATPase β-subunit. Effects on subunit assembly and intracellular transport. J. Biol. Chem. 266, 20491-20497.

Richter, H.P., Jung, D., & Passow, H. (1984). Regulatory changes of membrane transport and ouabain binding during progesterone-induced maturation of *Xenopus* oocytes. J .Membr. Biol. 79, 203-210.

Rose, A.M., & Valdes, R. (1994). Understanding the sodium pump and its relevance to disease. Clin. Chem. 40, 1674-1685.

Rose, J.K., & Doms. R.W. (1988). Regulation of protein export from the endoplasmic reticulum . Ann. Rev. Cell. Biol. 4, 257-288.

Sachs, A., & Wahle, E. (1993). Poly(A) tail metabolism and function in eucaryotes. J. Biol. Chem. 268, 22955-22958.

Sachs, G. (1994). The gastric H,K ATPase—regulation and structure/function of the acid pump of the stomach. Physiology of the Gastrointestinal Tract, 3rd edn. (Johnson, R., ed.), Vols. 1 and 2, pp. 1119-1138. Raven Press, New York.

Schmalzing, G., Kröner, S., & Pasow, H. (1989). Evidence for intracellular sodium pumps in permeabilized *Xenopus laevis* oocytes. Biochem. J. 260. 395-399.

Schmalzing, G., & Kröner, S. (1990). Micromolar free calcium exposes ouabain-binding sites in digitonin-permeabilized *Xenopus-laevis* oocytes. Biochem. J. 269, 757-766.

Schmalzing, G., Eckard, P., Kroner, S., & Passow, H. (1990). Downregulation of surface sodium pumps by endocytosis during meiotic maturation of *Xenopus-laevis* oocytes. Am. J. Physiol. 258 C179-C184.

Schmalzing, G., Mädefessel, K., Haase, W., & Geering, K. (1991). In: The sodium Pump: Recent developments (Kaplan, J.H. & De Weer, P., Eds) pp. 465-470, The Rockefeller University Press), New York.

Schmalzing, G., Kröner, S., Schachner, M., & Gloor, S. (1992). The adhesion molecule on glia (AMOG/β2) and α1 subunits assemble to functional sodium pumps in *Xenopus*-oocytes. J. Biol. Chem. 267, 20212-20216.

Schmidt, T.J., Husted, R.F., & Stokes, J.B. (1993). Steroid hormone stimulation of N$^+$ transport in A6 cells is mediated via glucocorticoid receptors. Am. J. Physiol. 264, C875-C884.

Schneider, B. (1992). Na$^+$,K$^+$-ATPase isoforms in the retina. Intl. Rev. Cytol. 133, 151-185.

Shin, J.M., & Sachs, G. (1994). Identification of a region of the H,K-ATPase α subunit associated with the β subunit. J. Biol. Chem. 269, 8642-8646.

Shull, G.E., Greeb, J., & Lingrel, J.B. (1986a). Molecular cloning of three distinct forms of the Na$^+$,K$^+$-ATPase α-subunit from rat brain. Biochemistry 25, 8125-8132.

Shull, G.E., Lane, L.K., & Lingrel, J.B. (1986b). Amino-acid sequence of the β-subunit of the (Na$^+$ + K$^+$)ATPase deduced from a cDNA. Nature 321, 429-431.

Shull, M. M., Pugh, D.G., & Lingrel, J.B. (1989). Characterization of the human Na$^+$-K$^+$-ATPase α2 gene and identification of intragenic restriction fragment length polymorphisms. J. Biol. Chem. 264, 17532-17543.

Shull, M.M., Pugh, D.G., & Lingrel, J.B. (1990). The human Na$^+$-K$^+$-ATPase-α-1 gene: Characterization of the 5'-flanking region and identification of a restriction fragment length polymorphism. Genomics 6, 451-460.

Sonenberg, N. (1994). mRNA translation: Influence of the 5' and 3' untranslated regions. Curr. Op. Gen. Dev. 4, 310-315.

Stewart, P.M., & Edwards, C.R.W. (1990). Specificity of the mineralocorticoid receptor—crucial role of 11-β-hydroxysteroid dehydrogenase. Trends Endocrinol. Metab. 1, 225-230.

Suzuki-Yagawa, Y., Kawakami, K., & Nagano, K. (1992). Housekeeping Na$^+$-K$^+$-ATPase α1 subunit gene promoter is composed of multiple cis elements to which common and cell type-specific factors bind. Mol. Cell. Biol. 12, 4046-4055.

Sweadner, K.J. (1985). Enzymatic properties of separated isozymes of the Na$^+$-K$^+$-ATPase. Substrate affinities, kinetic cooperativity, and ion transport stoichiometry. J. Biol. Chem. 260, 11508-11513.

Sweadner, K.J. (1992). Overlapping and diverse distribution of Na-K ATPase isozymes in neurons and glia. Can. J. Physiol. Pharmacol. 70, S255-S259.

Takeyasu, K., Tamkun, M.M., Renaud, K.J., & Fambrough, D.M. (1988). Ouabain-sensitive (Na$^+$K$^+$)-ATPase activity expressed in mouse L cells by transfection with DNA encoding the α-subunit of an avian sodium pump. J. Biol. Chem. 263, 4347-4354.

Tamkun, M.M., & Fambrough, D.M. (1986). The (Na$^+$ + K$^+$)-ATPase of chick sensory neurons: Studies on biosynthesis and intracellular transport. J. Biol. Chem. 261, 1009-1019.

Taormino, J.P., & Fambrough, D.M. (1990). Pre-translational regulation of the (Na$^+$-K$^+$)-ATPase in response to demand for ion transport in cultured chicken skeletal muscle. J. Biol. Chem. 265, 4116-4123.

Tsai, M.J., & O'Malley, B.W. (1994). Molecular mechanisms of action of steroid/thyroid receptor superfamily members. Ann. Rev. of Biochem. 63, 451-486.

Van Winkle, L.J., & Campione, A.L. (1991). Ouabain-sensitive Rb^+ uptake in mouse eggs and preimplantation conceptuses. Dev. Biol. 146, 158-166.

Vasilets, L.A., Schmalzing, G., Mädefessel, K., Haase, W., & Schwarz, W. (1990). Activation of protein kinase-C by phorbol ester induces down-regulation of the Na^+/K^+-ATPase in oocytes of Xenopus laevis. J. Membr. Biol. 118, 131-142.

Verrey, F. (1995). Transcriptional control of sodium transport in tight epithelia by adrenal steroids. J. Membr. Biol. in press.

Verrey, F., Schaerer, E., Zoerkler, P., Paccolat, M.P., Geering, K., Kraehenbuhl, J.P., & Rossier, B.C. (1987). Regulation by aldosterone of Na^+,K^+-ATPase mRNAs, protein synthesis, and sodium transport in cultured kidney cells. J. Cell Biol. 104, 1231-1237.

Verrey, F., Kraehenbuhl, J.P., & Rossier, B.C. (1989). Aldosterone induces a rapid increase in the rate of Na^+-K^+-ATPase gene transcription in cultured kidney cells. Mol. Endocrin. 3, 1369-1376.

Wang, Z.M., Yasui, M., & Celsi, G. (1994). Glucocorticoids regulate the transcription of Na^+-K^+-ATPase genes in the infant rat kidney. Am. J. Physiol. 267, C450-C455.

Watson, A.J., Kidder, G.M., & Schultz, G.A. (1992). How to make a blastocyst. Biochem. Cell Biol. 70, 849-855.

Watson, A.J., Pape, C., Emanuel, J.R., Levenson, R., & Kidder, G.M. (1990). Expression of Na^+-K^+-ATPase α-subunit and β-subunit genes during preimplantation development of the mouse. Dev. Genet. 11, 41-48.

Weder, A.B. (1994). Sodium metabolism, hypertension, and diabetes. Am. J. Med. Sci. 307, S53-S59.

Weinstein, S.P., Kostellow, A.B., Ziegler, D.H., & Morrill, G.A. (1982). Progesterone-induced down-regulation of an electrogenic Na^+,K^+-ATPase during the first meiotic division in amphibian oocytes. J. Membr. Biol. 69, 41-48.

Welling, P.A., Caplan, M., Sutters, M., & Giebisch, G. (1993). Aldosterone-mediated Na/K-ATPase expression is $\alpha1$ isoform specific in the renal cortical collecting duct. J. Biol. Chem. 268, 23469-23476.

Wiener, H., Nielsen, J.M., Klaerke, D.A., & Jørgensen, P.L. (1993). Aldosterone and thyroid hormone modulation of $\alpha1$-mRNA, $\beta1$-mRNA, and Na,K-pump sites in rabbit distal colon epithelium. Evidence for a novel mechanism of escape from the effect of hyperaldosteronemia. J. Memb. Biol. 133, 203-211.

Wiley, L.M., Kidder, G.M., & Watson, A.J. (1990). Cell polarity and development of the first epithelium. BioEssays 12, 67-73.

Wormington, M. (1993). Poly(A) and translation: Developmental control. Curr. Opin. Cell Biol. 5, 950-954.

Yagawa, Y., Kawakami, K., & Nagano, K. (1990). Cloning and analysis of the 5'-flanking region of rat Na^+/K^+-ATPase α-1 subunit gene. Biochim. Biophys. Acta 1049, 286-292.

Yamamoto, K., Ikeda, U., Seino, Y., Tsuruya, Y., Oguchi, A., Okada, K., Ishikawa, S.E., Saito, T., Kawakami, K., Hara, Y., & Shimada, K. (1993). Regulation of Na,K-adenosine triphosphatase gene expression by sodium ions in cultured neonatal rat cardiocytes. J. Clin. Invest. 92, 1889-1895.

Yamamoto, K., Ikeda, U., Okada, K., Saito, T., Kawakami, K., & Shimada, K. (1994). Sodium ion mediated regulation of Na/K-ATPase gene expression in vascular smooth muscle cells. Cardiovasc. Res. 28, 957-962.

Yiu, S.C., Lambert, R.W., Tortoriello, P.J., & Mircheff, A.K. (1991). Secretagogue-induced redistributions of Na^+-K^+-ATPase in rat lacrimal acini. Invest. Ophthalmol. Visual Sci. 32, 2976-2984.

Young, R.M., & Lingrel, J.B. (1987). Tissue distribution of mRNAs encoding the α isoforms and β subunit of rat Na^+,K^+-ATPase. Biochem. Biophys. Res. Commun. 145, 52-58.

Zamofing, D., Rossier, B.C., & Geering, K. (1988). Role of the Na^+-K^+-ATPase β-subunit in the cellular accumulation and maturation of the enzyme as assessed by glycosylation inhibitors. J. Membr. Biol. 104, 69-79.

Zamofing, D., Rossier, B.C., & Geering, K. (1989). Inhibition of N-glycosylation affects transepithelial Na^+ but not Na^+-K^+-ATPase transport. Am. J. Physiol. 256, C958-C966.

Zeidel, M.L. (1993). Hormonal regulation of inner medullary collecting duct sodium transport. Am. J. Physiol. 265, F159-F173.

Zlokovic, B.V., Mackic, J.B., Wang, L., Mccomb, J.G., & McDonough, A. (1993). Differential expression of Na^+-K^+-ATPase α-subunit and β-subunit isoforms at the blood-brain barrier and the choroid plexus. J. Biol. Chem. 268, 8019-8025.

STRUCTURE, MECHANISM, AND REGULATION OF THE CARDIAC SARCOLEMMA Na^+-Ca^{2+} EXCHANGER

Daniel Khananshvili

Advances in Molecular and Cell Biology
Volume 23B, pages 311-358.
Copyright © 1998 by JAI Press Inc.
All right of reproduction in any form reserved.
ISBN: 0-7623-0287-9

I. FUNCTIONAL LEVELS OF Na^+-Ca^{2+} EXCHANGE: FROM MOLECULAR STRUCTURE-FUNCTION RELATIONSHIPS TO GROSS PHYSIOLOGY

A. Short History and Overview

Na^+-Ca^{2+} exchange was first identified in guinea pig atria in tracer flux experiments carried out by Reuter and Seitz (1968) and in squid axons by Baker, Blaustein, Hodgkin, and Steinhardt (1969). Since then the subject of Na^+-Ca^{2+} exchange has aroused much interest. It appears that this system plays a key role (with Ca^{2+}-channels and Ca^{2+} pumps) in regulating Ca^{2+} homeostasis and electric currents in many cell types (Blaustein, 1974,1977; Mullins, 1977; Reeves, 1985; Carafoli, 1987; Powell and Nobel, 1989; Nobel et al., 1991; Bridge, 1991; Blaustein et al., 1991; Philipson and Nicoll, 1993). Because of the occurrence of large changes in intracellular calcium concentrations during the action potential, the role of cell membrane Na^+-Ca^{2+} exchange is especially important in excitable tissues (Kimura et al., 1987; Koch and Stryer, 1988; Powell and Nobel, 1989; Blaustein et al., 1991; Lipp and Niggli, 1994; Langer, 1994). Na^+-Ca^{2+} exchange represents a typical "exchange-only" system (Stein, 1986; Laüger, 1987,1991; Khananshvili, 1990a), otherwise known as an antiporter or counter transport system which catalyzes a coupled, stoichiometric exchange of substances at opposite sides of the membrane (Reeves and Hale, 1984; Kimura et al., 1986, 1987; Rasgado-Flores and Blaustein, 1987; Cervetto et al., 1989). The exchanger exhibits a carrier-type mechanism in which the ion-binding sites can be alternatively exposed at the opposite sides of the membrane along the transport cycle (Stein, 1986; Laüger, 1987,1991; Khananshvili, 1990a,b; Khananshvili et al., 1995 a). Nevertheless, the Na^+-Ca^{2+} exchanger can oscillate between functionally active and inactive states, a feature which resembles "channel-like" function (Hilgemann et. al., 1992; Matsuoka and Hilgemann, 1992; Hilgemann, 1996).

This chapter focuses on molecular, kinetic, cellular and regulatory aspects of Na$^+$-Ca^{2+} exchange. Excellent review articles are available, discussing the functional aspects of Na$^+$-Ca^{2+} exchange in integrated cardiac physiology (Hilgemann and Nobel, 1987; Powell and Nobel, 1989; Nobel et al., 1991; Langer, 1994), and electrophysiology (Bridge et al, 1991; Matsuoka and Hilgemann, 1992; Hilgemann et al., 1992). These topics will not be discussed in detail; rather the cardiac Na$^+$-Ca^{2+} exchanger, as a prototype will be the central theme of this chapter. This is because of the availability of considerable information about this exchanger and because of the author's bias. No attempt, however, is made to present the cardiac exchanger as a general model. It is fully understood that the molecular, kinetic and regulatory mechanisms might not be the same in various isoforms.

In the last few years significant advances have been made in the elucidation of the structure-function mechanisms of Na$^+$-Ca^{2+} exchange. Cloning, functional expression, and gene sequence analysis of the cardiac Na$^+$-Ca^{2+} exchanger (NCX1) in Philipson's laboratory (Nicoll et al., 1990) has made it possible to carry out functional and regulatory studies (Hilgemann et al., 1991). Although the major impetus molecular biology came from (Nicoll et al., 1990; Nicoll and Philipson, 1991; Reilander et al., 1992; Philipson and Nicoll, 1992; Shulze et al., 1993, 1996; Philipson et al., 1996), this triumph has in no way diminished the need for detailed functional analysis of the exchanger at different levels (Khananshvili, 1990a,b, 1991a,b, 1996a; Matsuoka and Hilgemann, 1992; Hilgemann, 1996, Khananshvili et al., 1996a). It is now clear that the Na$^+$-Ca^{2+} exchangers represent a family of isoforms with different molecular structure, tissue distribution, ion transport mechanisms, regulation and cellular biology (Nicoll et al., 1990; Reilander et al., 1992; Philipson and Nicoll, 1992; Reilly and Shurgue, 1992; Low et al., 1993; Kofuji et al., 1993; Lee et al., 1994; Shulze et al, 1996; Philipson et al, 1996; Rahamimoff et al., 1996). We are only at the very beginning of understanding the genomic organization of Na$^+$-Ca^{2+} exchangers (Kraev and Carafoli, 1996; Philipson et al., 1996; Shulze et al., 1996). Specific isoforms can arise from the splicing of two mutually exclusive exons and four cassette exons in the regulatory region of the Na$^+$-Ca^{2+} exchanger (Kofuji et al., 1993, 1994; Shulze et al., 1993, 1996). Distinct promoters may drive tissue-specific expression of the exchanger prototypes, suggesting that alternative splicing may also be involved in producing promoter diversity (Lee et al., 1994; Lytton et al., 1996). Site-directed mutagenesis and segment deletions suggest that specific domains and amino acids are involved in cation binding-translocation and in secondary regulation of the exchanger (Philipson et al., 1992; Philipson and Nicoll, 1992; Matsuoka et al., 1993; Nicoll et al., 1994,1996). Application of molecular approaches may have potential biomedical relevance.

Mechanistically, Na$^+$-Ca^{2+} exchanger proteins are especially interesting, because they can catalyze a carrier-type mechanism without formation and breakdown of covalent bonds (Stein, 1986; Lauger, 1987,1991; Krupka, 1989; Khananshvili,

1990a,b). The ion-exchange pumps, like the Na^+-K^+-ATPase, can also exhibit an alternative exposure of ion-binding sites (the carrier type mechanism), but to achieve this, the pump has first to undergo additional steps of phosphorylation and dephosphorylation (Stein, 1986; Lauger, 1987,1991; Glynn and Karlish, 1990; Sachs, 1991). It is not yet clear how the exchanger system can accomplish alternative exposure of ion-binding sites without covalent catalysis (Lauger, 1987, 1991; Khananshvili, 1990a,b, 1991a,b; Hilgemann et al, 1991,1992; Niggli and Lederer, 1991; Khananshvili et al., 1995a, 1996b). Recent kinetic studies suggest that Na^+-Ca^{2+} exchange can be described as separate movements of Na^+ and Ca^{2+} ions through the exchanger, while the Na^+-movement accounts for a rate-limiting and voltage-sensitive step of the exchange cycle (Khananshvili, 1990a, 1991a; Hilgemann et al., 1991; Matsuoka and Hilgemann, 1992; Khananshvili and Weil-Maslansky, 1994; Khananshvili et al., 1996b). This is encouraging, because the specific charge-carrying properties of ion-protein interactions may determine ion transport specificities along the transport cycle in ion pumps (Goldshlegger et al., 1987; Gadsby and Nakao, 1989; Glynn and Karlish, 1990; Laüger,1991; Sachs, 1991) and in ion-exchange systems (Laüger, 1987, 1991; Khananshvili, 1991a,b; Hilgemann et al., 1991, 1992). Despite the progress already made the mechanisms underlying partial reactions in the Na^+-Ca^{2+} exchanger (and in similar proteins) are still poorly understood. A main reason for this is the lack of specific inhibitors and absence of adequate techniques for detecting unstable intermediates.

The Na^+-Ca^{2+} exchanger is modulated by many regulatory modes such as ATP, methylation, proteolytic enzymes, lipid environment, pH, Ca_i, Na_i, redox potential, phospholipase C, and so forth. (Philipson, 1990; Hilgemann, 1990; DiPolo and Beauge, 1991; Hilgemann et al, 1992; Philipson and Nicoll, 1992,1993; Shigekawa et al., 1996; Hilgemann and Ball, 1996). Cellular and biochemical mechanisms underlying these cellular factors are poorly understood. In this chapter, I will discuss only the regulatory mechanisms that are of common interest nowadays. For example, substantial progress has been made in elucidating the mechanisms of secondary modulation induced by intracellular calcium and sodium (Hilgemann et al, 1992a,b, Matsuoka et al., 1993 and by ATP (Hilgemann and Ball, 1996). Despite this progress, practically nothing is known about some important regulatory aspects of the exchanger regulation. For example, we have no idea how the cardiac exchanger responses to rapid oscillation of calcium during the action potential. Application of new time-resolved techniques are necessary for elucidating these issues. Another missing link is how the specific hormones can modulate the exchanger activity in different tissues. No signal transduction pathways have been yet described for regulating the exchanger activity.

Many attempts at finding a selective and potent inhibitor of Na^+-Ca^{2+} exchange (either natural or synthetic) have met with limited success. The lack of a potent pharmacophore seriously hampers the development of new drugs and molecular probes. Recent findings suggest that the XIP-peptides (Li et. al., 1991; Kleiboeker et al., 1992; Chin et al., 1993; Hale, 1996), FMRFa-like peptides (Khananshvili et al, 1993; DiPolo and Beauge, 1994), and conformationally constrained cyclic hexapeptides (Khananshvili et al., 1995a) inhibit Na$^+$-Ca^{2+} exchange in different preparations. Nevertheless, exploration of these peptide inhibitors in most physiological experiments is still wanting.

B. Na$^+$-Ca^{2+} Exchange in Cardiac and Other Tissues

It is a fundamental feature of the living cell to maintain resting levels of intracellular free calcium as low as 0.1-0.2 μM (Carafoli, 1987; Wier, 1990). For fulfilling the physiological needs of a particular cell type, the oscillations of intracellular calcium have to take place in the right place and the right time (Lederer et al., 1990; Tsien and Tsien, 1990; Wier, 1990; Lipp and Niggli, 1994; Langer, 1994). This is a very sophisticated process that requires coordinated operation and fine tuning of carrier-type transport systems (Ca^{2+}-ATPases and Na$^+$-Ca^{2+} exchangers) and Ca^{2+}-channels (Tsien and Tsien, 1990; Bers and Bridge, 1989; Bers, 1991; Bassani and Bers, 1994). A major role of Na$^+$-Ca^{2+}exchange is to control intracellular calcium according to the physiological needs of specific cell types. The Na$^+$-Ca^{2+} exchanger is particularly important in cells in which it is required: (1) to extrude a large amount of intracellular calcium rapidly; (2) to regulate a balance between calcium entry and exit and control calcium levels in intracellular stores; (3) to regulate transient levels of cytosolic calcium during the action potential. Since the action potential is Ca^{2+}-dependent itself, Na$^+$-Ca^{2+} exchange can also control the Ca^{2+}-sensitive currents (for example, K$^+$ and Cl$^-$ channels) (Hilgemann and Nobel, 1987; Powell and Nobel, 1989; Nobel et al., 1991; Bridge et al., 1991). Na$^+$-Ca^{2+} exchange has a vital physiological role in cardiac contractility-relaxation, visual dark adaptation, exocytosis of synaptic vesicles for neurotransmitter release and regulating mitochondrial bioenergetics (Hilgemann and Nobel, 1987; Cervetto et al., 1989; Bridge et al., 1990; Philipson, 1990; Blaustein et al., 1991; Matsuoka and Hilgemann, 1992; Brierley et al., 1994; Reuter, 1995). Na$^+$-Ca^{2+} exchange activity has been detected in most tissues. Nevertheless, the contribution of the exchange system to physiological activities in many cell types is not yet well understood (Blaustein et al., 1991; Philipson and Nicoll, 1993).

In various types of muscle cells (and neurons) the intracellular calcium is modulated by a primary excitation of the membrane. However, excitation-contraction coupling in skeletal and cardiac muscle operates by different mechanisms. In cardiac muscle, the action potential activates voltage-dependent Ca^{2+}-channels in the sarcolemma through which external calcium enters the cell. Calcium entry triggers a larger release of calcium from the sarcoplasmic reticulum via "Ca-induced Ca-release" (Fabiato, 1985; Beukelmann and Wier, 1988; Adams and Beam, 1990). This mechanism reflects a very well-known physiological principle that calcium has to be added to external solutions for a repetitive heart contraction to occur. In *contrast, excitation-contraction coupling in skeletal muscle does not depend on the* entry of extracellular calcium, because a voltage sensor in the T-tubular membrane directly regulates the release of calcium from the sarcoplasmic reticulum (Schneider and Chandler, 1973).

Although it is widely accepted that the Na^+-Ca^{2+} exchanger plays a key role in excitation-contraction coupling (Hilgemann, 1986; Kimura et al., 1987; Bers and Bridge, 1989; Bridge et al., 1990; Nobel et al., 1991; Niggli and Lederer, 1993), the main question to be asked is, How is the Na^+-Ca^{2+} exchanger involved in coordinating the transient levels of calcium, the action potential currents, and muscle contraction-relaxation? Since the Na^+-Ca^{2+} exchanger controls Ca^{2+} extrusion, it may also control the Ca^{2+}-dependent Ca^{2+}-release from the sarcoplasmic reticulum and hence, muscle relaxation. However, the Na^+-Ca^{2+} exchange current can generate net Ca^{2+}-entry into the cell (outward exchange current) when Em > -40 mV and thus, may cause the triggering of Ca^{2+}-release from the sarcoplasmic reticulum, which, in turn, may trigger muscle contraction (Lablanc and Hume, 1990; Lederer et at al., 1990; Levi et al., 1992). One possible mechanism is that a transient increase in internal sodium can drive net calcium entry into the cardiac cell via the Na^+-Ca^{2+} exchanger during the early phase of the action potential (Lablanc and Hume, 1990). This calcium entry may elicit as much as half the normal contraction, as seen in rat myocytes whose membrane potential is reduced to values approaching the normal action potential (Levi et al., 1992). Although Na^+-Ca^{2+} exchange is capable of generating a calcium influx under fixed experimental conditions, the physiological significance of this mechanism remains unclear (Lopez-Lopez et al., 1995). However, it is possible that the mechanism of calcium entry via the exchanger may become relatively important under certain pathophysiological conditions; for example, when L-type Ca^{2+}-channels are suppressed for some reason or another (so, the exchanger may play a compensatory role). Therefore, a major physiological function of the exchanger (at least in excitable tissues) is to extrude the intracellular calcium.

Besides the crucial role of calcium in excitation-contraction coupling, the exchanger is responsible for the transient levels of calcium observed during the plateau phase of the action potential which control the duration and shape of the action potential (Hilgemann, 1986; Powell and Noble, 1989; Nobel et al., 1991; Le Guennec and Nobel, 1994). It was thus suggested that in cells with a relatively low or short plateau the extrusion of Ca^{2+} via the Na^+-Ca^{2+} exchanger may generate net Ca^{2+} extrusion early, and hence, generate a second slow plateau phase (Noble and Powell, 1989). In cells with a relatively high plateau, the time of onset of the exchange current and therefore of net Ca^2-extrusion is somewhat delayed. Presumably, this is because the membrane potential takes longer to reach the range of potential in which $I_{Na\text{-}Ca}$ is larger than I_{Ca} (Noble and Powell, 1989; Nobel et al., 1991).

As one would expect, the cell membrane Na^+-Ca^{2+} exchanger protein occurs in abundance in cardiac muscle (Cheon and Reeves, 1988; Philipson, 1990; Aceto et al, 1992; Philipson and Nicoll,1993) and retinal rods (Cook and Kaupp, 1988; Koch and Stryer, 1988; Schnetkamp et al., 1989,1991). However, the exchanger occurs in smaller amounts in skeletal and smooth muscle, neurons, kidney, intestinal epithelial cells, neutrophils, and various secretory cells (Blaustein et al., 1991; Philipson and Nicoll, 1993). In retinal rod cells, a specific type of exchanger controls visual dark adaptation by regulating cytoplasmic calcium (Koch and Stryer, 1988; Cook and Kaupp, 1988; Cervetto et al., 1989; Schnetkamp et. al., 1991; Reilander et al., 1992; Schnetkamp, 1995a). In neural tissue, the exchanger regulates the cytoplasmic calcium at the nerve terminals, which can, in turn, contribute to the termination of transmitter release (Blaustein et al., 1991; Low et al., 1993; Li et al., 1994; Fontana et al., 1995). Dynamic changes in intracellular calcium and exocytosis in synaptic boutons of hippocampal neurons are regulated by different types of Ca^{2+}-channels and Na^+-Ca^{2+} exchanger (Reuter, 1995). In the kidney, the Na^+-Ca^{2+} exchanger may play a specialized role. It can participate in the transepithelial reabsorption of calcium by extruding Ca^{2+} from the basolateral surface of the renal tubular cell (Smith et al., 1991; Lyu et al., 1991; Reilly et al., 1993; Lee et al., 1994). Therefore, in contrast to the cardiac and neuronal exchangers, the kidney Na^+-Ca^{2+} exchanger may drive the reabsorption of calcium and thus regulate extracellular calcium homeostasis. The physiological roles of Na^+-Ca^{2+} exchange in other cell types remain rather obscure (Reeves, 1992; Philipson and Nicoll, 1993).

C. Contribution of Na^+-Ca^{2+} Exchange to Integrated Ca^{2+} Transport

During excitation-contraction coupling in muscle tissue, the average calcium concentrations in the cytosol increase transiently from 0.1-$0.2\ \mu M$ to 1-$2\ \mu M$ (Wier, 1990; Noma et al, 1991; Ross, 1993) or higher (Cheng et al., 1996). To avoid unwanted Ca^{2+}-overload in intracellular Ca^{2+}-stores (sarcoplasmic reticulum and mitochondria) and to cause muscle relaxation, Ca^{2+}exit must equal Ca^{2+}entry (Bers and Bridge, 1989; Philipson, 1990; Philipson and Nicoll, 1993; Langer, 1994). Ca^{2+}-channels (either voltage- or ligand-activated) provide calcium entry from outside the cell and from intracellular Ca^{2+}-stores (sarcoplasmic reticulum), while the Ca^{2+}-ATPases and Na^+-Ca^{2+} exchanger extrude calcium from the cytoplasm (Carafoli, 1987, 1994; Tsien and Tsien, 1990; Philipson, 1990; Bers et al., 1996). Recent progress has clearly shown that the Na^+-Ca^{2+} exchanger rather than the sarcolemma Ca^{2+}-ATPase is by far the predominant system for intracellular Ca^{2+} extrusion, (Bers and Bridge, 1989; Bridge et al., 1990, 1991). Therefore, Na^+-Ca^{2+} exchange is the only available system that can use the Na^+-gradient for extruding cytosolic Ca^{2+} during the action potential.

In most tissues, the Ca^{2+}-ATPases (of either the cell membrane or endoplasmic reticulum) have a high affinity for calcium, while exhibiting a relatively low turnover rate of calcium transport (Bers and Bridge, 1989; Carafoli, 1987, 1994). Both Ca^{2+}-pumps are responsible for creating and maintaining a primary gradient of calcium in most cells (Stein, 1986; Carafoli, 1987; 1994; Laüger, 1987, 1991; Khananshvili, 1990b). In contrast, the cell membrane Na^+-Ca^{2+} exchanger has a relatively low affinity for calcium and exhibits a high turnover rate in response to rapid changes in membrane potential and ionic conditions (Cheon and Reeves, 1988; Lederer et al.,1990; Philipson, 1990; Matsuoka and Hilgemann, 1992; Shulze et al.,

1993; Hilgemann, 1996). Therefore, this carrier-type system can control the changes in transient levels of calcium in accordance with the physiological needs of the cell (Reeves, 1985; Hilgemann, 1986; Powell and Nobel, 1989; Khananshvili, 1990b; Reeves, 1992). In some exceptional cases (for example, when Ca^{2+}-ATPase activity is absent or suppressed), the Na^+-Ca^{2+} exchanger seems able to control the calcium gradient across the membrane. For example, the outer rod segment (ROS) exchanger is capable of reducing cytoplasmic calcium to very low levels, because it mediates K^+ co-transport and uses a transmembrane gradient of potassium (Cervetto et al., 1989; Schnetkamp and Szerencsei, 1991; Schnetkamp et. al., 1991; Schnetkamp 1995a,b). Interestingly, the ROS exchanger has a much lower turnover rate than the cardiac exchanger (Cook and Kaupp, 1988; Nicoll and Applebury, 1989; Rispoli, 1995, 1996).

For ventricular relaxation to occur intracellular calcium must be lowered by translocating calcium out of the cytosol, thereby allowing calcium to dissociate from troponin C (a trigger for activation of actomyosin complex). Four Ca^{2+}-transport systems can compete to remove intracellular calcium from the myoplasm in mammalian ventricular myocytes. Two of them (the sarcolemma Ca^{2+}-ATPase and the Na^+-Ca^{2+} exchanger) extrude calcium from the cell. Two other systems, the sarcoplasmic reticulum (SR) Ca^{2+}-ATPase and mitochondrial Ca^{2+}-uniport systems are responsible for Ca^{2+}-uptake into intracellular Ca^{2+}-stores (Bers, 1991). In general, the sarcolemma Na^+-Ca^{2+} exchanger and SR Ca^{2+}-ATPase are the most important quantitatively, while their relative roles vary in different species (Bassani and Bers, 1994, 1995; Bers et al., 1996). A physiological demand of the cell is that in the steady state the amount of calcium leaving the cell must equal the amount entering the cell during a cardiac cycle. This balance is properly controlled by Na^+-Ca^{2+} exchange, which removes nearly all the calcium (90-95%) that has entered the cell during the depolarization of the membrane (Bridge et al., 1990, 1991; Bouchard et al., 1996). However, most of the calcium that induces muscle contraction comes from the sarcoplasmic reticulum; this varies substantially in different species (Bassani and Bers, 1995; Bers et al., 1996). In fact, the relative roles of Na^+-Ca^{2+} exchange and sarcoplasmic reticulum Ca^{2+}-ATPase differ among species. For example, during a twitch in rat ventricle myocytes the SR Ca^{2+}-ATPase removes ~92 % of the calcium, while the Na^+-Ca^{2+} exchanger removes only 7%. In other species (rabbit, ferret, cat, and guinea-pig) the contribution is more or less in the range of 70% SR Ca^{2+}-ATPase and 25-30% Na^+-Ca^{2+} exchange (Bers et al., 1995).

D.   Subcellular Compartmentalization of Na^+-Ca^{2+} Exchange and Ca^{2+}-Pools

Besides the Na^+-Ca^{2+} exchange system of the plasma membrane, the Na^+-Ca^{2+} exchange also occurs in mitochondria, serving to remove calcium that enters the matrix of mitochondria via the ruthenium red-sensitive uniport system (Gunter and Pfieffer, 1990; Li et al., 1992b; Brierley et al., 1994). In other words, mitochondrial Na^+-Ca^{2+} exchange regulates calcium concentration in the mitochondrial matrix and by this means controls bioenergetics. The mitochondrial exchanger has been assumed to have an electroneutral stoichiometry of $2Na^+:Ca^{2+}$; however, recent thermodynamic studies suggest that the stoichiometry of mitochondrial exchange is not electroneutral (Brierley et al., 1994; Jung et al., 1995). Although the molecular structure of the mito-·chondrial exchanger is not known, it might be regarded as a distinct molecular entity (Li et al., 1992b; Nicoll and Philipson, 1993; Brierley, 1994; Jung et al., 1995).

Fluorescent antibodies raised against the cardiac sarcolemma Na^+-Ca^{2+} exchanger protein have been used to localize the exchanger in isolated cardiac myocytes (Frank et al., 1992). This work reveals that the sarcolemma of T-tubules consistently shows an intense immunofluorescence, while some regions of the sarcolemma surface show a relatively sparse density of exchangers. In contrast to this finding, the Na^+-Ca^{2+} exchanger protein has been detected in great amounts on the outer sarcolemma surfaces and on the intercalated discs of heart cells (Kieval et al., 1992). Apart from these differences, both reports are in agreement that the T-tubular system has an abundance of Na^+-Ca^{2+} exchanger protein. The distribution of the exchanger in neurons is different (Juhaszova and Blaustein, 1996) in that exchanger labeling occurs on neurites and soma, while the label is more concentrated at nerve terminals and growth cones. Recent evidence suggests that the cardiac exchanger binds to the cytoskeletal protein ankyrin, suggesting a possible interaction between the exchanger and cytoskeleton (Li et al., 1993). Co-localization of the exchanger with ankyrin has been examined in adult and neonatal myocardial cells (Frank et al., 1995). In adult cells, there is co-localization of these two proteins, while ankyrin occurs mainly at the Z-disk before the development of the T-tubular system in newborn rabbit cells. These data suggest that many exchanger sites are nonhomogeneously distributed in T-tubules close to Ca^{2+}-release channels, suggesting that ankyrin may immobilize the exchanger during the development of T-tubules (Frank et al., 1995). The view that there is a high site density of exchanger molecules in T-tubules may have considerable implications for excitation-contraction coupling in cardiac muscle. It has been proposed that the exchanger molecules are localized near the Ca^{2+}-release sites of the sarcoplasmic reticulum in the "diadic region," where the sarcoplasmic reticulum borders the T-tubules (Langer and Rich, 1992; Langer, 1994). According to this model, calcium concentrations can be held at high levels due to inner sarcolemma leaflet anionic phospholipid sites, which can optimize and prolong the function of the exchanger (Peskoff et al., 1992; Langer, 1994; Langer et al.,1995).

New methods have been introduced to study the "three-dimensional" distribution of the Na^+-K^+-ATPase, Na^+-Ca^{2+} exchanger and Ca^{2+}-storage proteins of the sarcoplasmic reticulum, as well as attachment of contractile filaments to the membrane in smooth muscle (Moore et al., 1993; Moore and Fay, 1993). These studies show that the Na^+-Ca^{2+} exchanger is largely co-distributed with the Na^+-K^+-pump in unique regions of the plasma membrane. These regions are close to the sarcoplasmic reticulum in sites distinct from the sites where contractile filaments attach to the membrane. The observed clustering of the Na^+-K^+-ATPase and Na^+-Ca^{2+} exchanger in regions closely apposed to the sarcoplasmic reticulum may have important implications for Ca^{2+}-homeostasis. It suggests that the Na^+-Ca^{2+} exchanger may have preferential access to calcium within the sarcoplasmic reticulum across a restricted diffusion space in which ion concentrations may vary significantly from their average cytosolic levels (Lederer et al., 1990; Langer and Rich, 1992; Langer, 1994). New findings strongly support local elevations in intracellular calcium ('Sparks'), reflecting perhaps the elementary calcium release events from the sarcoplasmic reticulum (Cheng et al., 1996).

II.   GENERAL PROPERTIES OF Na^+-Ca^{2+} EXCHANGE

A.   Experimental Approaches and Methodology

In isolated preparations of sarcolemma vesicles and reconstituted proteoliposomes, Na^+-Ca^{2+}, Na^+-Na^+ and Ca^{2+}-Ca^{2+} exchanges can be assayed by measuring the kinetics of ^{22}Na or ^{45}Ca (Reeves and Sutko, 1979; 1980; Reeves, 1988; Cheon and Reeves, 1988; Khananshvili, 1990a, 1991a; Khananshvili et al., 1993, 1995a,b; Khananshvili and Weil-Maslansky, 1994). The advantage of this approach is that the initial rates (as fast as 0.5-1.0 s) of exchange modes can be quantitatively estimated by using the semi-rapid mixing techniques (Khananshvili, 1990a, 1991a; Khananshvili and Weil-Maslansky, 1994; Khananshvili et al., 1995a). These techniques are still very useful for studying the kinetics of partial reactions; they also allow an estimation of the relative rates of ion movements through the exchanger under various conditions (Khananshvili and Weil-Maslansky, 1994; Khananshvili et al., 1995a). Unfortunately, Na^+-Ca^{2+} exchange activity in vesicular preparations (either in sarcolemma vesicles or reconstituted proteoliposomes) is insensitive to some regulatory factors. It is not clear whether these sarcolemma membrane preparations lack certain cellular factors and/or whether their regulatory abilities are impaired during the preparation procedures.

The cardiac sarcolemma vesicles contain a mixture of inside-out (55-70%), right-side out (15-25%) and leaky (5-30%) vesicles (Ambesi et al., 1991). But under fixed experimental conditions the inside-out vesicles contribute to most, if not all of the Na^+-Ca^{2+} exchange activity (Li et al., 1991; Ambesi et al., 1991; Philipson and Nicoll, 1993; Khananshvili et al., 1993, 1995a,b). Squid axons (DiPolo and Beauge, 1990,1991) and barnacle muscle cells (Rasgado-Flores et al., 1988, 1991) are especially useful for studying both the mechanistic and regulatory aspects of various exchange modes. This is because these preparations still have regulatory abilities and the exchange modes can be studied by ion fluxes (^{22}Na or ^{45}Ca) and electrophysiological techniques.

During the last few years electrophysiological approaches have been improved dramatically. Hilgemann has developed a new technique, the giant membrane patch method (Hilgemann, 1989; 1990, Hilgemann et. al., 1991). This technique has had a tremendous impact in the field, especially for mutant analysis (Philipson & Nicoll,

1992,1993; Philipson et al., 1992; Matsuoka et al., 1993; Nicoll et al., 1994, 1996). The giant patch technique offers many advantages as can be seen in work with the native exchanger in cardiac membrane and with the cloned exchanger in *Xenopus* oocytes (Hilgemann, 1990; Hilgemann et al., 1991; Matsuoka and Hilgemann, 1992; Matsuoka et al., 1993; 1995; Hilgemann, 1995). This approach allows high-resolution measurement of exchange current under well-controlled ionic conditions (20-150 pA with 4-12 pF and 2-10 GΩ seals) and resolution of current-voltage relationships (I-V curves) over a > 200 mV range. Hilgemann also introduced a voltage jump technique for direct measurement of charge movements through the cardiac exchanger (Hilgemann, 1996). Such techniques with microsecond time resolution could be especially useful for studying the mechanisms of partial reactions. Intracellular photorelease of Ca^{2+} from "caged calcium" (DM-nitrophen) was applied to investigate the Ca^{2+}-activated currents in ventricular myocytes by applying a patch clamp technique to whole-cell configurations (Niggli and Lederer, 1991, 1993). Similar techniques were also applied to giant excised membrane patches by using powerful laser flashes of 10 ns duration (Kappl and Hartung, 1995). These experimental approaches could be useful for studying transient reactions (conformations) of the exchanger and might help separate the Na$^+$-Ca^{2+} current from the other Ca^{2+}-activated currents in intact cells. Moreover, electrical currents have been obtained in black lipid membranes, containing a partially purified lobster exchanger (Eisenrauch et al., 1995). The Na$^+$-Ca^{2+} exchanger was first reconstituted into liposomes and then absorbed on black lipid membranes and the exchange reaction was activated by photolysis of caged Ca^{2+}.

More recently, Na$^+$-Ca^{2+} exchange reactions have been monitored using a combination of fast response optical probes and a multi syringe stopped-flow technique (Khanabshuili et al., 1996a). This permits the measurement of fast exchange reactions with a millisecond time resolution. The advantage of the stopped-flow technique is that a number of subsequent reactions can be carried out in succession, thus leading to the formation of specific intermediates that can be identified. The exchanger in intact cells has been difficult to investigate, because no specific inhibitor is yet available. It is noteworthy that specific inhibition of Na$^+$-Ca^{2+} exchange has been achieved in single cultured myocytes by using antisense oligodeoxynucleotide techniques (Lipp et al., 1995). Application of advanced laser-scanning confocal microscopy and related optical techniques could be useful for investigating the subcellular aspects of behavior of the exchanger under physiologically related conditions (Lipp and Niggli, 1994; Cannell et al., 1996; Lopez-Lopez et al., 1995; Cheng et al., 1996).

B. Stoichiometry and Electrogenicity of Na+-Ca2+ Exchange

Different experimental approaches provide evidence that the stoichiometry of ion-exchange in muscle cell membrane is 3Na$^+$:Ca^{2+} (Reeves and Hale, 1984; Kimura et al., 1987; Rasgado-Flores and Blaustein, 1987; Crespo et al., 1990; Le Guennec and Nobel, 1994). In contrast to muscle, in the rod outer segment exchanger is coupled to K^+ cotransport and the overall stoichiometry is $4Na^+:Ca^{2+},K^+$ (Cervetto et al., 1989; Schnetkamp et al., 1989). The differences in stoichiometry may reflect the fact that the cardiac exchanger (Nicoll et al. 1990; Philipson and Nicoll, 1992, 1993) and retinal exchanger (Reilander et al., 1992) are structurally and functionally related but different molecules. Although the ion-exchange stoichiometry is less established in other tissues, recent studies support a stoichiometry of $3Na^+:Ca^{2+}$ in neuronal cells (Blaustein et al., 1991; Fontana et al., 1995).

Ca^{2+} extrusion by the Na^+-Ca^{2+} exchanger is driven by the electrochemical gradient of Na^+. If E_{Na} is the sodium equilibrium potential and E_M the membrane potential, the overall driving force on Na^+ can be written as $n(E_{Na}\text{-}E_M)$, with $n = 3$. This force is used to overcome the electromotive force acting on Ca^{2+}, which is $2(E_{Ca}\text{-}E_M)$. No net movement of Na^+ and Ca^{2+} occurs under equilibrium conditions; that is, $3(E_{Na}\text{-}E_M) = 2(E_{Ca}\text{-}E_M)$. Both E_{Na} and E_{Ca} depend on $[Ca]_o/[Ca]_i$ and $[Na]_o/[Na]_i$, and they can be calculated using the Nerst equation. Substituting E_{Na} and E_{Ca} in the equilibrium equation the intracellular concentration of calcium can be calculated as:

$$[Ca]_i = [Ca]_o \; \{[Na]_o/ [Na]_i\}^3 \; exp(E_M F/RT) \tag{1}$$

Both $[Ca]_o$ and $[Na]_o$ change very little during the action potential, while $[Ca]_i$, $[Na]_i$ and E_M undergo very complicated time-dependent changes at various stages of the action potential, thereby affecting the exchange rate. On the basis of this equation, the exchange mechanism is in equilibrium at $E_M = -40mV$, while at more positive potentials, calcium enters the cell in exchange for sodium leaving the cell. At more negative potentials than -40 mV, calcium ions are extruded in exchange for extracellular sodium (Blaustein, 1974; Mullins, 1977, 1979; Reeves, 1985; Eisner and Lederer, 1985; Bers, 1987; Philipson, 1990). However, membrane depolarization leads to a rapid increase in intracellular calcium to about 1-2 μM via -Ca^{2+} -entry though the voltage-sensitive Ca^{2+}-channels and Ca^{2+} release from the sarcoplasmic reticulum. This allows the exchanger to extrude calcium. Thus, the thermodynamics of Na^+-Ca^{2+} exchange can favor net Ca^{2+}-efflux even during the plateau phase of the cardiac action potential. Although the thermodynamics of Na^+-Ca^{2+} exchange can determine the electrophysiological properties of the exchange, under most physiological conditions the exchanger does not reach equilibrium (Reeves, 1985; Eisner and Lederer, 1985; Philipson, 1990). Therefore, the kinetics may determine the voltage-sensitive response of the exchanger (Khananshvili, 1991a,b; Matsuoka and Hilgemann, 1992; Khananshvili and Weil-Maslansky, 1994). Practically nothing is known about the rapid modulatory responses of the exchanger during the resting and action potentials.

The stoichiometry of Na^+-Ca^{2+} exchange is a thermodynamic property of the system, which may or may not reflect the voltage sensitive response of the exchange rate. This is because the voltage response of different exchange modes is determined by a specificity of the rate-limiting pathway involving the charged or uncharged species (Khananshvili, 1991a,b; Khananshvili and Weil-Maslansky, 1994; Khananshvili et al., 1996a). For example, under certain conditions the system may show no response to voltage change; even if so, the stoichiometry is still electrogenic ($3Na^+$:Ca^{2+}). This could occur if the voltage-insensitive step (e.g., Ca^{2+}-transport) becomes rate-limiting; for example, at acidic pH when the Na$^+$-Ca^{2+} exchange loses voltage-sensitivity and the rates of Na$^+$-Ca^{2+} and Ca^{2+}-Ca^{2+} exchanges are equal (Khananshvili and Weil-Maslansky, 1994; Khananshvili et al., 1996a). It is thus clear that a separation of thermodynamics and kinetics is of primary importance not only from a mechanistic and regulatory point of view but also from a physiological point of view. This issue can be conceptually extended to include the mitochondrial Na$^+$-Ca^{2+} exchanger. It was assumed until recently that the mitochondrial Na$^+$-Ca^{2+} exchanger exhibits electroneutral exchange with a stoichiometry of $2Na^+$:Ca^{2+}. However, the recent work of Brierley and coworkers provides evidence that this stoichiometry in heart mitochondria is not electroneutral (Brierley et al., 1994; Jung et al., 1995). If the stoichiometry of mitochondrial Na$^+$-Ca^{2+} exchange is electrogenic, the rate of ion exchange may not respond to voltage, presumably because the rate-limiting step is voltage-insensitive.

C. Site Density and Turnover Rate of Na$^+$-Ca^{2+} Exchange

Na$^+$-Ca^{2+} exchange currents as high as 20-50 pmol.cm^{-2}.s^{-1} can be measured in mammalian heart cells by using the patch-clamp technique (Kimura et al., 1986, Hilgemann et. al., 1992). These values of ion-currents are comparable to a V_{max} = 5-20 nmol Ca.mg protein.s^{-1}, obtained by ion-flux studies in sarcolemma vesicles and reconstituted proteoliposomes (Reeves, 1985; Philipson, 1990; Khananshvili and Weil-Maslansky, 1994). Although these measurements were done by using ionic concentrations that are not physiological, it is obvious that Na$^+$-Ca^{2+} exchange represents a high-capacity system that is able to control intracellular calcium during the plateau phase of the action potential. The synaptic Na$^+$-Ca^{2+} exchanger is perhaps a high-capacity system too, but it remains uncertain whether the site density of this exchange system is as high as in cardiac tissue (Fontana et al., 1995). Based on estimates of the cardiac exchanger site density in reconstituted proteoliposomes, a turnover number of >1000 s^{-1} was reported (Cheon and Reeves, 1988). This lower limit estimate is consistent with the idea that the exchanger accounts for only 0.1-0.2% (or less) of the total sarcolemma protein. Niggli and Lederer (1991) who have used a combination of caged Ca^{2+} and patch-clamp techniques in cardiac cells found a transient current whose estimated value was 2500 s^{-1}. In patches derived from *Xenopus* oocytes expressing the cloned Na$^+$-Ca^{2+} exchanger, the measured turnover rate was 5000 s^{-1} (Hilgemann et al. 1991). In myocytes the site densities of the exchanger vary from 50 sites/μm^2 to 2500 sites/μm^2 (Niggli and Lederer, 1991; Li and Kimura, 1991; Hilgemann et al., 1991; Hilgemann, 1995). These estimates correspond to about 10^6-10^7 copies of exchanger molecules per cell, but they are to be taken with caution, since a turnover number of 1000-30,000 s^{-1} has been assumed. More recently, a direct charge movement was measured through the Na$^+$-Ca^{2+} exchanger, expressed in oocytes, by using the voltage jump technique

(Hilgemann, 1995). In these experiments the charge movements were nearly completed in 25 μs, showing a turnover rate of 20000-30000 s^{-1}. The resolution of the turnover rate is an important issue, because the limiting rate of the exchange cycle may determine the exchanger response modes during the action potential. For example, the exchanger can mediate the Ca^{2+}-entry at E_m>-40mV if the turnover rate is 'fast enough' the response the membrane potential changes during the action potential. The turnover number of an exchange cycle is even less obvious in other prototypes. In the mammalian ROS exchanger, for example, the reported values of site density vary from 200-450/μm^2 to 35000/μm^2 and the turnover rate from 10 s^{-1} to 115 s^{-1} (Cook and Kaupp, 1988; Nicoll and Applebury, 1989). Recent studies by Rispoli and collaborators show a lower turnover number of 2-3 s^{-1} and a high site density of 10000/μm^2 (Rispoli et al., 1995, 1996) for the lizard's ROS exchanger. It thus seems likely that the cardiac exchanger is 10^2-10^3 faster than the ROS exchanger. This is perhaps not surprising because the ROS exchanger, being only a Ca^{2+} extrusion system, is designed to bind intracellular Ca^{2+} in the nanomolar range. These concentrations of Ca^{2+} are one order of magnitude lower than the cytosolic calcium at which the muscle exchanger operates.

D.   Homogeneous Exchange Modes: Na^+- Na^+ and Ca^{2+}-Ca^{2+} Exchanges

Besides primarily catalyzing Na^+-Ca^{2+} exchange, the exchanger can also catalyze Na^+-Na^+ and Ca^{2+}-Ca^{2+} exchanges (Bartschat and Lindenmayer, 1980; Reeves, 1985; DiPolo and Beauge, 1991; Khananshvili, 1990b,1991a,b). These partial reactions can operate in intact cells, but their contribution to specific cellular activities and biochemical regulation is unclear. Homogeneous ion exchange reactions can be described as a separate movement of analogous ions (either Na^+ or Ca^{2+}) across the membrane but in opposite directions (Khananshvili, 1990a,1991a; Khananshvili and Weil-Maslansky, 1994; Khananshvili et al. 1995a, 1996b). Whereas Na^+ is not required for Ca^{2+}-Ca^{2+} exchange, Na^+ or other monovalent cations such as K^+, Li^+, and Rb^+ accelerate Ca^{2+}-Ca^{2+} exchange. There is no evidence that potassium can be co-transported in the cardiac exchanger (Slaughter et al., 1983; Yasui and Kimura, 1990). On the other hand, in cardiac cells and squid axons the intracellular calcium accelerates Na^+-Na^+ exchange (Reeves, 1985; Philipson, 1990; DiPolo and Beauge, 1991), suggesting that the regulatory Ca^{2+}-binding site can control the rate of the exchange. This is consistent with the view that the occupation of the regulatory Ca^{2+}-binding site(s) can accelerate the rate-limiting and voltage-sensitive movements of Na^+ during the Na^+-Ca^{2+} exchange (Khananshvili,1991a,b; Hilgemann et al., 1991; Matsuoka and Hilgemann, 1992). Although K^+ is co-transported in the ROS exchanger, Ca^{2+}-Ca^{2+} exchange can occur without the addition of K^+ to the medium (Schnetkamp et al., 1991).
We have only begun to understand the mechanisms of partial reactions (Khananshvili, 1991a,b; Khananshvili and Weil-Maslansky, 1994; Khananshvili et al., 1995a,1996b). Kinetic studies suggest that Ca^{2+}-Ca^{2+} exchange is an integral

part of Na$^+$-Ca^{2+} exchange; such independent evidence favors the consecutive (ping-pong) mechanism (Khananshvili and Weil-Maslansky, 1994). Under equilibrium exchange conditions, in which Na$^+$ or Ca^{2+} concentrations are subsaturating at both sides of the membrane, Ca^{2+}-Ca^{2+} and Na$^+$-Na$^+$ exchanges show some degree of voltage-sensitivity in reconstituted proteoliposomes (Khananshvili, 1991a,b). However, at saturating ionic conditions the steady-state exchange rates of both Ca^{2+}-Ca^{2+} and Na$^+$-Na$^+$ exchanges become apparently voltage-insensitive (Khananshvili, 1991a,b), suggesting that the binding of both ions is weakly voltage-sensitive. In the case of unidirectional Na$^+$-Ca^{2+} exchange Na$^+$ transport is a voltage-sensitive and rate-limiting step of the exchange cycle (Khananshvili, 1991a,b; Hilgemann et al., 1991; Matsuoka and Hilgemann, 1992). Recently, Hilgemann (1996) has obtained direct evidence for it by monitoring large electric signals under conditions in which Na$^+$-Na$^+$ exchange occurs; no such signal was detected for Ca^{2+}-Ca^{2+} exchange. In squid axons, Na$^+$-Na$^+$ exchange shows no voltage sensitivity while Ca^{2+}-Ca^{2+} exchange shows the same voltage dependence as Na$^+$-Ca^{2+} exchange (DiPolo and Beauge, 1990, 1991). It is possible that in this system the voltage-sensitive Ca^{2+}-transport step is rate-limiting.

III. STRUCTURE, ISOFORM DIVERSITY, AND REGULATION OF EXPRESSION

A. Structural Motifs of the Cardiac Na$^+$-Ca^{2+} Exchanger (NCX1)

The cardiac Na$^+$-Ca^{2+} exchanger (NCX1) gene (970 amino acids) was the first to be cloned and sequenced (Nicoll et al., 1990). The NCX1 gene is located on human chromosome 2p21-p23 (Sheih et al., 1992). The protein molecule contains six potential glycosylation sites (Nicoll et al., 1990), but only one of them is glycosylated in the mature protein (Hryshko et al., 1993). The NCX1 exchanger is abundant in cardiac muscle and can be detected in brain, kidney, lung, smooth muscle, and skeletal muscle. NCX1 contains an N-terminal sequence of 32 amino acids, which is cleaved during processing (Nicoll et al., 1990; Philipson and Nicoll, 1992; Durkin et al., 1991). The mature NCX1 protein (938 amino acids) has 11 putative transmembrane segments with a large hydrophilic domain (intracellular loop) of 520 amino acids between the membrane segments 5 and 6 (Philipson and Nicoll, 1992) (Figure 1). Therefore, NCX1 may fall into a class of transporter family that has a large hydrophobic domain and 11-12 transmembrane segments (Henderson, 1993).

An obvious similarity between NCX1 and other ion-transporters is to be found in the transmembrane segments. For example, the 23 amino acid sequences (amino acids 180-202) that are located in the transmembrane segments 4 and 5 exhibit ~ 50% identity to an appropriate portion (amino acids 308-330) of the α-subunit of Na$^+$-K$^+$-ATPase (Nicoll et al., 1990). At the C-terminal end of this region a 9 amino acid sequence (amino acids 194-202) shows 61% identity to the corresponding re-

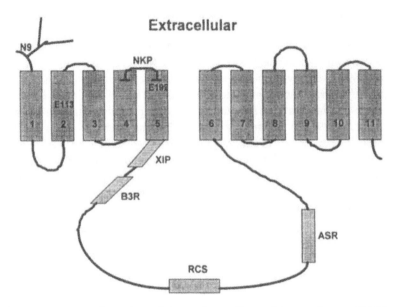

Figure. 1. Structural topology of the cardiac sodium-calcium exchanger (NCX1).

gions of the Na^+-K^+-ATPase, sarcoplasmic reticulum and plasma membrane Ca^{2+}-ATPase and K^+-H^+-ATPase (Nicoll et al., 1996; Philipson et al., 1996). Likewise, E199 is conserved in all these ATPases and is an important residue for Ca^{2+}-binding in the SR Ca^{2+}-ATPase (Clarke et al., 1989; Andersen and Vilsen, 1995). The transmembrane segments 2-3 and 7-8 of NCX1 share significant identity with an analogous region in the ROS exchanger, suggesting that these segments may involve ion transport pathways. Deletion of the intracellular loop does not alter basal activity of the exchanger but current-voltage relationships alter a secondary modulation induced by intracellular sodium or calcium (Matsuoka et al., 1993). These data provide evidence that the intracellular loop is not involved in ion-binding and transport, but can play a regulatory role for controlling exchanger activity. The putative membrane segments of NCX1 contain hydrophilic residues (especially those containing hydroxyl groups) (Nicoll and Philipson, 1991). Likewise, some of these hydrophilic residues in transmembrane segments 2,3,4, and 5 can be modeled to line one face of α-helixes and thus, may form the ion-transport pathway of the exchanger (Nicoll and Philipson, 1991; Philipson and Nicoll, 1992; Nicoll et al., 1996).

Thirty-seven mutations of 29 different amino acids were constructed and expressed in *Xenopus* oocytes (Philipson et al., 1992; Nicoll et al., 1994; 1996). Most of these mutations involve substitutions of either acidic or basic amino acids. They are located in one of two regions similar to the ROS exchanger involving transmembrane segments 2-3 or 7-8 (the RETX region), or in a region similar to the Na^+-K^+ pump (the NKP region). Amino acids in the RETX region were the most sensitive to mutagenesis and produce inactive exchangers (S109A, S11A, T203V, T810V, S818A, N842V). Among the seven serine residues, six are critical for exchange activity (Nicoll et al., 1994; 1996). Two of the mutants (G138A and G837A) exhibited altered voltage-current relationships (IV-curves), suggesting that the relevant transmembrane regions might be involved in a cation translocating process (Nicoll et al., 1994,1996).

The sequence analysis suggests that the putative transmembrane segments contain only four negatively charged residues (E113, E120, E196, and E199). Among the negatively charged amino-acids that belong to putative transmembrane segments, only two glutamic acids (E113 and E199) have been identified as essential for ion-exchange activities (Philipson et al., 1992; Nicoll et al., 1996). E199, which belongs to the NKP region is highly conserved in other pumps and plays a critical role in ion binding/translocation (Clarke, 1989; Nicoll et al., 1995; Andersen and Vilsen, 1995). These findings are important because they shed new light on the suggestion that the ion-translocating domain of the cardiac exchanger may contain two negative charges (Khananshvili, 1991a,b; Khananshvili et al., 1995a, 1996b). In this model, the ECa species have no charge, while the ENa$_3$ species are positively charged. Note that the charge-carrying properties of specific intermediates and the rate-limiting pathway (for example, the rate-limiting transport of sodium or calcium) could be different in various isoforms. Therefore, the charge-carrying properties of ion-protein intermediates may determine the relative rates of the exchange modes (for example, Na$^+$-Ca^{2+} vs. Ca^{2+}-Ca^{2+} exchange) and voltage-sensitivity of specific exchange modes (Khananshvili, 1991a,b; Khananshvili and Weil-Maslansky, 1994; Khananshvili et al., 1995a,1996b). In contrast to NCX1, the barnacle muscle exchanger may contain three negative charges on the ion-transport surface. This interesting possibility might be tested, when molecular biology methods make it possible to study this exchanger prototype.

B. Gene Family of Na$^+$-Ca^{2+} Exchangers

The ROS Na$^+$-Ca^{2+},K$^+$ exchanger (1199 amino acids) shows a very similar overall topology to the NCX1 exchanger and exhibits 11 transmembrane segments interrupted by a hydrophilic intracellular loop (408 amino acids) between transmembrane segments 5 and 6 (Reilander et al., 1992; Philipson and Nicoll, 1992). The ROS exchanger contains six potential asparagine-linked glycosylation sites, compared with one in NCX1 (Nicoll et al., 1990; Reilander et al., 1992). Despite the overall topological similarities of the NCX1 and ROS exchangers, the similarities at the amino-acid level are restricted to two short regions spanning transmembrane segments 2 and 3, and transmembrane segments 7 and 8 (Reilander et al., 1992). In the case of the rabbit kidney exchanger, it is very similar to that of NCX1, but contains an 80 amino acid sequence in the intracellular loop that is significantly different from that in the NCX1 loop (Reilly and Shugrue, 1992; Reilly et al., 1993). Another isoform of the exchanger, NCX2 (921 amino acids), has been identified and sequenced (Li, et al.,

1994; Philipson et al., 1996). In contrast to NCX1, the NCX2 is located on human chromosome 14 and its expression is limited to brain and skeletal muscle. NCX1 and NCX2 show 61% and 65% identity at the nucleotide and amino acid levels, respectively, and apparently are the products of distinct genes (Li, et al., 1994). Intracellular calcium secondarily regulates both NCX1 and NCX2, but with different affinity (Li et al., 1994; Matsuoka et al., 1995; Philipson et al., 1996). The Na^+-Ca^{2+} exchanger gene (Calx) of *Drosophila melanogaster* has been cloned and sequenced (Shwarz and Benzer, 1995). *Calx* has a very similar topological structure to NCX1 and NCX2 (Philipson et al., 1996; Shulze et al., 1996). *Calx* encodes two repeated motifs, *Calx*-α and *Calx*-β, that overlap domains required for ion transport and regulation (Shwarz and Benzer, 1995). Likewise, the sequence analysis of these structural motifs suggests that *Calx* was formed from several gene duplications in the Precambrian Age. One would expect the *Drosophila* exchanger to have the same regulatory properties with respect to intracellular calcium and sodium as that of other known isoforms (Philipson et al., 1996). A distant homolog of the mammalian NCX genes has been identified in *C. elegans* and completely sequenced (Kraev and Carafoli, 1996). It encodes 20% shorter protein with 55% homology to human NCX1, and lacks most of the region that undergoes alternative splicing in vertebrates. All these data suggest that a primary NCX gene had emerged in the primitive nervous system which has subsequently adapted to other species by using novel domains, encoded in additional exons.

C. Functional Domains on the Intracellular Loop

Electrophysiological analysis of Na^+-Ca^{2+} exchange shows that addition of calcium at the cytoplasmic side activates exchange, while addition of cytoplasmic sodium causes partial inactivation of exchange current (Hilgemann et al., 1992; Matsuoka and Hilgemann, 1994). This analysis suggests that secondary modulation by calcium is caused by a regulatory site that is different from transport sites. To identify the relevant regions on the intracellular loop, the regulatory capacity of the exchanger was examined by deletion mutagenesis (Matsuoka et al., 1993, 1995; Philipson et al., 1996). It was found that mutants with deleted residues 240-679 and 562-685 show no regulation by cytoplasmic calcium (Matsuoka et al., 1993). However, mutant 562-685 (but not mutant 240-679) exhibited sodium-dependent inactivation, suggesting that a putative regulatory site may reside in the region of 562-685 (Figure. 1). Mutations have parallel effects on the high affinity of the exchanger loop for calcium-binding and alter the kinetic parameters of calcium and sodium-dependent secondary modulation (Levitski et al., 1994; Matsuoka et al., 1995). The number of Ca^{2+} ions binding in this region was not quantified but could be more than one. A number of point mutations (for example, D447 and D498) and a short range deletion of 450-456 amino-acids alter dramatically the affinity of calcium-dependent activation (Matsuoka et al., 1995). It is noteworthy that mutations that eliminate sodium-dependent inactivation also accelerate the response to regulatory calcium (Matsuoka et al, 1995; Philipson et al., 1996). Surprisingly, the Na$^+$-Ca^{2+} exchanger cloned from *Drosophila* is anomalous; this exchanger is inhibited, rather than stimulated by regulatory calcium (Philipson et al., 1996). These peculiar differences in regulatory capacities may provide useful information on structure-function relationships in the intracellular loop.

Although Na$^+$-Ca^{2+} exchange can be activated by intracellular ATP in various tissues (Blaustein, 1977; DiPolo and Beauge, 1979, 1991; Hilgemann, 1990), the underlying biochemical mechanisms and relevant structure-function relationships are not yet known. The finding of a new splicing variant shows an extra sequence of 9 amino acids in the intracellular loop of the frog heart sarcolemma exchanger (Iwata and Carafoli, 1996). This short sequence is located at the site where most alternative splicing events occur (amino acids 570-645) in other species. It is possible that this 9 amino acid stretch is coded in a frog by a novel exon (so far not found in mammals) that involves a putative ATP-binding site. This insert completes a "Walkers A" nucleotide binding motif, not yet known to be present in any mammalian exchangers (Iwata and Carafoli, 1996).

A 20 amino acid segment (amino acids 219-238) of the intracellular loop (close to transmembrane segment 5) was identified as a possible calmodulin-binding domain with autoinhibitory potential (Nicoll et al., 1990; Li et al., 1991). This region contains mostly basic and hydrophobic residues and resembles the calmodulin-binding domains of various proteins (see Carafoli, 1994). In plasma membrane Ca^{2+}-ATPase, the relevant binding site becomes functional upon the binding of calmodulin, resulting in a change in the kinetic parameters of the pump (Carafoli, 1994). There is no evidence that the calmodulin-binding site is functional in any prototype of the Na$^+$-Ca^{2+} exchanger. The sequence of amino acid residues 263-279 shows 48-60% identity to erythrocyte anion antiporter proteins (band 3). The functional role of this region is not known. It was proposed that this segment may be involved in interactions with the cytoskeleton (Reeves et al., 1994), since some antiporters may interact with ankyrin (Li et al., 1993; Frank et al., 1995).

D. Alternative Splicing as a Mode for Isoform Production

The human, rat and cow heart exchangers are strikingly similar (>90% amino acid identity) to the canine heart exchanger (Kofuji et al., 1993,1994; Nakasaki et al., 1993; Shulze et al., 1993,1996). Likewise, it appears that they differ primarily within a limited region of the central domain of an intracellular loop (amino acids 570-645). Shulze and collaborators have suggested that alternative splicing has the potential of generating the various isoforms of the exchanger in different tissues (Kofuji et al., 1994; Shulze et al., 1996). To date, eight alternatively-spliced isoforms of NCX1 have been described (depicted as NACA 1-8), expressed in a tissue-specific manner (Reilly and Shurgue, 1992; Kofuji et al., 1993, 1994; Shulze et al., 1993; Reilly and Lattanzi, 1995). NACA1 and 8 are expressed in the heart, NACA2,3, and 7 in the kidney, and NACA4,5, and 6 in the brain. According to

Shulze et al., the isoforms arise from the splicing of two mutually exclusive exons (A,B), and four cassette exons (C,D,E,F) at the carboxyl end of the intracellular regulatory loop (Kofuji et al., 1993, 1994; Shulze et al., 1993, 1996). Further diversity is then generated by alternative splicing of the cassette exons. This unusual arrangement of exons could allow for the generation of up to 32 different isoforms and accounts for the isoforms identified to date (Kofuji et al., 1994; Shulze et al., 1996). Interestingly enough, all known isoforms contain exon D, suggesting that 6 amino acids serve a critical function. In regard to exon A it is expressed only in the heart and brain, and exon B only in the kidney. The alternatively-spliced isoform NACA8 (A,C,D,E) is similar to NACA1(A,C,D,E,F), but exon F is missing (Reillly and Lattanzi, 1995). NACA3 and NACA7 isoforms have been identified in pancreatic B cells that are responsible for insulin production (Van Eylen et al., 1995b). Alternative splicing may also occur in the C-terminal hydrophobic domains of the exchanger (Gabellini et al., 1995).

Lytton and coworkers found that the NCX1 transcripts from different tissues have unique 5'-end sequences, suggesting the possibility that different promoters drive expression of the exchanger in different tissues (Lee et al., 1994; Lytton et al., 1996). These sequences have been identified as distinct exons, and mapped as the transcriptional starting points for the exchanger in heart and kidney (Lytton et al., 1996). Therefore, the tissue-specific expression of the exchanger can be regulated by independent promoters in different tissues. It is thus possible that promoter diversities, in specific tissues, are the result of alternative splicing.

E.   Biogenesis, Processing, and Related Structure-Function Relationships

Hydropathy analysis of NCX1 led to an initial model with 12 transmembrane regions and a large intracellular domain (Nicoll et al., 1990). However, the NH_2-terminal region (32 amino acids that corresponds to the first transmembrane segment in the original model) possesses the characteristics of a signal sequence. Subsequent studies show that the N-terminal sequence of the purified bovine cardiac Na^+-Ca^{2+} exchanger begins at the predicted cleave site (Durkin et al., 1991). Similarly, the ROS exchanger may also possess a cleaveable signal sequence (Reilander et al., 1992). However, the results obtained with the cloned NCX1 in HeLa cells do not support the idea that the cleaved N-terminal segment is a signal sequence (Rahamimoff et al., 1996; Furman et al., 1995). This work shows that the exchange activity of a mutated gene, in which the initial 21, 26, or 31 amino acids beyond the initiating Met were deleted, was similar to that of the wild-type exchanger gene. Deletion of the C-terminal extramembranous tail (12 amino acids) does not impair significantly transport activity whereas further deletion of one or more amino acids of the last transmembrane segment, severely alter transport activity, (Rahamimoff et al., 1996; Furman et al., 1995).

A general motif among transmembrane proteins is the presence of glycosylation at the extracellular surface. Although the role of glycosylation remains to be clarified, it may be important for detecting certain proteins on the extracellular surface.

According to the primary sequence of NCX1, there are 6 potential N-linked glycosylation sites (Nicoll et al., 1990; Philipson and Nicoll, 1992), while hydropathy analysis places three of them at intracellular locations. Since the reliability of hydropathy analysis is limited, the identification of glycosylation sites provides important topological information. Recent results suggest that glycosylation occurs at a single site (N-9), which is located after the N-cleaved segment before the first transmembrane segment in the protein (Philipson and Nicoll, 1992; Hryshko et al., 1993). Glycosylation does not affect the exchange activity expressed at the cell surface in oocytes. Likewise, the elimination of glycosylation does not modify voltage-sensitivity of the exchanger (Hryshko et al., 1993).

F. Cellular Expression and Regulation

The cloned cardiac Na$^+$-Ca^{2+} exchanger has been expressed in various systems including *Xenopus laevis* oocytes (Nicoll et al., 1990; Komuro et al., 1992; Aceto et al., 1992), insect (Li et al., 1992a), COS-7 cells (Aceto et al., 1992), CHO cells (Pijuan et al., 1993), and HeLa cells (Rahamimoff et al., 1996). It appears that exchanger expression is exquisitely sensitive to environmental stimuli such as hormones and growth factors (Smith and Smith, 1994; Smith et al., 1996; Shigekawa et al., 1996). For example, activation of adenylyl cyclase by forskolin down-regulates NCX1 activity, protein and mRNA in the rat aortic myocytes (Smith et. al., 1994, 1996). Activation of protein kinase C by phorbol esters markedly down-regulates mRNA, protein and exchange activity in renal epithelial cells (Smith et al., 1996). It was suggested by Smith and collaborators that transcription factors that regulate myocyte growth may mediate the opposing influences on expression and hence, may contribute to upregulation. It was found recently that the steady-state levels of mRNA were 6-fold higher in the late fetus and new newborn rabbit hearts, while the mRNA levels declined postnatally to adult levels (Boerth et al., 1995). Thus, the possibility exists that some specific hormones can downregulate exchanger expression that occurs postnatally.

Cardiac hypertrophy is characterized by the selective up and down regulation of distinct genes (Menick et al., 1996). One of the genes up-regulated very early in response to a hemodynamic load (cardiac hypertrophy) is the Na$^+$-Ca^{2+} exchanger (Menick et al., 1996). Further analysis suggests that the induction of the exchanger occurs at the transcriptional level, while a number of isoforms can be involved in this process. Enhanced expression of mRNA and protein was also detected in failing versus nonfailing hearts (Flesch et al., 1995; Reinecke et al., 1995). These data are consistent with the proposal that under certain pathophysiological conditions the Ca^{2+}-extruding systems, sarcolemmal Na$^+$-Ca^{2+} exchanger and Ca^{2+}-ATPase are up-regulated. Under the same conditions the sarcoplasmic Ca^{2+}-ATPase might be down-regulated. This adaptive mechanism may serve to prevent Ca^{2+}-overload in the sarcoplasmic reticulum, and it may alter calcium homeostasis and transient levels of calcium during the action potential. This is to say, upregulation of the exchanger may contribute to delayed after-depolarization, and therefore, may contribute to the increased risk of arrhythmias. The mechanism of expression regulation could be different during ischemia. For example, mRNA of NCX1 is down-regulated within 10-60 min of cardiac ischemia, suggesting that regulation may occur at the transcriptional level (Bersohn, 1995).

IV. KINETIC MECHANISMS AND STRUCTURE-FUNCTION RELATIONSHIPS

A.   Consecutive (Ping-Pong) Mechanism of Ion-Exchange

Ion-exchange is a bireactant reaction which can be classified as either consecutive(ping-pong) or simultaneous (sequential) (Stein, 1986; Lauger, 1987, 1991; Khananshvili, 1990a,b). In the case of the consecutive (ping-pong) mechanism, the first substrate (for example, Na^+ or Ca^{2+}) binds at one side and is released on the other side of the membrane, before binding the second substrate. This means that the translocation of either Na^+ or Ca^{2+} ion is a separate event as indicated by Scheme 1 shown below. In the case of the simultaneous mechanism, both substrates bind

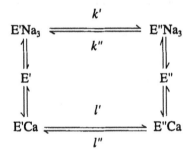

Scheme 1.

before the products are released and ion-translocation occurs in a single step. A distinction between the two basic mechanisms has been a controversial issue for many years and a simultaneous mechanism was suggested as a favored mechanism. However, recent kinetic analyses provide strong support for a consecutive (ping-pong) mechanism (Khananshvili,1990a; Niggli and Lederer, 1991; Hilgemann et al., 1991; Li and Kimura, 1991; Matsuoka and Hilgemann, 1992; Khananshvili and Weil-Maslansky, 1994, Khananshvili et al., 1995a,1996b; Rispoli et al., 1996; Schnetkamp, 1995a). The initial rates of Na_i-dependent ^{45}Ca-uptake have been measured in EGTA-entrapped reconstituted proteoliposomes by using the semi-rapid-mixing techniques (Khananshvili, 1990a). These experiments have been done under "zero-trans" conditions, in which the $[^{45}Ca]_o$ was varied at fixed $[Na]_i$

levels. It was found that $V_{max}(Ca)$ and $K_m(Ca)$ values increase with increasing $[Na]_i$, while $V_{max}(Ca)/K_m(Ca)$ values are still $[Na]_i$-independent (Khananshvili, 1990a). Likewise, at very low, fixed $[Ca]_o$, and varying $[Na]_i$ the initial rates of Na^+-Ca^{2+} exchange are independent of $[Na]_i$. These studies have provided basic kinetic evidence for a consecutive mechanism (Scheme I). Hilgemann's studies (Hilgemann et al., 1991; Matsuoka and Hilgemann, 1992) have shown that in giant excised patches of cardiac sarcolemma or cloned cardiac exchanger the apparent affinity of the exchanger for one ion increased as the concentration of the trans-ion decreases. Similar conclusions have been made by measuring the exchange currents in whole-cell voltage clamped ventricular cells (Li and Kimura, 1991). Consistent with the consecutive mechanism, Niggli and Lederer (1991) observed a transient signal, which was obtained during flash photolysis of caged calcium in myocytes. A transient signal was obtained in the cardiac exchanger during the rapid addition of cytoplasmic sodium, showing voltage-sensitive transport of Na^+ (Hilgemann et al., 1991). Although the ion-exchange mechanism in the ROS-type exchanger might be more complicated, recent studies support the consecutive mechanism of ion transport in this system, too (Rispoli, 1995, 1996; Schnetkamp. 1995a).

Kinetic analysis of Ca^{2+}-Ca^{2+} exchange in cardiac sarcolemma vesicles has provided an independent evidence that this homogeneous exchange mode can be described in terms of separate partial reactions of Ca^{2+}-efflux and Ca^{2+}-influx (Khananshvili and Weil-Maslansky, 1994). Likewise, it has been shown that the Ca^{2+}-Ca^{2+} exchange reaction can be presented as an integral part of the Na^+-Ca^{2+} exchange cycle, as can be predicted from the consecutive mechanism (Khananshvili, 1991b; Khananshvili and Weil-Maslansky, 1994). Under fixed conditions the ratios of Na^+-Ca^{2+}/Ca^{2+}-Ca^{2+} exchanges (the R-values) are much higher than unity $(R = 3$–$4)$, meaning that the Ca^{2+}-influx is at least 5-10 times faster than the Ca^{2+}-efflux (Khananshvili and Weil-Maslansky, 1994; Khananshvili et al., 1995a,1996b). Thus, it is quite possible that the opposite movements of Ca^{2+} through the exchanger might be due to some highly asymmetric-process. For example, factors such as acidic pH, low temperature and/or high concentrations of potassium decrease the R values from 3.0–4.0 to 0.8–1.0 (Khananshvili and Weil-Maslansky, 1994; Khananshvili et al., 1995a,b). It is therefore likely that such factors may alter the asymmetry of bidirectional Ca^{2+}-movements.

B. Rate-Limiting and Voltage-Sensitive Mechanisms of Exchange Modes

Genetic deletion or proteolytic digestion of an intracellular loop (520 amino acids) does not seem to considerably affect current-voltage relationships of the cardiac Na^+-Ca^{2+} exchanger (Matsuoka et al., 1993, 1995). These data suggest that the ions interacting with the specific ion-binding sites within transmembrane segments may determine the voltage-response of exchange modes. Depending on the magnitude of the charge-carrying parameters of ion-protein intermediates, the voltage-dependence of ion fluxes can differ widely (Stein, 1986; Lauger, 1987; 1991). The crucial question is what is the charge-carrying specificity of ENa_3 or ECa species and which species are involved in the rate-limiting pathway (Khananshvili, 1991a,b; Hilgemann et al., 1991; Niggli and Lederer, 1991; Matsuoka and Hilgemann, 1992, Khananshvili and Weil-Maslansky, 1994; Khananshvili et al., 1995b, 1996b). A significant feature of Na^+-Ca^{2+} exchange is that, under zero-trans conditions and saturating $[Ca]_o$ and $[Na]_i$ it shows a linear I-V curve (and not a bell-shape curve) from -100mV to +130 mV (Khananshvili, 1991a,b; Hilgemann et al., 1992; Hilgemann and Matsuoka, 1992). This makes unlikely the charge (z) values of 0, -1 or -4 for unliganded protein, since in these cases both the ENa_3 and ECa species would carry either a positive charge (for $z = 0$ or -1) or negative charge (for $z = -4$), thus resulting in a bell-shaped voltage-curve (Lauger, 1987). This is because the extreme values of voltage (either negative or positive) would inhibit a rate-limiting partial reaction (no matter whether Na^+ or Ca^{2+} transport is involved) causing a bell-shaped voltage-dependence of Na^+-Ca^{2+} exchange. Since no bell-shaped I-V-curves could be obtained, an alternative, possibility had to be considered. It was thus reasoned that a model with a single charge-carrying step would reproduce a monotonic flux-voltage relation which leads to a plateau value at extremes of voltage (Lauger, 1987, 1991). The observed linear voltage-curves are most simply explained as a part of such a curve in the range of tested voltage (Khananshvili, 1991a,b; Hilgemann et al., 1991,1992; Matsuoka and Hilgemann, 1992). This would indicate a charge value of unloaded protein of $z = -2$ (the $E.Na_3$ species are positively charged and $E.Ca$ are electroneutral) or of $z = -3$ (the $E.Ca$ species carry a negative charge and $E.Na_3$ are electroneutral). To distinguish between these two possibilities, $z = -2$ and $z = -3$, one may ask: Which step in Na^+ or Ca^{2+} transport, is rate-limiting for Na^+-Ca^{2+} exchange? In general, the rate-limiting step may or may not be voltage-sensitive. But in any case the rate-limiting intermediate(s) might determine the voltage-sensitive property of Na^+-Ca^{2+} exchange. In other words, if the rate-limiting step is voltage-sensitive, then Na^+-Ca^{2+} exchange would have to be voltage-sensitive. However, if the rate-limiting step involves voltage-insensitive species (for example, $E.Ca$), then Na^+-Ca^{2+} exchange would not be expected to respond to voltage change, regardless of whether the stoichiometry of ion exchange is electrogenic or not (Khananshvili and Weil-Maslansky, 1994; Khananshvili et al., 1995a, 1996b).

An outward current transient was seen during rapid cytoplasmic application of sodium in the absence of trans calcium in giant excised patches with native or cloned cardiac exchanger (Hilgemann et al. 1991). These data were interpreted as showing movement of positive charge through the membrane field during outward Na^+-translocation, because no transient signal was observed by adding cytoplasmic calcium in the absence of trans sodium. On the basis of this finding it was suggested that the Na^+-induced transient signal accounts for a movement of one positive charge per cycle, when most, if not all the voltage-dependence of the exchange cycle lies in the Na^+-transport pathway (Hilgemann et al., 1991, 1992; Matsuoka and Hilgemann, 1992).

In order to elucidate the rate-limiting and voltage-sensitive mechanisms of the exchange reactions, the steady-state kinetics of Na$^+$-Ca^{2+}, Ca^{2+}-Ca^{2+} and Na$^+$-Na$^+$ exchanges have been systematically investigated in the isolated cardiac sarcolemma vesicles and reconstituted proteoliposomes (Khananshvili, 1990a, 1991a,b; Khananshvili and Weil-Maslansky, 1994; Khananshvili et. al., 1995a, 1996b). It was found that with saturating concentrations of ions on both sides of the membrane, the rate of Ca^{2+}-Ca^{2+} exchange was at least 3-5 times faster than the rate of Na$^+$-Na$^+$ exchange (Khananshvili, 1991a,b). Such data are consistent with the concept that the Na$^+$-transport step is rate-limiting for unidirectional Na$^+$-Ca^{2+} exchange. However, the observation that with saturating concentrations of ions both Ca^{2+}-Ca^{2+} and Na$^+$-Na$^+$ exchanges were voltage-insensitive was puzzling, because if Na$^+$-transport is a rate-limiting and voltage-sensitive step, then a voltage-sensitive Na$^+$-Na$^+$ exchange would be expected (Khananshvili, 1991a,b; Matsuoka and Hilgemann, 1992). One possible explanation for this is that Na$^+$-transport during Na$^+$-Na$^+$ exchange at equilibrium may involve multiple steps in which the dielectric distances of charge movements are vanishing (Laüger, 1987, 1991; Khananshvili, 1991a,b; Matsuoka and Hilgemann, 1992; Hilgemann, 1996). Data obtained by time-resolved (microsecond) voltage-jump techniques support this possibility, since very little charge movements were monitored during Ca^{2+}-Ca^{2+} exchange, while large signals were observed for charge movements during Na$^+$-Na$^+$ exchange (Hilgemann, 1996). Together, these data support the idea that Na$^+$-transport is rate-limiting and a voltage-sensitive step occurs in the cardiac exchanger, although this ion-translocating process may involve multiple species and conformational transitions.

Both the rate-limiting pathway and voltage-sensitive step(s) might not be the same in various exchanger prototypes. For example, at saturating concentrations of ions Ca^{2+}-Ca^{2+} exchange in squid axon and barnacle muscle is voltage-sensitive, while Na$^+$-Na$^+$ exchange is not (DiPolo and Beauge, 1991). This is in agreement with the proposal that the voltage-sensitive Ca^{2+} transport step is rate-limiting. In this particular instance one may assume that the ion-binding sites contain three negative charges ($z = -3$) and thus, the E.Ca species are negatively charged when the E.Na$_3$ species are electroneutral. At saturating concentrations of ions, the ROS exchanger rate becomes voltage-insensitive (see Lagnado and McNaughton, 1990). It has also been reported that voltage-insensitive Ca^{2+}-transport might be a rate-limiting step of the ROS exchanger transport cycle (Rispoli, 1995, 1996b). Although the ion-transport mechanism of the ROS exchanger is rather complicated, one may speculate that the E.Ca.K species are "electroneutral," and the E.Na$_4$ species are positively charged. This would be in keeping with a view that the ion-binding domain of the ROS exchanger contains three negative charges ($z = -3$). On this view, voltage-sensitivity of the ROS exchanger cannot be seen at saturating ionic conditions, because the putative charged E.Na$_4$ species may not limit the exchange cycle.

Kinetic analysis suggests that the rate-limiting and voltage-sensitive steps in the cardiac exchanger can be changed by various factors (Khananshvili and Weil-Maslansky, 1994; Khananshvili et al., 1995a, 1996b). For example, in isolated cardiac sarcolemma vesicles the ratio of Na^+-Ca^{2+} and Ca^{2+}-Ca^{2+} exchange rates are comparable ($R\approx1$) at pH 6.0. By increasing pH, however, the Na^+-Ca^{2+} exchange increases, while Ca^{2+}-Ca^{2+} exchange decreases and exhibits R values of 3–4. Likewise, at acidic pH the rate of Na^+-Ca^{2+} exchange becomes apparently voltage-insensitive (Khananshvili and Weil-Maslansky, 1994; Khananshvili et al., 1995a, 1996b). At present there is no evidence that the exchange stoichiometry is changed by acidic pH, suggesting that at pH>6.5 the different reactions, the Na^+ efflux and Ca^{2+} efflux, limit the Na^+-Ca^{2+} and Ca^{2+}-Ca^{2+} exchanges, respectively. In contrast, at low pH, the voltage-insensitive Ca^{2+} influx can limit the rates of both Na^+-Ca^{2+} and Ca^{2+}-Ca^{2+} exchanges (Figure 2). Likewise, at pH >6.5 potassium accelerates Ca^{2+}-Ca^{2+} exchange, but not Na^+-Ca^{2+} exchange (Khananshvili et al., 1995a, 1996b). This would be expected if potassium affects the rate-limiting step (Ca^{2+} efflux) of Ca^{2+}-Ca^{2+} exchange, but not the rate-limiting step (Na^+ efflux) of Na^+-Ca^{2+}-exchange (the effect of potassium on Ca^{2+}-influx cannot be seen, because it does not limit either exchange mode at pH >6.5).

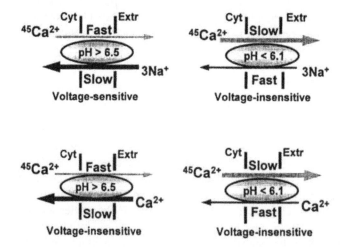

Figure. 2. Proton-dependent changes of rate-limiting pathways for sodium-calcium and calcium-calcium exchanges. This scheme describes the relative rates of exchange modes in cardiac sarcolemma vesicles as a function of extravesicular (cytosolic) pH. At pH > 6.5 the Na^+-efflux and Ca^{2+}-efflux can limit Na^+-Ca^{2+} and Ca^{2+}-Ca^{2+} exchanges, respectively. At pH < 6.1, the same partial reaction, ^{45}Ca-influx, determines the rates of both Na^+-Ca^{2+} and Ca^{2+}-Ca^{2+} exchanges. At pH > 6.5 the Na^+-Ca^{2+} exchange can be voltage-sensitive because the rate-limiting Na^+-efflux involves the charge carrying species. At pH<6.1 Na^+-Ca^{2+} exchange becomes voltage-insensitive, because the electroneutral ^{45}Ca-influx can determine the exchange rate.

C. Inactive States, Ground-State Intermediates, and Parallel Reactions

Although the consecutive (ping-pong) mechanism provides a framework for a primary description of Na$^+$-Ca^{2+} exchange and its partial reactions (Khananshvili, 1990a; Niggli and Lederer, 1991; Hilgemann et al. 1991; Li and Kimura, 1991; Matsuoka and Hilgemann, 1992; Hilgemann, 1996), the actual mechanism of the exchange cycle might be much more complex than a simple consecutive model, described earlier in Scheme I (Khananshvili, 1990a, 1991a,b; Matsuoka and Hilgemann, 1992; Khananshvili et al., 1995a,b; Hilgemann, 1996). For example, voltage-dependent shifts of apparent affinity for Na$^+$ and Ca^{2+} are much smaller and the current-voltage relations (especially of inward current) are not as steep as those predicted by the simple consecutive mechanism (Matsuoka and Hilgemann, 1992). Secondary modulation by intracellular calcium was suggested as a "gating" factor that may switch between fully active and fully inactive exchanger forms (Hilgemann et al., 1992; Matsuoka and Hilgemann, 1992; Hilgemann, 1996). According to this model, the exchanger can occur in two inactive states, I_1 and I_2. The I_1 state represents an interaction of cytoplasmic Ca^{2+} to the ENa$_3$ species, while the I_2 state depicts a binding of cytoplasmic Ca^{2+} to the unloaded exchanger.

Voltage-sensitivity of Na$^+$-Ca^{2+} exchange can be considered on the basis of an alternating access (exposure) mechanism, which focuses on two general principles for describing a voltage-response of the carrier-type transporter (Stein, 1986; Lauger, 1987, 1991; Khananshvili, 1990b). The first principle states that the net charge carried by transporter binding sites and loaded ions may be translocated through the electric field during the occlusion and/or movement of ions by protein. The second assumes that the ions freely diffuse to the binding site without the aid of a transport-related conformational change in the protein (an "access channel"). One therefore expects that in a high-field access channel (ion well) the affinity of ion binding sites is strongly voltage-dependent (Lauger, 1987,1991). However, the available data on the cardiac exchanger do not support this possibility (Khananshvili, 1991a,b; Matsuoka and Hilgemann, 1992), suggesting that the high-resistance access channel is not an intrinsic property of the exchanger. In the absence of direct evidence, one may postulate that the voltage-sensitivity of the exchange cycle is related to the occlusion of ions associated with conformational changes of ion-binding domains.

An occluded state of ions is known to exist in ion pumps, such as the Na$^+$-K$^+$-ATPase, Ca^{2+}-ATPase and H$^+$-K$^+$-ATPase (Post et al., 1965; Dupont, 1980; Stein, 1986; Jencks, 1989; Glynn and Karlish, 1990; Sachs, 1991), but its structure-function relationships are still unclear even in these systems (Andersen and Vilsen, 1995). It is not entirely clear whether catalytic phosphorylation of the pump is absolutely necessary for the occurrence of the occluded state, or whether ion-occlusion may occur without covalent catalysis. Extensive proteolytic digestion of the Na$^+$-K$^+$-ATPase results in a complete loss of ATP-dependent reactions, while the protein can still occlude Na$^+$ and Rb$^+$ (Karlish et al., 1990). These findings suggest that the phosphorylation step of the enzyme is not absolutely necessary for alternative exposure of the ion-binding sites in the pump.

A bell-shaped curve of temperature dependence of Ca^{2+}-Ca^{2+} exchange has been found in isolated cardiac sarcolemma vesicles, exhibiting a maximum at 27-29 °C (Khananshvili et al., 1995a, 1996b). This effect is not caused by irreversible inactivation (for example, thermal denaturation) of the exchanger and cannot be explained by a single reversible exchange reaction involving two rate-constants. Therefore, for the description of the bell-shaped temperature-curve of Ca^{2+}-Ca^{2+} exchange it is essential to assume that Ca^{2+}-transport involves at least two steps with the three ground-state species (Khananshvili et al., 1995a, 1996b). Three possible mechanisms can be considered (Khananshvili et al., 1995a, 1996b): (1) A ground-state intermediate may involve Ca^{2+}-transport, reflecting the "occluded" state (or other conformation) of the transporter. (2) Inactive conformations of the exchanger may be in rapid equilibrium with active species of the exchange cycle. (3) Two (or more) interconvertable conformers may be in dynamic equilibrium, catalyzing parallel reactions with different rates. These proposed mechanisms are not mutually exclusive and they may exist, in principle, at various stages of the transport cycle (Khananshvili et al, 1996b). Time-resolved measurements of charge movements through the exchanger suggest that Na^+ transport occurs in multiple steps and may involve parallel reactions (Hilgemann, 1996). This is perhaps one main reason why voltage-sensitive Na^+-Na^+ exchange was not observed previously under steady-state conditions, although this kind of behavior was expected in terms of a simple consecutive mechanism (Khananshvili, 1991a; Matsuoka and Hilgemann, 1992).

D. Proposed Ion-Protein Interactions

If sodium and calcium ions interact with putative binding sites in transmembrane segments, one could suggest that the actual motion ("conformational change") of ion-transport protein may be associated with the reorientation of a relatively small number of amino-acids and functional groups for ion occlusion. Still, the question which remains is, How can the specific structure–function relationships are reflected in the properties of ECa and ENa_3 species? For a discussion of this issue it is essential to separate the ion-binding events from the charge movement processes through the exchanger. Kinetic analysis of exchange currents and ion-fluxes suggest that the binding of either Na^+ or Ca^{2+} to the cardiac exchanger is a weakly voltage-sensitive process (Khananshvili, 1991a,b; Hilgemann et al., 1992; Matsuoka and Hilgemann, 1992; Hilgemann, 1996). For example, $K_m(Ca)$ changed only twice over the range of 200 mV, thus supporting the view that the calcium-binding site is not situated deeply in the membrane (Khananshvili, 1991a,b; Matsuoka and Hilgemann, 1992). Likewise, at nonsaturating concentrations of ions both Na^+-Na^+ and Ca^{2+}-Ca^{2+} exchanges are weakly voltage-sensitive (Khananshvili, 1991a,b), suggesting that Na^+ and Ca^{2+} may interact with a negatively charged domain. In the case of the ROS exchanger, a change in membrane potential does not affect the affinity of calcium, but hyperpolarization increases the apparent binding affinity of external sodium (Lagnado et al., 1988). Since the effect of membrane potential disappears at saturating concentrations of sodium, it was proposed that the voltage-dependence of ROS exchange is due to the external Na$^+$-binding site being sensitive to voltage (presumably because it is located within the membrane electric field) (Lagnado et al., 1988; Lagnado and McNaughton, 1990). These data suggest again that the ion-binding and charge carrying (charge-translocating) properties of the cardiac and ROS exchangers are fundamentally different.

As already discussed, a primary response of the cardiac Na$^+$-Ca^{2+} exchange to membrane potential can be determined by a rate-limiting movement of Na$^+$ ions through the exchanger ($z = -2$). Thus, the voltage-sensitive and rate-limiting properties of Na$^+$-Ca^{2+} exchange can be determined by rate-limiting and charge-carrying intermediates (presumably E. Na$_3$ species), while the properties of uncharged intermediates (presumably E.Ca species) are not predominant (this is because they involve fast reactions). A model with -2 charges on the ion-binding site had been put forward for the Na$^+$-K$^+$-ATPase (Goldshleger et al., 1987). However, one obvious difference between the Na$^+$-Ca^{2+} exchanger and the Na$^+$-K$^+$-ATPase is that the Na$^+$-transport step is not rate-limiting in the transport cycle, catalyzed by the pump (Post et al., 1965; Sachs, 1977; Gadsby et al., 1985; Gadsby and Nakao, 1989; Glynn and Karlish, 1990; Sachs, 1991). Interestingly, both Na$^+$-K$^+$-ATPase and Na$^+$-Ca^{2+} exchangers are the only known proteins that bind 3Na$^+$ ions in a highly cooperative manner (Hill coefficient n=2-3) and have a homologous sequence of 23 amino acids (a putative domain for ion binding/transport) (Nicoll et al., 1990; Philipson and Nicoll, 1992). Recent studies on site-directed mutagenesis of the cardiac exchanger (Philipson et al., 1992; Nicoll et al., 1994, 1996) show that among the negatively charged residues of putative transmembrane segments only the two (Glu-113 and Glu-199) are essential for Na$^+$-Ca^{2+} exchange activity. Glu-199 is known to be highly conserved in the Na$^+$-K$^+$-ATPase, Ca^{2+}-ATPase and H$^+$-K$^+$-ATPase (Nicoll and Philipson, 1991; Reeves, 1992). Likewise, this glutamic acid is essential for Ca^{2+}-transport in the SR Ca^{2+}-ATPase (Clarke et al., 1989; Andersen and Vilsen, 1995).

Ten years ago Reeves described a model of the cardiac exchanger with two classes of cation binding-translocating sites carrying three negative charges: a divalent site (A) with two negative charges which binds either a single Ca^{2+} ion or one to two Na$^+$ ions, and a second monovalent site (B) which has one negative charge and binds the third Na$^+$ (Reeves, 1985). Here I offer an alternative model, in which the cation-binding domain of the cardiac exchanger contains three ion-binding sites; two of them are negatively charged (the α and β sites) and one is "electroneutral" (the γ-site). In the ECa species the Ca^{2+} ion may interact with the α-site (Glu-113) and β-site (Glu-199), thereby generating the "electroneutral" and "voltage-insensitive" E.Ca species (Figure 3). In the E.Na$_3$ species two Na$^+$ ions may interact with the negatively charged α- and β-sites, while the third Na$^+$ ion may interact with

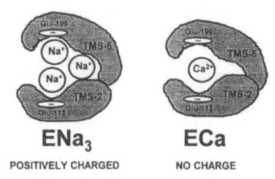

ENa₃

POSITIVELY CHARGED

ECa

NO CHARGE

Figure. 3. Three-site model describing the ion-protein interactions in the E.Ca and E.Na$_3$ species. It is proposed that in the E.Ca species Ca^{2+} ion interacts with the α-site (Glu-113, located on the transmembrane segment 2) and β-site (Glu-113, located on the transmembrane segment 5), yielding an uncharged ion-protein intermediate. In the E.Na$_3$ species two Na^+ ions interact with negatively charged α- and β-sites, while the third Na^+ ion interacts with an uncharged domain (the γ-site). Therefore, the E.Na$_3$ species represent a positively charged intermediate, which may involve a voltage-sensitive and rate-limiting step.

the uncharged γ-site (Figure 3). As a result, the E.Na$_3$ species can carry a net positive charge and therefore, can respond to voltage within the rate-limiting translocation of Na^+ ions. It needs to be explained that a manifestation of E.Ca species as being "electroneutral" is considered as phenomenological, meaning that a net charge of total E.Ca species is "neutral." If there are multiple E.Ca species, then some of them could be charged. For example, the Ca^{2+} ion may not interact simultaneously with the negatively charged α- and β-sites, yielding the intermediate E_α.Ca and E_β.Ca species. In this case, Ca^{2+} movement between the α- and β-sites can generate transient electric movements with charged E_α.Ca and E_β.Ca species. This hypothesis is in accord with the recent findings, suggesting that: (i) Ca^{2+}-transport involves multiple steps and intermediates (Khananshvili et al., 1995a, 1996b); and (ii) Ca^{2+}-translocation is associated with a small (but detectable) transient signal, reflecting Ca^{2+} charge movements through the cardiac Na^+-Ca^{2+} exchanger (Hilgemann, 1996).

The Glu-113 (the α-site) and Glu-199 (the β-site) amino acids are located on the two transmembrane segments, TMS-2 and TMS-5 (Nicoll et al., 1990; Philipson and Nicoll, 1992). One would therefore expect that Ca^{2+}-binding can induce "hinge bending" or adequate conformational change, while Na^+ ions may not (Figure 3). Although the transmembrane segments contain a number of positively charged amino acids (Nicoll et al., 1990; Philipson et al., 1992), the question of their involvement in ion-transport events is not yet resolved. Even though the positively charged groups of transmembrane segments interact with the translocated cations, they might reject both Na^+ and Ca^{2+} ions (this process should be fast and may not contribute to the rate-limiting of the exchange cycle). Besides the charged α and β sites, the Ca^{2+} ion may interact with some "electroneutral" (polar) ligation centers (for example, carbonyl or hydroxyl oxygens) generating a coordination number of 6-8. In general, at least one carboxyl and carbonyl oxygen are involved in Ca^{2+}-binding to soluble proteins (Strynadka and James, 1989; Glusker, 1991). A putative γ site may contain only uncharged functional groups with 6 (or less) ligation centers and thus, may favor binding of Na$^+$ over Ca^{2+}. The hydrophilic residues of transmembrane segments 2,3,4, and 5 can be modeled in such a way that they form a putative ion-transport pathway of the exchanger (Nicoll and Philipson, 1991; Philipson and Nicoll, 1992). The fact that among the seven mutated transmembrane serines the six are essential for the exchanger activity is in line with this hypothesis (Nicoll et al., 1994, 1996).

The present model for ion-protein interactions is applicable for the cardiac Na$^+$-Ca^{2+} exchanger (NCX1) and relevant isoforms. It is possible that the specific structure-function relationships are somewhat different in other exchanger prototypes.

E.   Organic Inhibitors

In the past, amiloride and its derivatives were recognized as relatively effective inhibitors of Na$^+$-Ca^{2+} exchange (see Kaczorowski et al., 1989), but their application in most biomedical experiments is strictly limited. For example, the most popular amiloride analogs (for example, benzamil, dichlorobenzamil, or benzobenzamil) inhibit Na$^+$-Ca^{2+} exchange with an IC$_{50}$ of 10^{-4}-10^{-5} M. However, the main problem is that amiloride derivatives inhibit a number of other Na$^+$-transport systems including the Na$^+$-K$^+$-ATPase, Na$^+$-H$^+$ exchange and Na$^+$-channels. This is not altogether surprising, because amiloride derivatives are known to mimic Na$^+$ in many Na$^+$-dependent transport systems, reflecting perhaps the fact that these transport systems may contain some similar tertiary structures (Kaczorowski et al., 1989; Or et al., 1993; Garty, 1994). It is thus unlikely that selective ligands for ion-transport systems that bind to the ion-binding sites will be developed.

A 20 amino-acid sequence on the intracellular loop (amino-acids 219-238) has been identified as a possible calmodulin-binding domain which shows auto-inhibitory potency (Li et al., 1991). This is a very positively charged segment containing five arginines and three lysines. Similar sequences were found in a number of calmodulin-binding proteins (see Carafoli, 1994). Based on this information, the XIP peptide has been synthesized and tested for inhibition of Na$^+$-Ca^{2+} exchange activities. The XIP peptide inhibits most of the exchanger activity with an IC$_{50}$ of 1.5 μM in the preparation of isolated cardiac sarcolemma vesicles and 0.1 μM in the giat patches of myocytes (Li et al., 1991; Kleiboeker et al., 1992). Although the XIP peptide is more potent and specific than dichlorobenzamil (a most potent amiloride derivative), it may also interact with other calmodulin-binding proteins. In addition, since the XIP-binding site is situated at the intracellular surface, it is inaccessible for most physiological experiments. The suggestion has been made that the XIP domain may interact with either the negatively charged phospholipid bilayer or another domain of the transporter protein, thus inhibiting or increasing the exchanger activity (Kleiboeker et al., 1992; Hale, 1996). At least one lysine among the three is functionally important, while not all arginines are perhaps obligatory for retaining XIP-induced inhibition (Gatto et al., 1995; Hale, 1996).

Studies with isolated cardiac sarcolemma vesicles have led to evidence that the Phe-Met-Arg-Phe-$CONH_2$ (FMRFa) tetrapeptide and its analogs are able to inhibit Na^+-Ca^{2+} and Ca^{2+}-Ca^{2+} exchanges with an IC_{50} of 10^{-6}-10^{-3} M (Khananshvili et al., 1993). The inhibitory FMRFa-peptides behave as noncompetitive inhibitors with respect to extravesicular calcium and like the XIP-peptide interact with the intracellular surface. Both XIP and FMRFa peptides inhibit Na^+-Ca^{2+} exchange in squid axons, suggesting that the putative XIP and FMRFa-sites are also present in neuronal tissue (DiPolo and Beauge, 1994). The FMRFa- peptides also inhibit Na^+-Ca^{2+} exchange in insulin producing β-cells (Van Eylen et al., 1995a).

In general, there are common structural disadvantages with linear peptide-inhibitors, which are difficult to overcome without applying alternative approaches. For example, both the XIP and FMRFa-peptides contain positively charged amino acids such as Arg and/or Lys that makes them attractive targets for proteolytic enzymes. Likewise, linear peptides undergo many conformational transitions that may decrease the specificity and affinity of peptide-receptor interaction (Hruby et al., 1990; Gilon et al., 1991). Chemical studies show that intramolecular cyclization may restrict the conformational flexibility of a peptide, resulting in improved affinity, selectivity, stability, and resistance to proteolytic enzymes (Hruby et al., 1990; Tonioli, 1990; Gilon et al., 1991). Taking this into account, a number of positively charged cyclic hexapeptides (with an intramolecular disulfide bond) have been designed, synthesized, and tested for their inhibitory activity in cardiac sarcolemma vesicles (Khananshvili et al., 1995b). The most potent cyclic hexapeptide, FRCRCFa (Figure 4), inhibits both Na^+-Ca^{2+} and Ca^{2+}-Ca^{2+} exchanges, with an IC_{50} of 2-10 μM. FRCRCFa is a noncompetitive inhibitor with respect to extravesicular calcium for Na^+-Ca^{2+} or Ca^{2+}-Ca^{2+} exchange (Khananshvili et al., 1995b, 1996a). In intact myocytes, FRCRCFa shows the IC_{50} of 20 nM, suggesting that some unknown cytosolic factors may cause enhancement of the inhibitory potency (Hobai et al., 1997). Interestingly, the inhibitory interaction of FRCRCFa to the cytosolic side of the membrane is very rapid (<20 msec), although the identity of the peptide binding site is still unknown (Khananshvili et al., 1996a, 1997). Application of more sophisticated molecular approaches (Khananshvili et al., 1997) may produce new peptidomimetic blockers with improved pharmacokinetics and therapeutic potency.

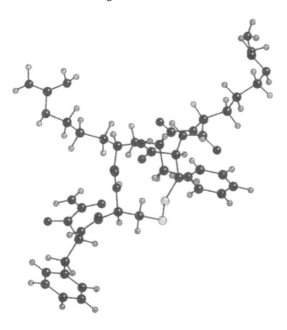

Figure. 4. Calculated structure of cyclic hexapeptide FRCRCFa. The three-dimensional structure of FRCRCFa was computed on the basis of "energy minima" criteria (computer program Nemesis).

V. REGULATORY MECHANISMS OF EXCHANGE REACTIONS

A. Secondary Modulation by Intracellular Calcium and Sodium

An interaction of cytosolic calcium with a regulatory site on the exchanger (secondary modulation by calcium) has been observed in internally dialyzed squid axons, barnacle muscle and cardiac myocytes (see Hilgemann et al., 1991; DiPolo and Beauge, 1991; Hilgemann et al., 1992; Matsuoka and Hilgemann, 1992). The use of giant excised patches has greatly advanced our knowledge of mechanisms of secondary calcium regulation in myocytes (Hilgemann, 1989, 1990; Matsuoka and Hilgemann, 1992; Matsuoka et al., 1993, 1994). These techniques allow the measurement of the "reverse mode" of Na$^+$-Ca^{2+} exchange (the intracellular Na$^+$-dependent Ca^{2+}-influx currents) and have provided evidence that the cytoplasmic regulatory site and the Ca^{2+}-transport site are not the same (Hilgemann et al., 1991; Matsuoka and Hilgemann, 1992, 1994; Matsuoka et al., 1993). However, it is not yet clear if the same Ca^{2+}-regulatory site affects the opposite mode of exchange. This is because it is impossible at the present time to distinguish between the effects occurring on regulatory and transport sites. Although the rate of Ca^{2+}-dissociation may determine the exchanger activity (e.g.. when the cytosolic calcium drops during the action potential), this kinetic parameter has not been yet measured by any technique. In contrast to the activatory effect of calcium, cytoplasmic sodium causes rapid inactivation of Na^+-Ca^{2+} exchange currents (Hilgemann et al., 1991, 1992; Matsuoka et al., 1993,1994). This Na^+-dependent inactivation process takes place on a multisecond time scale and the kinetics of inactivation and recovery from Na^+-dependent inactivation are affected by various factors (for example, pH). The secondary modulatory effect of sodium can be explained by a consecutive (ping-pong) mechanism, when the exchanger can oscillate between the fully active and fully inactive state (Hilgemann et al., 1992; Matsuoka & Hilgemann, 1992). This secondary regulatory effect by sodium (or calcium) can be functionally abolished by limited proteolysis from the cytoplasmic side or by genetic deletion of the intracellular loop, leaving the exchanger in a highly stimulated state (Hilgemann, 1990; Hilgemann et al., 1992; Matsuoka et al., 1993). Surprisingly, the *Drosophila* exchanger is an exception, since it is inhibited rather than stimulated by regulatory calcium, and is stimulated rather than inhibited, by regulatory sodium (Philipson et al., 1996).

Patch clamp and molecular biology techniques have shed new light on the problem of structure-function relationships (Matsuoka et al., 1993, 1995; Levitsky, 1994; Hryshko et al., 1994; Philipson et al., 1996). There is evidence which suggests that Na^+-dependent inactivation and Ca^{2+}-dependent activation are interrelated (Matsuoka et al., 1996; Philipson et al., 1996). It appears that the site(s) of secondary Ca^{2+}-activation is located on the intracellular loop between amino-acid residues 371-508. Two acidic segments with amino acids, 446-455 and 498-510 are essential for high-affinity Ca^{2+}-binding. The segment 498-510 is closely similar to the Ca^{2+}-binding domains of EF hand structures. However, the removal of the C-terminal glutamic acid (which is expected to be critical for EF hand structure), fails to affect Ca^{2+}-binding to the exchanger (Matsuoka et al., 1996; Philipson et al., 1996). The amino acid residues involving Na^+-dependent inactivation are now under investigation (Philipson et al., 1996).

B. Regulation by Intracellular ATP

Although the Na^+-Ca^{2+} exchanger does not form an acylphosphate intermediate as some ion pumps do (for example, Na^+-K^+-ATPase or Ca^{2+}-ATPase), ATP affects the kinetic properties of the exchange modes in squid axons, barnacle muscle, vascular smooth muscle cells, and cardiac myocytes (Baker and Glitsch, 1973; Blaustein, 1974, 1977; Rasgado-Flores et al., 1988; Hilgemann, 1990; DiPolo and Beauge; 1991, Collins et al., 1992; Hilgemann et al., 1992; Shigekawa et al., 1996). Some of these effects induced by ATP are also seen in CHO cells transfected with NCX1 (Condrescu et al., 1995; Reeves et al., 1995). In squid axons and barnacle muscle, ATP decreases the K_m for cytosolic calcium and for extracellular sodium (Blaustein, 1974, 1977; Rasgado-Flores et al., 1988; Beauge and

DiPolo, 1991). In giant cardiac patches, ATP reduces the effects of cytosolic Na$^+$ in promoting Na$^+$-dependent inactivation (Collins et al., 1992; Hilgemann et al., 1992).

The mechanisms of ATP-induced regulation of Na$^+$-Ca^{2+} exchange are still poorly understood. In giant cardiac patches stimulation by MgATP is not blocked by protein kinase and phosphatase inhibitors, nor is it mimicked by MgATPγS; it is also not reversed by phosphatase (Hilgemann et al., 1991, 1992; Collins et al., 1992; Hilgemann, 1995). This led to the suggestion that stimulation of Na$^+$-Ca^{2+} exchange by MgATP in cardiac sarcolemma is not likely to involve protein kinase(s). It was also postulated, but not confirmed that MgATP can stimulate Na$^+$-Ca^{2+} exchange by aminophospholipid translocase which "pumps" phosphatidylserine from the extracellular to the cytoplasmic side of the bilayer leaflet (Hilgemann and Collins, 1992). Recently Hilgermann described a new pathway for regulating the cardiac Na$^+$-Ca^{2+} exchanger by PIP$_2$ (Hilgermann and Ball, 1996). According to this pathway, the stimulation of ATP probably involves the generation of PIP$_2$ by lipid kinases that phosphorylate PI and PIP.

Direct evidence that the Na$^+$-Ca^{2+} exchanger can be phosphorylated comes from experiments with ^{32}P in cultured cells of aortic smooth muscle (Shigekawa et al., 1995). The extent of phosphorylation correlates with an increase in diacylglycerol produced by platelet-derived growth factor-BB (PDGF-BB). At least four phosphopeptides were detected by tryptic digestion of the exchanger. Analysis of these phosphopeptides revealed that the phosphorylated amino acids were exclusively serine residues in both quiescent and stimulated cells. It was therefore concluded that the Na$^+$-Ca^{2+} exchanger is activated by protein kinase C in response to growth factors in vascular smooth muscle cells (Shigekawa et al., 1996). However, no phosphorylation of the exchanger has been found in CHO cells transfected with cDNA of NCX1 (Reeves et al., 1995).

C. Modulation by Cytoplasmic Protons

It is widely accepted that Na$^+$-Ca^{2+} exchange is considerably modulated by cytoplasmic pH in cardiac, neuronal, and visual tissues (Baker and McNaughton, 1977; Wakabayashi and Goshima, 1981; DiPolo and Beauge, 1982; Philipson et al., 1982; Hodgkin and Nunn, 1987; Doering and Lederer, 1993, 1996; Khananshvili and Weil-Maslansky, 1994; Khananshvili et al., 1995a). Although the pH-titration profiles are not quite the same in various tissues and preparations, the general rule is that Na$^+$-Ca^{2+} exchange is inhibited by acidic pH and activated by alkaline pH. This phenomenon, otherwise known as "proton block," was first described in squid axon (Baker and McNaughton, 1977; DiPolo and Beauge, 1982) and later characterized in cardiac sarcolemma vesicles, giant patches and myocytes (Wakabayashi and Goshima, 1981; Philipson et al. 1982; Hilgemann et al., 1992; Doering and Lederer, 1993, 1996). Since intracellular pH can change in many pathological conditions (for example, acidosis, ischemia), effects of pH on Na^+-Ca^{2+} exchange may have some clinical relevance. In contrast to the intracellular pH effects, the exchange currents are stimulated by extracellular acidification and inhibited by extracellular alkalinization (Doering and Lederer, 1996).

The regulatory mechanisms underlying pH-dependent modulation of the exchanger remain poorly understood. An increase in cytoplasmic pH from 6.8 to 7.8 attenuates current decay in giant sarcolemma patches and shifts the apparent dissociation constant of secondary calcium activation from 9.6 μM to < 0.3 μM (Hilgemann et al., 1992a). These data support the idea that deprotonation of the regulatory Ca^{2+}-binding domain is associated with an increase in calcium-binding affinity at the regulatory site, suggesting the possibility that H^+ may compete with Ca^{2+} for regulatory site(s). This mechanism should be distinguished from potential competition of H^+ with Ca^{2+} at the ion-transport domain. Using giant patches of cardiac sarcolemma membranes, it was shown that a step decrease in cytoplasmic pH from 7.2 to 6.4 produced a biphasic but monotonic decrease in Na^+-Ca^{2+} exchange currents, while alkalinization of cytoplasmic pH from 7.2 to 8.0 exhibited a large, biphasic increase in exchange currents (Doering and Lederer, 1993, 1996). In addition, rapid (perhaps, diffusion controlled) inhibition of the exchanger follows cytoplasmic acidification without cytoplasmic sodium, while slow (within the second time scale) inhibition of the exchanger by protons requires cytoplasmic sodium. Doering and Lederer concluded that there are at least two types of proton blocks: the "primary" proton block that is rapid and Na^+-independent, and the "secondary" block that is slow and Na^+-dependent. Addition of proteolytic enzyme α-chymotrypsin (partial digestion) on the cytoplasmic side largely abolished the sensitivity of the exchanger to protons (Doering and Lederer, 1993, 1996). These data suggest that proteolytic treatment removes (or inactivates) a specific domain involved in protein-dependent regulation or one that "deregulates" other modulatory sites (for example, the secondary calcium, sodium and/or ATP binding domains). The finding that a specific histidine amino acid, His-226, plays a role in the pH-sensor of the *E. coli* Na^+-H^+ exchanger (Gerchman et al., 1993) raises the possibility that a similar "pH-sensor" may also exist in the Na^+-Ca^{2+} exchanger.

The effects of extravesicular (intracellular) pH on Na^+-Ca^{2+} and Ca^{2+}-Ca^{2+} exchanges have been examined in isolated cardiac sarcolemma vesicles by using the semi-rapid-mixing techniques (Khananshvili and Weil-Maslansky, 1994; Khananshvili et al., 1995a, 1996b). Without monovalent ions the pH-titration curve of Ca^{2+}-Ca^{2+} exchange follows a bell shape in the acidic range (pK_{a1}=5.1±0.1 and pK_{a2} =6.3±0.1) followed by activation of the exchange in the alkaline range (pK_{a3}=8.6±0.2). In contrast, Na^+-Ca^{2+} exchange shows a monotonic increase from pH 5.0 to 10.0, exhibiting three apparent pK_a values (pK_{a1}=5.2±0.1, pK_{a2} =6.4±0.1 and pK_{a3} =8.7±0.1). Although the observed pK_a values do not necessarily represent the pK_a values of specific functional groups involving ion-transport and/or modulatory activities, three apparent phases (see Scheme 2 below) in exchanger deprotonation can be described as follows: (i) At pH < 6.1 both exchange modes have a very similar rate while deprotonation with $pK_{a1} \approx 5.1$ accelerates both Na$^+$-Ca^{2+} and Ca^{2+}-Ca^{2+} exchanges to a similar extent (R _ 1). This phase of deprotonation may reflect a function of carboxyl residues that involve ion biding/translocation (for example, Glu-113 and Glu-199). (ii) Further deprotonation ($pK_{a2} \approx 6.3$) of the exchanger accelerates Na$^+$-Ca^{2+} exchange but decelerates Ca^{2+}-Ca^{2+} exchange, suggesting that this phase of deprotonation has opposite effects on rate-limiting reactions. That is, it accelerates Na$^+$-efflux during Na$^+$-Ca^{2+} exchange and decelerates Ca^{2+}-efflux during Ca^{2+}-Ca^{2+} exchange. A histidine and/or carboxyl group(s) might be suitable candidates to account for these effects. (iii) Deprotonation in the alkaline range ($pK_{a3} \approx 8.7$) accelerates both Ca^{2+}-Ca^{2+} and Na$^+$-Ca^{2+} exchanges, suggesting that deprotonation of some functional hydroxyl and/or amino group may result in accelerating both Na$^+$ and Ca^{2+}-movements.

$$pK_{a1} = 5.1 \qquad pK_{a2} = 6.3 \qquad pK_{a3} = 8.7$$
$$H_3E \rightleftharpoons H_2E \rightleftharpoons H_1E \rightleftharpoons E$$

Scheme 2

It was also found that potassium induces a prominent shift in pK_{a1} and pK_{a2} values on the pH-titration curve of Ca^{2+}-Ca^{2+} exchange so that the difference between these two pK_a values declines dramatically (there is little or no effect on the pK_{a1}) (Khananshvili et al., 1995a). These data support the idea that potassium interacts with deprotonated species, thereby affecting the rate limiting step of Ca^{2+}-Ca^{2+} exchange; this is not the case with Na$^+$-Ca^{2+} exchange, presumably because potassium does not affect the rate-limiting transport of sodium (Khananshvili and Weil-Maslansky, 1994; Khananshvili et al., 1995a). Taking into account all the available data, it seems reasonable to conclude that protons can interact with a number of regulatory and transport sites. This is why the meaning of "a proton block" (that is, competition of protons with Na$^+$ and/or Ca^{2+} for transport sites) needs to be reconsidered.

SUMMARY

Cell membrane Na$^+$-Ca^{2+} exchange plays a key rolel in Ca^{2+} homeostasis in most cell types. Cardiac contractility, vision, neurotransmitter and hormone secretion and bioenergetics are highly dependent on Na$^+$-Ca^{2+} exchange. In cardiac cells, Na$^+$-Ca^{2+} exchange (3Na$^+$:Ca^{2+}) has a turnover rate of 10^3-10^4 s^{-1}, thereby providing a voltage-sensitive extrusion of nearly all the calcium entering the cell via Ca^{2+}-channels. The exchanger can be modulated by various factors (ATP, Ca_i, Na_i, pH, lipids, etc.), but the mechanisms underlying its operation have not yet been elucidated. The cardiac Na^+-Ca^{2+} exchanger (NCX1) is a large polypeptide (970 amino acids) with 11 putative transmembrane segments and a large intracellular loop (520 amino acids) containing a number of regulatory sites. Specific functional domains involving ion transport, regulation, and glycosylation have been identified and characterized by using site-directed and deletion techniques. A putative ion-transport domain of NCX1 shows some similarities to the Na^+-K^+-ATPase and other ion-pumps. Besides NCX1, prototypes have also been identified and characterized in brain, kidney, and photoreceptor cells, indicating that distinct promoters may control tissue-specific expression of exchanger isoforms. Specific mechanisms have been described for alternative splicing that can explain isoform diversity in different species and tissues. The XIP-peptide (20 amino acids) and positively charged cyclic hexapeptide FRCRCFa have been designed to inhibit exchanger activity. These peptide blockers interact with the cytosolic side of the membrane, exhibiting the $IC_{50}=20$-$100\mu M$ in intact myocytes. The Na^+-Ca^{2+} exchange cycle can be described in terms of separate movements of Na^+ and Ca^{2+} through the membrane segment of the exchanger (the ping-pong or consecutive mechanism). However, ion-translocation is not a simple reaction and may involve ground state intermediates and inactive states. Although the binding of both Na^+ and Ca^{2+} is weakly voltage-sensitive, a primary response of Na^+-Ca^{2+} exchange to voltage arises within the rate-limiting and voltage-sensitive movement of Na^+ ions through the exchanger. Protons can change the rate-limiting pathway (Ca^{2+}-transport becomes rate limiting instead of rate-limiting Na^+-transport at physiological pH), yielding a voltage-insensitive Na^+-Ca^{2+} exchange. A model is described, in which the $E.Na_3$ species are positively charged, while the E.Ca species have "no charge" (for example, the unloaded cation-binding domain may contain -2 charges). It is proposed that Glu-113 (the α-site) and Glu-199 (the β-site) bind either $2Na^+$ or Ca^{2+}, while the third Na^+ interacts with an uncharged domain (the γ-site).

ACKNOWLEDGMENTS

I am grateful to my colleagues L. Beauge, D. M. Bers, M. P. Blaustein, G. P. Brierley, R. DiPolo, A. Doering, F. S. Fay, C. C. Hale, D. W. Hilgemann, T. Iwata, S. J. D. Karlish, J. Kimura, G. A. Langer, J. Lederer, W. J. Lytton, M. A. Milanic, E. Niggli, D. Nobel, K. D. Philipson, H., Rahamimoff, H. Rasgado-Flores, J. P. Reeves, R. F. Reilly, H. Reuter, G. Rispoli, P. P. M. Shnetkamp, M. Shigekawa, D. Shulze, and J. B. Smith for sending me reprints and for their comments.

This work was supported by the U.S.A.-Israel Binational Foundation (BSF)-Grant # 9300096, the Israeli Science Foundation, Administrated by the Israel Academy of Sciences and Humanities-Grant # 196/93 and the Ministry of Science and the Arts. Financial support from the Shlezak foundation is greatly acknowledged.

REFERENCES

Aceto, J. F., Condrescu, M., Kroupis, C., Nelson, H, Nelson N., Nicoll, D., Philipson, K. D. & Reeves, J. P. (1992). Cloning and expression of the bovine cardiac sodium-calcium exchanger. Arch. Biochem. Biophys. 298, 553-560.

Adams, B. A., & Beam, K. G. (1990) Muscular dysgenesis in mice: a model system for studying excitation-contraction coupling. FASEB J. 4, 2809-2816.

Ambesi, A., Van Alstyne, E. L., Bagwell, E. E., & Lindenmayer, G. E. (1991). Effect of polyclonal antibodies on the cardiac sodium-calcium exchanger. Ann. N.Y. Acad. Sci. 639, 245-247.

Andersen, J. P., & Vilsen, B. (1995). Structure-function relationships of cation translocation by Ca^{2+}-and Na$^+$,K$^+$-ATPases studied by site-directed mutagenesis. FEBS Lett. 359, 101-106.

Baker, P. F., Blaustein, M. P. Hodgkin, L., & Steinhardt (1969). The influence of calcium on sodium efflux in squid axons. J. Physiol. 200, 431-438.

Baker, P. F., & Glitsch, H. G. (1973). Does metabolic energy participate directly in the Na$^+$-dependent extrusion of Ca^{2+} ions from squid giant axons? J. Physiol. 233, 44P-46P.

Baker, P. F., & McNaughton, P. A. (1977). Selective inhibition of the Ca-dependent Na efflux from intact squid axons by a fall in intracellular pH. J. Physiol. 269, 78P-79P.

Bartschat, D. K., & G. E. Lindenmayer (1980). Calcium movements promoted by vesicles enriched sarcolemma preparation from canine ventricles. J. Biol. Chem. 255, 9626-9634.

Bassani, R. A., & Bers, D. M. (1994). Na-Ca exchange is required for rest-decay but not for rest-potentiation of twitches in rabbit and rat ventricular myocytes. J. Mol. Cell Cardiol. 26, 1335-1347.

Bassani, R. A., & Bers, D. M. (1995). Rate of diastolic Ca release from the sarcoplasmic reticulum of intact rabbit and rat ventricular myocytes. Biophs. J. 68, 2015-2022.

Beukelmann, D. J., & Wier, W. G. (1988). Mechanism of release of calcium from sarcoplasmic reticulum of guinea-pig cardiac cells. J. Physiol. 405, 233-255.

Bers, D. M. (1987). Ryanodine and Ca content of cardiac SR assessed by caffeine and rapid cooling contractures. Am. J. Physiol. 253, C408-415.

Bers, D. M. (1991). Excitation-Contraction Coupling and Cardiac Contractile Force. (Single Author Monograph) pp. 1-258, Kluwer Academic Press, Dordrecht, The Netherlands.

Bers, D. M., & Bridge, J.H.B. (1989). Relaxation of rabbit ventricular muscle by Na-Ca exchange and sarcoplasmic reticulum Ca-pump: Ryanodine and voltage-sensitivity. Circ. Res. 65, 334-342.

Bers, D. M., Bassani, J. W. M., & Bassani, R. (1996). Na/Ca exchange and Ca fluxes during contraction and relaxation in mammalian ventricular muscle. Ann. N.Y. Acad. Sci. 779, 430-442.

Bersohn, M. M. (1995). Sodium-calcium exchange expression in ischemic rabbit hearts. Third International Conference on Sodium-Calcium Exchange, Woods Hole, Mass., April 23-26, 1995, Abstract P4.

Blaustein, M. P. (1974). The interrelationship between sodium and calcium fluxes across cell membranes. Rev. Physiol. Biochem. Pharmacol. 70, 33-82.

Blaustein, M. P. (1977). Effects of internal and external cations and of ATP on sodium-calcium and calcium-calcium exchange in squid axons. Biophys. J. 20, 79-111.

Blaustein, M. P., Goldman, W. F., Fontana, G., Krueger, B. K., Santiago, E. M., Steele, T. D., Weiss, D. N., & Yarawsky, P. J. (1991). Physiological roles of the sodium-calcium exchanger in nerve and muscle. Ann. N.Y. Acad. Sci. 639, 254-274.

Boerth, S. R., Coetzee, W. A., & Artman, M. (1995). Ontogenic and normal regulation of cardiac Na-Ca exchanger expression in rabbit ventricles. Third International Conference on Sodium-Calcium Exchange, Woods Hole, Mass., April 23-26, 1995, Abstract P1.

Bouchard, R. A., Clark, W. R., & Giles, W. R. (1996). Action potential voltage clamp measurements of E-C coupling in a rat ventricle. Ann. N.Y. Acad. Sci. 779, 417-429.

Bridge, J. H. B, Smolley, J. R., & Spitzer, K. W. (1990). The relationship between charge movements associated with I_{Ca} and I_{Na-Ca} in cardiac myocytes. Science 248, 376-378.

Bridge, J. H. B, Smolley, J. R., Spitzer, K. W., & Chin T. K. (1991). Voltage dependence of sodium-calcium exchange and the control of calcium extrusion in the heart. Ann. N.Y. Acad. Sci. 639, 34-47.

Brierley, G. P., Baysal, K., & Jung, D. W. (1994). Cation transport systems in mitochondria: Na^+ and K^+ uniports and exchangers. J. Bioenerg. Biomembr. 26, 519-526.

Cannell, M. B., Main, M. J., & Evans, A. M. (1996). The roles of intracellular sodium, sodium current and L-type calcium current in triggering SR calcium release. Ann. N.Y. Acad. Sci. 779, 443-450.

Carafoli, E. (1987). Intracellular calcium homeostasis. Ann. Rev. Biochem. 56, 395-433.

Carafoli, E. (1994). Biogenesis: Plasma membrane calcium ATPase: 15 years of work on the purified enzyme. FASEB J. 8, 993-1002.

Cervetto, L., Lagnado, L., Perry, R. J., Robinson, D. W., & McNaughton, P. A. (1989). Extrusion of calcium from rod outer segments is driven by both sodium and potassium gradients. Nature 337, 740-743.

Cheon, J., & Reeves, J. P. (1988). Site-density of the sodium-calcium exchange carrier in reconstituted vesicles from bovine cardiac sarcolemma. J. Biol. Chem. 263, 2309-2315.

Cheng, H., Lederer, M. R., Xiao, R. P., Gomez, A. M., Zhou, Y. Y., Ziman, B., Spurgeon, H., Lakatta, E. G., & Lederer, W. J. (1996). Cell Calcium 20, 129-140.

Chin, T., K., Spitzer, K. W., Philipson, K. D., & Bridge, H. B. (1993). The effect of exchanger inhibitory peptide (XIP) on sodium-calcium exchange current in guinea pig ventricular cells. Circ. Res. 72, 497-503.

Clarke, D.M., Loo, T. W., Inesi, G., & MacLennan, D. H. (1989). Location of high-affinity Ca-binding sites within the predicted transmembrane domain of the sarcoplasmic reticulum Ca-ATPase. Nature 339, 476-478.

Collins, A., Somlyo, A. V., & Hilgemann, D. W. (1992). The giant cardiac membrane patch method: stimulation of outward Na^+-Ca^{2+} exchange current by MgATP. J. Physiol. 454, 27-57.

Condrescu, M., Gardner, J. P., Chernaya, G., Aceto, J. F., Kroupis, C., & Reeves, J. P. (1995). ATP-dependent regulation of sodium-calcium exchange in CHO cells transfected with the bovine cardiac sodium-calcium exchanger. J. Biol. Chem. 270, 9137-9146.

Cook, N. J., & Kaupp, U. B. (1988). Solubilization, purification and functional reconstitution of the sodium-calcium exchanger from bovine rod outer segments. J. Biol. Chem. 263, 11382-11388.

Crespo, L. M., Grantham, C. J., & Cannel, M. B. (1990). Kinetics, stoichiometry and role of the Na-Ca exchange mechanism in isolated cardiac myocytes. Nature 345, 618-621.

DiPolo, R., & Beauge, L. (1979). Physiological role of ATP-driven calcium pump in squid axons. Nature 278, 271-273.

DiPolo, R., & Beauge, L. (1982). The effect of pH on Ca^{2+}-extrusion mechanisms in dialyzed squid axons. Biochem. Biophys. Acta 688, 227-245.

DiPolo, R., & Beauge, L. (1990). Asymmetrical properties of the Na-Ca exchanger in voltage-clamped, internally dialyzed squid axons under symmetrical ionic conditions. J. Gen. Physiol. 95, 819-835.

DiPolo, R., & Beauge, L. (1991). Regulation of Na-Ca exchange: An overview. Ann. N.Y. Acad. Sci. 639, 100-111.

DiPolo, R., & Beauge, L. (1993). Effects of some metal-ATP complexes on Na^+-Ca^{2+} exchange in internally dialysed squid axons. J. Physiol. 462, 71-86.

Di Polo R., & L. Beauge (1994). Cardiac sarcolemmal Na/Ca-inhibiting peptides XIP and FMRF-amide also inhibit Na/Ca exchange in squid axons. Am. J. Physiol. 267, C307-C311.

Doering, A. E., & Lederer, W. J. (1993). The mechanism by which cytosolic protons inhibit the sodium-calcium exchange in guinea pig heart cells. J. Physiol. 466, 481-499.

Doering, A. E., & Lederer, W. J. (1996). Cardiac Na/Ca exchange and pH. Ann. N.Y. Acad. Sci 779, 182-198.

Durkin, J. T., Ahrens, D. C., Pan, Y. C. E., & Reeves, J. P. (1991). Purification and amino-terminal sequence of the bovine cardiac sodium-calcium exchanger. Arch. Biochem. Biophys. 290, 369-375.

DuPont, Y. (1980). Occlusion of divalent cations in the phosphorylated calcium pump of sarcoplasmic reticulum. Eur. J. Biochem. 109, 231-238.

Eisenrauch, A., Juhaszova, M., Ellis-Davies, G.C.R., Kaplan, J. H., Bamberg, E., & Blaustein, M.P. (1995). Electrical currents generated by a partially purified Na/Ca exchanger from lobster muscle reconstituted into liposomes and adsorbed on black lipid membranes: activation by photolysis of caged Ca^{2+}. J. Membr. Biol. 145, 151-164.

Eisner, D. A., & Lederer, W. J. (1985). Na-Ca exchange: stoichiometry and electrogenicity. Am. J. Physiol. 248, C189-C202.

Fabiato, A. (1985). Simulated calcium current can both cause calcium loading in and trigger calcium release from the sarcoplasmic reticulum of a skinned canine cardiac Purkinje cell. J. Gen. Physiol 85, 291-320.

Fontana, G., Rogowski, R. S., & Blaustein, M. P. (1995). Kinetic properties of the sodium/calcium exchanger in rat brain synaptosomes. J. Physiol. 485, 349-364.

Flesch, M., Schwinger, R. H. G., Muler-Ehmsen, Putz, F., & Bohm, M. (1995). Enhanced expression of the Na-Ca exchanger and its functional relevance in idiopathic dilated cardiomyopathy. Third International Conference on Sodium-Calcium Exchange, Woods Hole, Mass., April 23-26, 1995, Abstract P14.

Frank, J. S., Mottino, G., Molday R. S., & Philipson, K. D. (1992). Distribution of the Na$^+$-Ca^{2+} exchange protein in mammalian cardiac myocytes: an immunofluorescence and immmonocolloidal gold-labeling study. J. Cell. Biol. 117, 337-345.

Frank, J. S., Chen, F., Garfinkel, E., Moore, E., & Philipson, K. D. (1995). Immunolocalization of the Na$^+$-Ca^{2+} exchanger in cardiac myocytes. 3rd International Conference on Sodium-Calcium Exchange, Woods Hole, Mass., April 23-26, 1995, Abstract P12.

Furman, I., Cook, O., Kasir, J., Low, W., & Rahamimoff, H. (1995). The putative amino-terminal signal peptide of the cloned rat brain Na$^+$-Ca^{2+} exchanger Gene (RBE-1) is not mandatory for functional expression. J. Biol. Chem. 270, 19120-19127.

Gabellini, N., Iwata, T., & Carafoli, E. (1995). Alternative splicing in the C-terminal hydrophobic domains of the Na$^+$-Ca^{2+} exchanger. 3rd International Conference on Sodium-Calcium Exchange, Woods Hole, Mass., April 23-26, 1995, Abstract P15.

Gadsby, D. C., Kimura, J., & Noma, A. (1985). Voltage-dependence of Na/K pump current in isolated heart cells. Nature 315, 63-65.

Gadsby, D. C., & Nakao, M. (1989). Steady-state current-voltage relationship of the Na/K pump in guinea pig ventricular myocytes. J. Gen. Physiol. 94, 511-537.

Gatto, C., Xu, W.-Y., Denison, H. A., Hale, C. C., & Milanic, M. A. (1995). Modification of XIP, the autoinhibitory regions of the Na/Ca exchanger alter its ability to inhibit the Na/Ca exchanger in bovine sarcolemmal vesicles. 3rd International Conference on Sodium-Calcium Exchange, Woods Hole, Mass., April 23-26, 1995, Abstract P17.

Garty, H.(1994). Molecular mechanisms of epithelial, amiloride-blockable Na$^+$-channels. FASEB J. 8, 522-528.

Gerchman, Y., Olami, Y., Rimon, A., Taglight, D., Shuldiner, S., & Padan, E. (1993). Histidine-226 is part of the pH sensor of NhaA of Na$^+$/H$^+$ antiporter in *Escherichia coli*. Proc. Natl. Acad. Sci. 90, 1212-1216.

Gilon, C., Halle D., Chorev, M., Selinger, M., & Byk, G. (1991). Backbone cyclization: A new method for conferring conformational constraint on peptides. Biopolm.31, 745-750.

Glusker, J. P. (1991). Structural aspects of metal liganding to functional groups in proteins. Adv. Prot. Chem. 42, 1-75.

Glynn, I. M., & Karlish, S. J. D. (1990). Occluded cations in active transport. Ann. Rev. Biochem. 59, 171-205.

Goldshlegger, R., Karlish, S. J. D., Rephaeli, A., & Stein, W. D. (1987). The effect of membrane potential on the mammalian sodium-potassium pump reconstituted into phospholipid vesicles. J. Physiol. 387, 331-355.

Gunter, T. E., & Pfieffer, D. R. (1990). Mechanisms by which mitochondria transport calcium. Am. J. Physiol. 258, C755-C786.

Hale, C. C. (1996). Mechanism of XIP in cardiac sarcolemmal vesicles. Ann. N.Y. Acad. Sci.779, 171-181.

Henderson, P. J. F. (1993). The 12-transmembrane helix transporters. Curr. Opin. Cell Biol. 5, 708-721.

Hilgemann, D. W. (1986). Extracellular calcium transients at single excitations in rabbit atrium. J. Gen. Physiol. 87, 707-735.

Hilgemann, D.W., & Nobel, D. (1987). Excitation-contraction coupling and extracellular calcium transients in rabbit atrium. Reconstitution of basic cellular mechanisms. Proc. R. Soc. London B 230, 163-205.

Hilgemann, D. W. (1989). Giant excised cardiac sarcolemmal membrane patches:sodium and sodium-calcium exchange currents. Pflügers Arch., 415, 247-249.

Hilgemann, D. W. (1990). Regulation and deregulation of cardiac Na^+-Ca^{2+} exchange in giant excised sarcolemmal membrane patches. Nature 344, 242-245.

Hilgemann, D. W., Nicoll, D. A., & Philipson, K. D. (1991). Charge movement during Na^+-translocation by native and cloned Na^+/Ca^{2+} exchanger. Nature 352, 715-718.

Hilgemann, D. W., Matsuoka, S., Nagel, G.A., & Collins, A. (1992). Steady-state and dynamic properties of cardiac sodium-calcium exchange: Sodium-dependent inactivation. J. Gen. Physiol. 100, 905-931.

Hilgemann, D. W., & Collins, A. (1992). Mechanism of cardiac Na^+-Ca^{2+} exchange current stimulation by MgATP: possible involvement of aminophospholipid translocase. J. Physiol. 454, 59-82.

Hilgemann, D. W. (1995). Third International Conference on Sodium-Calcium Exchange, Woods Hole, Mass, April 23-26, 1995, Abstract A9.

Hilgemann, D. W. (1996). Function and regulation of the cardiac sodium-calcium exchanger. Ann. N.Y. Acad. Sci.779, 136-158.

Hilgemann, D. W., & Ball, R. (1996). Regulation of cardiac Na, Ca exchange and K_{ATP} channels by PIP_2. Science, 273, 956-959.

Hobai, I. A., Khananshvili, D., & Levi, A. J. (1997). The peptide "FRCRCFa", dialysed intracellularly, inhibits the Na/Ca exchange in rabbit ventricular myocytes with high affinity. Pflügers Arch. 433, 455-463.

Hodgkin, A. L., & Nunn, B. J. (1987). The effects of ions on sodium-calcium exchange in salamander rods. J. Physiol. 391, 371-398.

Hruby, V. J., Al-Obeidi F., & Kazmierski, W. (1990). Emerging approaches in the molecular design of receptor-selective peptide ligands: conformational, topological and dynamic considerations. Biochem. J. 268, 249-262.

Hryshko, L. V., Matsuoka, S., Nicoll, D. A., Weiss, J. N., & Philipson, K. D. (1993). Biosynthesis and initial processing of the cardiac sarcolemmal Na^+-Ca^{2+} exchanger. Biochim. Biophys. Acta 1151, 32-42.

Hryshko, L. V., Matsuoka, S., Nicoll, D. A., Levitsky, D., Hilgemann, D. W., Weis, J. N., & Philipson, K. D. (1994). Ca_i regulation of the cardiac Na-Ca exchanger. Biophys. J. 66, A331.

Iwata, T., & Carafoli, E. (1996). A new splicing variant in the frog heart sarcolemmal Na/Ca exchanger. Ann. N.Y. Acad. Sci. 779, 37-45.

Jencks, W.P. (1989). How does a calcium pump pump calcium? J. Biol. Chem. 264, 18855-18858.

Juhaszova, M., & Blaustein, M. P. (1996). Localization of the Na/Ca exchange in vascular smooth muscle, neurons, and astrocytes. Ann. N.Y. Acad. Sci. 779, 318-335.

Jung, D. W., Baysal, K., & Brierley, G. P. (1995). The sodium-calcium antiporter of heart mitochondria is not electroneutral. J. Biol. Chem. 270, 672-678.

Kappl, M., & Hartung, K. (1995). Transient Na-Ca exchange currents in giant patches induced by photolysis of caged Ca. Third International Conference on Sodium-Calcium Exchange, Woods Hole, Mass., April 23-26, 1995, Abstract P26.

Karlish, S. J. D., Goldshleger, R., & Stein, W. D. (1990). Identification of a 19kD C-terminal tryptic fragment of the alpha chain of Na/K-ATPase, essential for occlusion and transport of cations. Proc. Nat. Acad. Sci. 87, 4566-4570.

Kaczorowski, G. I., Slaughter, R. S., King, V. F., & Garcia, M. L. (1989). Inhibitors of sodium-calcium exchange: identification and development of probes of transport activity. Biochim. Biophys. Acta. 988, 287-302.

Khananshvili, D. (1990a). Distinction between the two basic mechanisms of cation transport in the cardiac Na$^+$-Ca^{2+} exchange system. Biochemistry 29, 2437-2442.

Khananshvili, D. (1990b). Cation antiporters. Curr. Opin. Cell Biol. 2, 731-734.

Khananshvili, D. (1991a). Voltage-dependent modulation of ion-binding and translocation in the cardiac Na$^+$-Ca^{2+} exchanger. J. Biol. Chem. 266, 13764-13769.

Khananshvili, D. (1991b). Mechanism of partial reactions in the cardiac Na$^+$-Ca^{2+} exchanger. Ann. N.Y. Acad. Sci. 639, 85-98.

Khananshvili, D., Price, D. C., Greenberg, M.J., & Sarne, Y. (1993). Phe-Met-Arg-Phe-NH$_2$ (FMRFa)-related peptides inhibit Na$^+$-Ca^{2+} exchange in cardiac sarcolemma vesicles. J. Biol. Chem. 268, 200-205.

Khananshvili, D., & Weil-Maslansky, E. (1994). The cardiac Na$^+$-Ca^{2+} exchanger: Relative rates of calcium and sodium movements and their modulation by protonation-deprotonation of the carrier. Biochemistry 33, 312-319.

Khananshvili, D., Shaulov, G., & Weil-Maslansky, E. (1995a). Rate-limiting mechanisms of exchange reactions in the cardiac sarcolemma Na$^+$-Ca^{2+} exchanger. Biochemistry 33, 10290-10297.

Khananshvili, D., Shaulov, G., Weil-Maslansky, E., & Baazov, D. (1995b). Positively charged cyclic hexapeptides, novel blockers for the cardiac sarcolemma Na$^+$-Ca^{2+} exchange. J. Biol. Chem. 270, 16182-16188.

Khananshvili, D., Baazov, D., Weil-Maslansky, E., Shaulou, G., & Mester, B. (1996a). Rapid interaction of FRCRCFa with cytosolic side of the cardiac sarcolemma Na$^+$-Ca^{2+} exchanger blocks the ion transport without preventing the binding of either sodium of calcium. Biochemistry 35, 15933-15940.

Khananshvili, D., Weil-Maslansky, E., & Baazov, D. (1996b). Kinetics and mechanism: Modulation of ion-transport in the cardiac sarcolemma sodium-calcium exchanger by protons, monovalent ions and temperature. Ann. N.Y. Acad. Sci. 779, 217-235.

Khananshvili, D., Mester, B., Saltown, M., Wang, X., Shaulov, G., & Baazov, D. (1997). Inhibition of the cardiac sarcolemma Na$^+$/Ca^{2+} exchanger by conformationally constrained small cyclic peptides. Molec. Pharmacol. 51, 126-131.

Kieval, R. S., Bloch, R. J., Lindenmayer, G. E., Ambesi, A., & Lederer, W. J. (1992). Immunofluorescence localization of the Na-Ca exchanger in heart cells. Am. J. Physiol. 263, C545-550.

Kimura, J., Noma, A., & Irisawa, H (1986). Na-Ca exchange current in mammalian heart cells. Nature 319, 596-597.

Kimura, J., Miyamae, S., & Noma, A. (1987). Identification of sodium-calcium exchange current in single ventricular cells of guinea-pig. J. Physiol. 384, 199-222.

Kleiboeker, S. B., Milanick, M. A., & Hale, C. C. (1992). Interaction of the exchange inhibitory peptide with Na-Ca exchange in bovine cardiac sarcolemmal vesicles and ferret red cells. J. Biol. Chem. 267, 17836-17841.

Koch, K.-W., & Stryer, L. (1988). Highly cooperative feedback control of retinal rod guanylate cyclase by calcium ions. Nature 334, 64-66.

Kofuji, P., Lederer, W. J., & Shulze, D. H. (1993). Na/Ca exchanger isoforms expressed in kidney. Am. J. Physiol. 265, F598-F603.

Kofuji, P., Lederer, W. J., & Shulze, D. H. (1994). Mutually exclusive and cassette exons underlie alternatively spliced isoforms of the Na/Ca exchanger. J. Biol. Chem. 269, 5145-5149.

Komuro, I., Wenninger, K. E., Philipson. K. D., & Izomo, S. (1992). Molecular cloning and characterization of the human cardiac Na+/Ca2+ exchanger cDNA. Proc. Natl. Acad. Sci. USA 89, 4769-4773.

Kraev, A., & Carafoli, E. (1995). Molecular biological studies on the cardiac sodium-calcium exchanger. Ann. NY Acad. Sci. 779, 103-109.

Krupka, R. M. (1989). Role of substrate binding forces in exchange-only transport systems: I. Transition state theory. J. Membr. Biol. 109, 151-158.

Lagnado, L., Cervetto, L., & McNaughton, P. A. (1988). Ion transport by the Na-Ca exchange in isolated rod outer segment. Proc. Natl. Acad. Sci. USA 85, 4548-4552.

Lagnado, L., & McNaughton, P. A. (1990). Electrogenic properties of the Na:Ca exchange. J. Membr. Biol. 113, 117-191.

Langer, G. A. (1994). Myocardial calcium compartmentation. Trends Cardiovas. Med. 4, 103-109.

Langer, G. A., & Rich, T. L. (1992). A discrete Na-Ca exchange-dependent Ca compartment in rat ventricular cells: exchange and localization. Am. J. Physiol. 262, C1149-C1153.

Langer, G. A., Wang, S. Y., & Rich, T. L. (1995). Localization of the Na/Ca exchange-dependent Ca compartment in cultured neonatal rat heart cells. Am. J. Physiol. 268, C119-C126.

Lauger, P. (1987). Voltage dependence of sodium-calcium exchange: predictions from kinetic models. J. Membr. Biol. 99, 1-11.

Lauger, P. (1991) In:Electrogenic Ion Pumps, pp. 3-135, Sinauer, Sunderland, MA.

Leblanc, N., & Hume, J. R. (1990). Sodium current-induced release of calcium from cardiac sarcoplasmic reticulum. Science 248, 372-375.

Lederer, W. J., Niggli, E., & Hadley, R. W. (1990). Sodium-calcium exchange in excitable cells. Science 248, 283.

Le Guennec, J.-Y., & Nobel, D. (1994). Effect of rapid changes of external Na+ concentration at different moments during the guine-pig myocytes. J. Physiol. 478, 493-504.

Lee, S.-L., Yu, A. S. L., & Lytton, J. (1994).Tissue-specific Expression of Na+-Ca2+ exchanger isoforms. J. Biol. Chem. 269, 14849-14852.

Levi, A. J., Brooksby, P., & Hancox, J. (1992). Depolarization of rat cardiac myocytes induces Ca entry on Na/Ca exchange which triggers intracellular Ca release and contraction. J. Molec. Cell Cardiol. 24, Suppl. IV, S.14.

Levitsky, D., O., Nicoll, D. A., & Philipson, K. D. (1994). Identification of the high affinity Ca2+ exchanger-binding domain of the cardiac Na+-Ca2+ exchanger. J. Biol. Chem. 269, 22847-22852.

Li, J., & Kimura, J. (1991). Translocation mechanism of cardiac Na-Ca exchange. Ann. N.Y. Acad. Sci. 639, 48-60.

Li, Z., Nicoll, D. A., Collins, A., Hilgemann, D. W., Filoteo, A. G., Penniston, J. T., Tomich, J. M., & Philipson, K. D. (1991). Identification of a peptide inhibitor of the cardiac sarcolemmal Na+-Ca2+ exchanger. J. Biol. Chem. 266, 1014-1020.

Li, Z., Smith, C. D., Smolley, J. R., Bridge, J. H. B., Frank, J. S., & Philipson, K. D. (1992a). Expression of the cardiac Na+-Ca2+ exchanger in insect cells using a baculovirus vector. J. Biol. Chem. 267, 7828-7833.

Li, W., Shariat-Madar, Z., Powers, M., Sun, X., Lane, R. D., & Garlid, K. D. (1992b). Reconstitution, identification, purification, and immunological characterization of the 110-kDa Na+/Ca2+ antiporter from beef heart mitochondria. J. Biol. Chem. 267, 17983-17989.

Li, Z., Burke, E. P., Frank, J. S., Bannett, V., & Philipson, K. D. (1993). The cardiac Na+-Ca2+ exchanger binds to the cytoskeletal protein ankyrin. J. Biol. Chem. 268, 11489-11491.

Li, Z., Matsuoka, S., Hryshko, L. V., Nicoll, D. A., Bersohn, M. M., Burke, E. P., Lifton, R. P., & Philipson, K. D. (1994). Cloning of the NCX2 isoform of the plasma membrane Na+/Ca2+ exchanger. J. Biol. Chem. 269, 17434-17439.

Lipp, P & Niggli, E. (1994). Modulation of Ca2+ releases in cultured neonatal rat cardiac myocytes: insights from subcellular release patterns revealed by confocal microscopy. Circ. Res. 74, 979-990.

Lipp, P., Swaller, B., & Niggli, E. (1995). Specific inhibition of Na-Ca exchange function by antisence oligodeoxynucleotides. FEBS Lett. 364, 198-202.

Lopez-Lopez, J. R., Shacklock, P. S., Balke, C. W., & Wier, W. G. (1995). Local calcium transients triggered by single L-type calcium channel currents in cardiac cells. Science 268, 1042-1045.

Low, W., Kasir, J., & Rahamimoff, H. (1993). Cloning of the rat heart Na⁺-Ca²⁺ exchanger and its functional expression in HeLa cells. FEBS Lett. 316, 63-67.

Lytton, J., Lee, S.-L., & Lee, W.-S. (1996). The kidney sodium-calcium exchanger. Ann. NY Acad. Sci. 779, 58-72.

Lyu, R.-M., Smith, L., & Smith, J. B. (1991). Sodium-calcium exchange in renal epithelial cells: dependence on cell sodium and competitive inhibition by magnesium. J. Membr. Biol. 124, 73-83.

Matsuoka, S., & Hilgemann, D. W. (1992). Steady-state and dynamic properties of cardiac sodium-calcium exchange: ion and voltage dependencies of the transport cycle. J. Gen. Physiol. 100, 963-1001.

Matsuoka, S., Nicoll, D. A., Reilly, R. F., Hilgemann, D. W., & Philipson, K. D. (1993). Initial localization of regulatory regions of the cardiac sarcolemmal Na⁺-Ca²⁺ antiporter. Proc. Natl. Acad. Sci. USA 90, 3870-3874.

Matsuoka, S., & Hilgemann, D. W. (1994). Inactivation of outward Na⁺-Ca²⁺ exchange current in guinea-pig ventricular myocytes. J. Physiology (London), 476, 443-458.

Matsuoka, S., Nicoll, D. A., Hryshko, L., V., Levitsky, D. O., Weis, J. N., & Philipson, K. D. (1995). Regulation of the cardiac Na⁺-Ca²⁺ exchanger by Ca²⁺: mutational analysis of the Ca²⁺-binding domain. J. Gen. Physiol. 105, 403-420.

Menick, D. R., Barnes, K. V., Thacker, U. F., Dawson, M. M., McDermott, D. E., Rozich, J. D., Kent, R. L., & Copper, G. (1996). The exchanger and cardiac hypertropy. Ann. N.Y. Acad. Sci.779, 489-501.

Moore, E. D. W., & Fay, F. S. (1993). Isoproterenol stimulates rapid extrusion of sodium from isolated smooth muscle cells. Proc. Natl. Acad. Sci. USA 90, 8058-8062.

Moore, E. D. W., Etter, E. F., Philipson, K. D., Carrington, W. A., Fogarty, K. E, Lifshitz, L. M., & Fay, F. S. (1993). Coupling of the Na⁺/Ca²⁺ exchanger, Na⁺/K⁺ pump and sarcoplasmic reticulum in smooth muscle. Nature 365, 657-660.

Mullins, L. J. (1977). A mechanism for Na-Ca transport. J. Gen. Physiol. 70, 681-695.

Mullins, L. J. (1979). The generation of electric currents in cardiac fibers by Na/Ca exchange. Am. J. Physiol. 236, C103-C110.

Nakasaki, Y., Iwamoto, T., Hanada, H., Imagawa, T., & Shigekawa, M. (1993). Cloning of the rat aortic smooth muscle Na⁺/Ca²⁺ untiporter and tissue-specific expression of isoforms. J. Biochem. Tokyo, 114, 528-534.

Nicoll, D. A., & Applebury, M. L. (1989). Purification of the bovine rod outer segment Na⁺/Ca²⁺ exchanger. J. Biol. Chem. 264, 16207-16213.

Nicoll, D. A., Longoni, S., & Philipson, K. D. (1990). Molecular cloning and functional expression of the cardiac sarcolemmal Na⁺-Ca²⁺ exchanger. Science 250, 562-564.

Nicoll, D. A., & Philipson, K. D. (1991). Molecular studies of the cardiac sarcolemma sodium-calcium exchanger. Ann. NY Acad. Sci. 639, 181-188.

Nicoll, D. A., Hryshko, L. V., Matsuoka, S, Wu, R.-Y., Hilgemann, D. W., & Philipson, K. D. (1994). Mutations in the putative transmembrane segments of the canine cardiac sarcolemmal Na/Ca exchanger. Biophys. J. 66, A330.

Nicoll, D. A., Hryshko, L. V., Matsuoka, S, Frank, J. S., & Philipson, K. D. (1996). Mutagenesis studies of the cardiac Na⁺-Ca²⁺ exchanger. Ann. N. Y. Acad. Sci. 779, 86-92.

Niggli, E., & Lederer W. J. (1991). Molecular operations of the sodium-calcium exchanger revealed by conformational currents. Nature 349, 621-624.

Niggli, E., & Lederer W. J. (1993). Activation of Na-Ca exchange current by photolysis of "caged calcium." Biophys. J. 65, 882-891.

Niggli, E., & Lipp, P. (1993). Subcellular restricted spaces: significance for cell signaling and excitation-contraction coupling. J. Muscle Res. Cell Motil. 14, 288-291.

Nobel, D., Nobel, S. J., Bett , G. C. L., Earm, Y. E., Ho, W. K., & So, I. K. (1991). The role of sodium-calcium during the cardiac action potential. Ann. N.Y. Acad. Sci. 639, 334-354.

Noma, A., Shioya, T., Paver, L. F. C., Twist, V. W., & Powell, T. (1991). Cytosolic free Ca^{2+} during operation of sodium-calcium exchange in guinea-pig heart cells. J. Physiol. 442, 257-276.

Or, E., David, P., Shainskaya A., Tal, D. M., & Karlish, J. D. (1993). Effects of competitive sodium-like antagonists on Na,K-ATPase suggest that cation occlusion from the cytoplasmic surface occurs in two steps. J. Biol. Chem. 268, 16929-16937.

Peskoff, A., Post, J. A., & Langer, G. A. (1992). Sarcolemma calcium binding sites in heart. II. Mathematical model for diffusion of calcium released from the sarcoplasmic reticulum into the diadic region. J. Membr. Biol. 129, 59-69.

Philipson, K. D., Bersohn, M. M., & Nishimoto, A. Y. (1982). Effect of pH on Na^+-Ca^{2+} exchange in canine cardiac sarcolemma vesicles. Circ. Res. 50, 287-293.

Philipson, K. D. (1990). The cardiac Na^+-Ca^{2+} exchanger. In: Calcium in the Heart (Langer, G. A, Ed.), pp. 85-108, Raven Press, New York.

Philipson, K. D., Nicoll, D. A., Li, Z., Frank, J. S., & Hilgemann, D. W. (1992). Molecular aspects of cardiac Na^+-Ca^{2+} exchange, 23S. 8th International Symposium on Calcium-Binding Proteins and Calcium Function in Health and Disease. August, 23-27 1992, Davos, Switzerland.

Philipson, K. D., & Nicoll, D. A. (1992). Sodium-calcium exchange. Curr. Opin. Cell Biol. 4, 678-683.

Philipson, K. D., & Nicoll, D. A. (1993). Molecular and kinetic aspects of sodium-calcium exchange. Intl. Rev. Cytol. 137C, 199-226.

Philipson, K. D., Nicoll, D. A., Matsuoka, S., Hrysko, L. V., Levitsky, D. O., & Weiss, J. N. (1996). Molecular regulation of the Na^+-Ca^{2+} exchanger. Ann. N.Y. Acad. Sci. 779, 20-28.

Pijuan, V., Zhuang, Y., Smith, L., Kroupis, C., Condrescu, M., Aceto, J. F., Reeves, J. P., & Smith, J. B. (1993). Stable expression of the cardiac sodium-calcium exchanger in CHO cells. Am. J. Physiol. 264, C1066-C1074.

Post, R. L., Sen, A. K., & Rosenthal, A. S. (1965). The phosphorylated intermediate in the ATP-dependent sodium and potassium transport across kidney membranes. J. Biol. Chem. 240, 1437-1445.

Powell, T., & Nobel, D. (1989). Calcium movement during each heart beat. Molec. Cell. Biochem. 89, 103-108.

Rahamimoff, H. Cook, O., Furman, I., Low, W., & Kasir, J. (1996). The role of different structural domains in functional expression of Na^+-Ca^{2+} exchange activity. Ann. NY Acad. Sci. 779, 29-36.

Rasgado-Flores, H., & Blaustein, M. P. (1987). Na/Ca exchange in barnacle muscle cells has a stoichiometry of 3 Na^+/1Ca^{2+}. Am. J. Physiol. 252, C499-C504.

Rasgado-Flores, H., Santiago, E. M., & Blaustein, M. P. (1988). Kinetics and stoichiometry of coupled Na efflux and Ca influx (Na/Ca exchange) in barnacle muscle cells. J. Gen. Physiol. 93, 1219-1241.

Rasgado-Flores, H., DeSantiago, J., & Espinosa-Tanguma, R. (1991). Stoichiometry and regulation of the Na-Ca exchanger in barnacle muscle cells. Ann. N.Y. Acad. Sci. 639, 22-33.

Reeves, J. P., & Hale, C.C. (1984). The stoichiometry of the cardiac sodium-calcium exchange system. J. Biol. Chem. 259, 7733-7739.

Reeves, J. P. (1985). The sarcolemmal sodium-calcium exchange system. Curr. Top. Membr. Transp. 25, 77-127.

Reeves, J. P. (1988). Measurement of sodium-calcium exchange activity in plasma membrane vesicles. Meth. Enzymol. 157, 505-510.

Reeves, J. P. (1992). Molecular aspects of sodium-calcium exchange. Arch. Biochem. Biophys. 292, 329-334.

Reeves, J. P., Condrescu, M., Chernaya, G., & Gardner, J. P. (1994). Na^+-Ca^{2+} antiport in the mammalian heart. J. Exp. Biol. 196, 375-388.

Reeves, J. P., Chernaya, G., & Condrescu, M. (1996). Sodium-calcium exchange and calcium homeostasis in transfected chinese hamster ovary cells. Ann. N.Y. Acad. Sci. 779, 73-85.

Reilander, H., Achillesb, A., Friedel, U., Maul, G., Lottspeich, F., & Cook, N. J. (1992). Primary structure and functional expression of the Na/Ca,K-exchanger from bovine rod photoreceptors. EMBO J. 11, 1689-1695.

Reilly, R. F., & Shurgue, C. A. (1992). cDNA cloning of a renal Na^+-Ca^{2+} exchanger. Am. J. Physiol. 262, F1105-F1109.

Reilly, R. F., Shurgue, C. A., Lattanzi, D., & Biemesderfer, D. (1993). Immunolocalization of the Na^+/Ca^{2+} exchanger in the rabbit kidney. Am. J. Physiol. 265, F327-F332.

Reilly, R. F., & Lattanzi, D. (1995). Identification of a novel alternatively-spliced isoform of the Na-Ca exchanger (NACA 8) in heart. Third International Conference on Sodium-Calcium Exchange, Woods Hole, Mass., April 23-26, 1995, Abstract, P33.

Reinecke, H., Studer, R., Vetter, R., Just, H., Holtz, J., & Drexler, H. (1995). Role of the sarcolemmal Na^+/Ca^{2+}-exchanger in end-stage human heart failure. Third International Conference on Sodium-Calcium Exchange, Woods Hole, Mass., April 23-26, 1995, Abstract, P36.

Reuter, H., & Seitz, N. (1968). The dependence of calcium efflux from cardiac muscle on temperature and external ion composition J. Physiol., 195, 451-470.

Reuter, H. (1995). Regulation of exocytosis in presynaptic boutons by Ca-channels and Na/Ca-exchange. Third International Conference on Sodium-Calcium Exchange, Woods Hole, Mass., April 23-26, 1995, Abstract A8.

Rispoli, G., Navangione, A., & Vellani, V. (1995). Transport of K^+ by Na^+-Ca^{2+},K^+ exchanger in isolated rods of lizard retina. Biophys. J. 69, 74-81.

Rispoli, G., Navangione, A., & Vellani, V. (1996). Turnover rate and number of Na^+:Ca^{2+},K^+ exchange sites in retinal photoreceptors. Ann. N.Y. Acad. Sci. 779, 346-355.

Ross, W. N. (1993). Calcium on the level. Biophys. J. 64, 1655-1656.

Sachs, J. R. (1977). Kinetic evaluation of the Na-K pump reaction mechanism. J. Physiol. 273, 489-514.

Sachs, J. R. (1991). Successes and failure of the Alberts-Post model in predicting ion-flux kinetics. In: The Sodium Pump: Structure, Mechanism, and Regulation (Kaplan, J. H., & De Weer, P. eds.) pp. 249-266. The Rockefeller University Press, New York.

Schneider, M., & Chandler, W. K. (1973). Voltage-dependent charge movement in skeletal muscle: a possible step in excitation-contraction coupling. Nature 242, 244-246.

Schnetkamp, P. P. M., Basu, D. K., & Szerencsei, R. T. (1989). Na^+-Ca^{2+} exchange in bovine rod segments requires and transports K^+. Am. J. Physiol. 257,C153-157.

Schnetkamp, P. P. M., & Szerencsei, R. T. (1991). Effect of potassium ions and membrane potential on the Na-Ca-K exchanger in isolated intact bovine rod outer segments. J. Biol. Chem. 266, 189-197.

Schnetkamp, P. P. M., Li, X. B., Basu, D. K., & Szerencsei, R. T. (1991). Regulation of free cytosolic Ca^{2+} concentration in the outer segments of bovine retinal rods Na-Ca-K exchange measured with fluo-3. J. Biol. Chem. 266, 22975-22982.

Schnetkamp, P. P. M. (1995a). How does the retinal rod Na-Ca+K exchanger regulate cytosolic free Ca^{2+}? J. Biol. Chem. 270, 13231-13239.

Schnetkamp, P. P. M. (1995b). Chelating properties of the Ca^{2+} transport site of the retinal Na-Ca+K exchanger: evidence for a common Ca^{2+} and Na^+ binding sites. Biochemistry, 34, 7282-7287.

Schwarz, E. M., & Benzer, S. (1995). Expression and evolution of Calx, a sodium-calcium exchanger of *Drosophila melanogaster*. Third International Conference on Sodium-Calcium Exchange, Woods Hole, Mass., April 23-26, 1995, Abstract P35.

Shieh, B. H., Xia, Y., Sparkers, R. S., Klisak, I., Lusis, A. J., Nicoll, D. A., & Philipson, K. D. (1992). Mapping of the gene for the cardiac sarcolemmal Na^+-Ca^{2+} exchanger to human chromosome 2p21-p23. Genomics, 12, 616-617.

Shigekawa, M. Iwamoto, T., & Wakabayashi (1995). Phosphorylation and modulation of Na^+-Ca^{2+} exchanger in vascular smooth muscle cells. Ann. NY Acad. Sci. (in press).

Shulze, D., Kofuji, P., Hadley, R., Kirby, M. S., Kieval, Doering, A., Niggli, E., & Lederer, W. J. (1993). Sodium/calcium exchanger in heart muscle: molecular biology, cellular function, and its special role in excitation-contraction coupling. Cardiovasc. Res. 27, 1726-1734.

Slaughter, R. S., Sutko, J. L., & Reeves, J. P. (1983). Equilibrium calcium-calcium exchange in cardiac sarcolemmal vesicles. J. Biol. Chem. 258, 3183-3190.

Smith, J. B., Lyu, R.-M., & Smith, L. (1991). Sodium-calcium exchange in aortic myocytes and renal epithelial cells. Ann. NY. Acad. Sci. 639, 505-520.

Smith, L., & Smith, J. B. (1994). Regulation of sodium-calcium exchanger by glucocorticoids and growth factors in vascular smooth muscle: dependence on metabolic energy and intracellular sodium. J. Biol. Chem. 269, 27527-27531.

Smith, J. B., Lee, H.-W., & Smith, L. (1996). Regulation of sodium-calcium exchanger expression. Ann. NY Acad. Sci.

Stein, W. D. (1986). Transport and Diffusion Across Cell Membranes, pp. 55-120, Academic Press, NY.

Strynadka, N. C. J., & James, M. N. G. (1989). Crystal structures of the helix-loop-helix calcium-binding proteins. Ann. Rev. Biochem. 58, 951-958.

Tonioli, C. (1990). Conformationally restricted peptides through short-range cyclizations. Int. J. Peptide Protein Res. 35, 287-300.

Tsien, R. W., & Tsien, R.Y. (1990). Calcium channels, stores, and oscillations. Ann. Rev. Cell Biol. 6, 715-760.

Van Eylen, F., Gourlet, P., Valdermeers, A., Lebrun, P., & Herchelz, A. (1996a). Effects of Phe-Met-Arg-Phe-NH$_2$ (FMRFa)-related peptides on Na/Ca exchange and ion fluxes in rat pancreatic ß-cells. Third International Conference on Sodium-Calcium Exchange, Woods Hole, Mass., April 23-26, 1995, Abstract P37.

Van Eylen, F., Svoboda, M., Bollen, A., & Herchelz, A. (1996b). Presence of NACA3 and NACA7 exchanger isoforms in insulin producing cells. Third International Conference on Sodium-Calcium Exchange, Woods Hole, Mass., April 23-26, 1995, Abstract P38.

Yasui, K., & J. Kimura (1990). Is potassium co-transported by the cardiac Na-Ca exchanged. Pflügers Arch. 415, 513-515.

Wakabayashi, S., & Goshima, K. (1981). Comparision of kinetic characteristics of Na-Ca exchange in sarcolemma vesicles and cultured cells from chick heart. Biochem. Biophys.Acta 645, 311-317.

Wier, W. G. (1990). Cytosolic [Ca^{2+}] in mammalian ventricle: dynamic control by cellular processes. Ann. Rev. Physiol. 52, 467-485.

STRUCTURE AND DYNAMICS OF THE F_1F_0-TYPE ATPase

Roderick A. Capaldi and Robert Aggeler

I. Introduction. 360
II. Subunit Composition 360
III. Organization of the F_1F_0 Complex from Electron Microscopy 361
IV. The High Resolution Structure of F_1 Shows the Overall Fold
    of the α and β Subunits 363
V. Features of Nucleotide Binding Sites 364
VI. Structure of the γ Subunit. 365
VII. Structure of the ε Subunit of ECF_1 and Interaction of this
    Subunit with the Core $\alpha_3\beta_3\gamma$ Complex 365
VIII. Structure of the δ Subunit of ECF_1 or CF_1, and OSCP of MF_1 366
IX. Asymmetry of the F_1 Structure. 367
X. Arrangement of Subunits in the F_0 Part. 368
XI. Unisite Catalysis, ATP Hydrolysis, and ATP Synthesis
    in a Single Catalytic Site 368
XII. Multisite or Cooperative Functioning of the F_1 and F_1F_0 Complexes 369
XIII. Studies of the Dynamics of the F_1F_0 Complex during Functioning. 370

Advances in Molecular and Cell Biology
Volume 23B, pages 359-380.
Copyright © 1998 by JAI Press Inc.
All right of reproduction in any form reserved.
ISBN: 0-7623-0287-9

I. INTRODUCTION

F_1F_0-type ATPases are found in the plasma membrane of prokaryotes, the mitochondrial inner membrane of animal cells, and in both the chloroplast thylakoid membrane and mitochondrial inner membrane of plants. This key enzyme synthesizes ATP in response to an electron-transfer-generated proton gradient, and in the reverse direction, generates an ATP hydrolysis-driven proton gradient for use in ion and substrate transport processes. With more than thirty years of study, a considerable amount is known about the structure and functioning of F_1F_0-ATPases. Several reviews have focused on mechanistic aspects of this complex enzyme (Senior, 1988, 1990; Boyer, 1989, 1993; Futai et al., 1989; Penefsky and Cross, 1991; Allison et al., 1992; Hatefi, 1993). Here, we focus on new data on the structure of the F_1F_0-ATPases and relate these to the cooperativity of catalysis that is a key feature of this enzyme, as well as to the still poorly understood mechanism of coupling between catalytic sites and the proton channel.

II. SUBUNIT COMPOSITION

As the name implies, F_1F_0-type ATPases are composed of two parts, which can be separated and studied independently. The F_1 part, once released from the F_0, is a water-soluble complex that contains the nucleotide binding sites. The F_0 part of the complex is membrane-intercalated and contains the proton channel.

The F_1 part from bacteria, chloroplasts and mitochondria (BF_1, CF_1, and MF_1, respectively) contains five different subunits α, β, γ, δ and ε, in the molar ratio 3:3:1:1:1. The α, β, and γ subunits, respectively, of the enzyme from all sources are equivalent and homologous (Walker et al., 1982a, 1985; Falk et al., 1985). The δ subunit of the bacterial enzyme, such as from *Escherichia coli* (ECF_1), is the equivalent of the δ subunit in CF_1 and of the polypeptide called the oligomycin sensitivity conferring protein (OSCP) in MF_1 (which does not usually co-purify with MF_1) (Walker et al., 1985). The ε subunit of ECF_1 and CF_1 is the equivalent of the δ subunit of MF_1 (Hamasur and Glaser, 1992; Giraud and Velours, 1994). The so-called ε subunit in MF_1 has no known counterpart in the bacterial or chloroplast enzyme (Walker et al., 1985; Guelin et al., 1993).

The F_0 complexes from bacteria, for example, ECF_0, are simplest in that they contain only 3 different polypeptides—a, b, and c (Walker et al., 1982b; Cox et al., 1984; Fillingame, 1990). In chloroplasts, there are four different subunits usually numbered I-IV (Fromme et al., 1987), while the mitochondrial enzyme contains 9 different polypeptides, first resolved by Ludwig et al. (1980) and, more recently, isolated and characterized (Walker et al., 1991; Collinson et al., 1994a).

An as yet undecided issue is the stoichiometry of subunits in the F_0 part. The molar ratio of subunits a and b in ECF_0 is 1:2 (Fillingame, 1990). In CF_0, subunits I and II, which are both equivalents structurally of subunit b (Berzborn et al., 1990), and subunit IV which is the homologue of a in bacteria (Cozens et al., 1986), are present in 1:1:1 stoichiometry (Gräber et al., 1990). Less certain at present is the number of c subunits in the F_0 complex, and most reviews give a range of 9-12 to accommodate the various determinations made. The most careful work on the number of c subunits is by Foster and Fillingame (1982) for ECF_1F_0. These workers isolated the intact complex from cells grown in radiolabeled amino acids, and obtained a value of 10 ± 1 for the number of c subunits. We have recently been working with mutants of ECF_1F_0 which contain a Cys residue genetically engineered into the c subunits. When enzyme isolated from such mutants is labeled with [^{14}C]NEM, after first denaturing and unfolding the polypeptides with sodium dodecyl sulfate, the molar ratio of NEM incorporated is $\alpha12$, $\beta3$, $\gamma2$, $\delta2$, and c10-12 for subunits containing 4, 1, 2, 2, and 1 Cys residues, respectively (Watts, S.D. and Capaldi, R.A. unpublished data). Therefore, these data support a value of 10 for the number of c subunits in the F_0 complex.

In MF_1, the equivalent subunits of a, b, and c are a or subunit 6, b and c respectively (Walker et al., 1991). Of the remaining polypeptides, OSCP (equivalent of δ: see above), F_6 and subunit d are hydrophilic and water soluble components, and appear to be a part of the stalk region (Collinson et al., 1994b; Belogrudov et al., 1995) (see below). At present, there is disagreement on the stoichiometry of b, OSCP, d and F_6. According to Collinson et al. (1994b), these are present in a 1:1:1:1 stoichiometry. In contrast, Hekman et al. (1991) have determined a ratio of 2:1:1:2 for these four subunits.

III. ORGANIZATION OF THE F_1F_0 COMPLEX FROM ELECTRON MICROSCOPY

There have been numerous studies of the structure of the F_1 part and the F_1F_0 complex using negative-staining electron microscopy (e.g., Fernandez-Moran, 1962; Lien and Racker, 1971; Tsuprun et al., 1989; Boekema et al., 1990; Gräber et al., 1990). More recently, cryoelectron microscopy has been used (Gogol et al., 1987, 1989a; Lücken et al., 1990; Wilkens and Capaldi, 1994), in which samples are examined after rapid freezing in a thin layer of amorphous ice and, hence, at close to physiological conditions.

Earlier electron microscopy studies, as well as low resolution X-ray analysis, had been interpreted to show that the F_1 part was arranged as six masses (the α and β subunits) in two layers, one above the other (with separate layers of α and β subunits) (Amzel et al., 1982; Tiedge et al., 1983; Tsuprun et al., 1984). Our three-dimensional reconstruction of ECF_1 from two dimensional arrays and concomitant cryoelectron microscopy studies gave a very different picture. We found that the α and β subunits were interdigitated for most of their length with the two different subunits alternating around the periphery of a central cavity, in which was located the γ subunit (Gogol et al., 1989a, b). The recent high-resolution X-ray structure of MF_1 (see later) confirms this arrangement of the F_1 part (Abrahams et al., 1994).

Earlier negatively stained images of F_1F_0 had shown the F_1 and F_0 parts separated by a relatively narrow stalk (Fernandez-Moran, 1962; Lien and Racker,

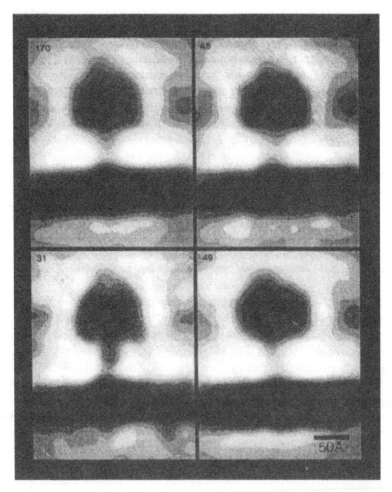

Figure 1. Averaged images of side views of the ECF_1F_0. Top left is an average of 170 images, while the other three panels are subclasses of this average showing different features of the F_1 part. (Reproduced from Lücken et al., 1990, with permission).

1971). However, such an arrangement had been largely discounted as an artifact of the negative stain dislodging the F_1 during sample preparation. The idea that the F_1 was separated from the bilayer intercalated part of the F_0 by a stalk-like structure received strong support when the F_1F_0 was imaged in ice (Gogol et al., 1987; Lücken et al., 1990), thereby avoiding the potential artifacts of staining. Figure 1 shows different classes of images of ECF_1F_0 obtained in cryoelectron microscopy, each with different features of the F_1 part. In each case, the stalk is clearly evident, 40-45 Å in length, and relatively narrow. It is now clear that views such as those in Figure 1 are themselves averages. For instance, the view down the cleft of adjacent α and β subunits is the combination of the views through three such clefts. As a result, asymmetric features in the stalk (and in the F_0) are averaged out and the stalk is likely wider in some projections, thereby allowing for the amount of protein that must be in this region (that is, in ECF_1F_0, the stalk must contain all of the δ and ϵ, part of the γ subunit, and part of the two b subunits see later). The images shown in Figure 1 have protein extending from the F_0 at the F_1 side of the membrane but there is apparently very little protein outside the bilayer on the opposite side of the membrane. There have been attempts to image the F_0 part alone by electron microscopy (Birkenhäger et al., 1995), and these are discussed later in relation to the organization of this subcomplex.

IV. THE HIGH RESOLUTION STRUCTURE OF F_1 SHOWS THE OVERALL FOLD OF THE α AND β SUBUNITS

A major advance in the understanding of F_1F_0 ATPases has come with the recently published high-resolution structure of MF_1 (Abrahams et al., 1994). This structure shows almost all of the α and β subunits and segments of the γ subunit involving residues 1-45, 82-99, and 223-286 (numbering for ECF_1). The remaining half of the γ subunit, along with the δ and ϵ subunits, were not resolved, either because they are disordered, or because they had become dissociated during crystallization. MF_1 is roughly globular, as previously indicated by electron microscopy, with dimensions of 80 Å top to bottom and 100 Å wide, as shown in Figure 2.

As expected from the significant homology in their amino acid sequences, the alternating α and β subunits have almost identical folds. Both subunits consist of three domains. At the top is an amino-terminal six-stranded β barrel. These barrels form a continuous β sheet linking all six subunits. There is a central domain containing the nucleotide binding sites. In both the α and β subunits, the nucleotide binding domain is a nine-stranded β sheet with nine associated α helices. This "core" nucleotide site has an almost identical overall fold to that of the nucleotide binding domain of RecA (Story et al., 1992). The third and bottom domain consists of a carboxyl-terminal bundle containing seven α helices in the α subunit and six α helices in the β subunit.

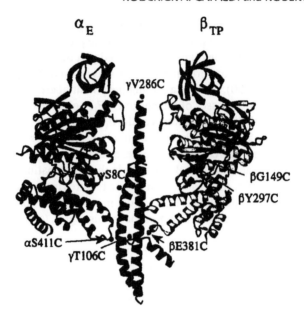

Figure 2. Arrangement and structural features of the α, β, and γ subunits of F_1. The α and β subunits shown are α_E and β_{TP}. This figure identifies sites on the α, β, and γ subunits at which cys residues have been introduced in our studies to date.

V.   FEATURES OF NUCLEOTIDE BINDING SITES

There are a total of six nucleotide binding sites on F_1, three called catalytic sites and three called noncatalytic sites, based on functional studies (Cross and Nalin, 1982; Perlin et al., 1984; Issartel et al., 1986; Pedersen et al., 1995). Earlier biochemical studies and mutagenesis experiments had indicated that the catalytic sites are located predominantly on β subunits, while the non-catalytic sites are mainly located on α subunits (reviewed in Allison et al., 1992; Futai et al., 1992; Pedersen et al., 1995; Senior et al., 1995); this is confirmed in the X-ray structure of MF_1.

All six nucleotide-binding sites are at the interfaces between α and β subunits with an average distance between adjacent catalytic and noncatalytic sites of 27 Å between β phosphates. In catalytic sites, the adenine part of the substrate nucleotide is in a hydrophobic pocket provided by β Tyr331, Phe404, and Phe410. The binding of the phosphates and associated Mg^{2+} ion is via the so-called "glycine rich" loop involving particularly β Lys155 and β Thr156. There is at least one water molecule in the catalytic site, providing hydrogen-bonding between side chains and substrate. One is bound to the carboxylate of β Glu181, and this residue is appropriately positioned to activate the water molecule, thereby promoting an in-line nucleophilic attack on the terminal phosphate. The guanidinium of Arg376 of the α subunit is positioned such that it may stabilize the negative charge of the terminal phosphate, with ATP bound in a penta-coordinate transition state.

The overall structure of noncatalytic sites is similar to that of the catalytic sites. The major differences between the two types of sites are that the adenine pocket on the α subunit is less hydrophobic. Also, β Glu181 in the catalytic sites is replaced by a Gln, which would explain why the noncatalytic sites are unable to hydrolyze or synthesize ATP. Finally, in noncatalytic sites, β Tyr354 is close to the 3 position of the adenine ring.

VI. STRUCTURE OF THE γ SUBUNIT

The X-ray determination shows three α-helical segments of the γ subunit. There is a C-terminal helix involving residues 223-286, the very C terminus of which is located in the "pit" at the top of the molecule. This helix then extends through the length of the $\alpha_3\beta_3$ barrel into a stem which is the stalk region. The N-terminal α helix, composed of amino acids 1-45, forms a left-handed antiparallel coiled coil with the lower half of the C-terminal helix, as shown in Figure 2. In the particular conformation of the enzyme being observed, the C-terminal α helix is twisted so the top of its axis is coincident with the pseudo-threefold axis relating α and β subunits. At the lower end, in the stalk region, the C-terminal helix is displaced from this axis by 7 Å. A third short α helix, provided by residues between 82 and 99 of the γ subunit, binds to β (and α subunits see below) at the bottom of the structure, that is, in the C-terminal domain of the β (and α) subunits. Secondary structure prediction algorithms (which had predicted abundant α helical structure of the N-terminal and C-terminal parts), indicate β sheet conformation of much of the γ subunit within the stalk. In chloroplasts, the γ subunit has an additional loop region not found in the mitochondrial or bacterial enzyme, which contains a disulfide bridge that is the site of light-dependent regulation of ATPase activity by the protein thioredoxin (Nalin and McCarty, 1984; Ketcham et al., 1984).

We have begun to explore the interactions of the γ subunit within the stalk region (Tang, C. and Capaldi, R.A. unpublished data). One face interacts with the ε subunit and this includes residues around 70, along with 106 and a region from residue 202 to around 240. The γ subunit also interacts with the c subunits of the F_0. In collaborative studies, we have shown that a thioether bond can be formed between Cys44 (introduced by mutagenesis) in the loop region of the c subunit and a Tyr or Trp in the region of residues 202-223 of the γ subunit (Watts et al., 1995).

VII. STRUCTURE OF THE ε SUBUNIT OF ECF$_1$ AND INTERACTION OF THIS SUBUNIT WITH THE CORE $\alpha_3\beta_3\gamma$ COMPLEX

A structure of the ε subunit of ECF$_1$ in solution has been obtained by nuclear magnetic resonance methods (Wilkens et al., 1995). These studies show that the ε

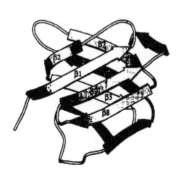

Figure 3. Two views of the polypeptide fold of the ε subunit of ECF₁. (Reproduced from Wilkens et al., 1995, with permission).

subunit is composed of two domains. The N-terminal 84 residues are a flattened 10-stranded β barrel or "sandwich", the inside of which contains predominantly hydrophobic residues. The C-terminal domain of 48 residues forms an α-helix-turn-α-helix structure. In solution, the C-terminal domain folds back to the β barrel structure, interacting with one end of this β sandwich, as shown in Figure 3.

A combination of genetic, cross-linking and chemical-labeling experiments have helped place the ε subunit in the overall F_1F_0 structure. The C-terminal domain interacts with (both) α and β subunits, an interaction which involves the loop region of ε residues (106-111, particularly residue 108), the DELSEED region of β (residues 380-386), and the equivalent region (including residue 411) in the α subunit (Dallmann et al., 1992; Aggeler et al., 1995a; Capaldi et al., 1995). The opposite end of the β sandwich, involving residues 31 and 38 interacts with F_0 (Aggeler et al., 1995b) via the c subunits (Zhang et al., 1994; Zhang and Fillingame, 1995a), while the face of the β sandwich that contains residues 10, 38, and 43 binds to the γ subunit (Aggeler et al., 1992, 1995b; Tang, C. and Capaldi, R.A., unpublished data). This face contains a stripe of hydrophobic residues including Val9, Met15, Leu42, Ile68, Thr77, and Leu79, which may contribute to the linkage of ε and γ.

VIII. STRUCTURE OF THE δ SUBUNIT OF ECF₁ OR CF₁, AND OSCP OF MF₁

The δ subunit of the bacterial and chloroplast enzyme, and the equivalent polypeptide, OSCP of MF₁, is the least characterized of the subunits of the F₁ part. Binding studies (Mendel-Hartvig and Capaldi, 1991a) and cross-linking experiments (Beckers et al., 1992) place the δ subunit in contact with the α subunit of F₁ and the b subunits of F₀ (subunit I of chloroplasts). Circular dichroism measurements indicate that the δ subunit is predominantly α helical (Engelbrecht et al., 1991). Whether the polypeptide is an α helical hairpin or a 3-4 α helix bundle, should be clearer when our ongoing NMR structure determination is complete.

IX. ASYMMETRY OF THE F_1 STRUCTURE

The crystal structure shows F_1 as a highly asymmetrical molecule with the 3 α-β pairs in different conformations, induced by either a different nucleotide occupancy of the catalytic sites they contain, or to the different interactions they have with the γ subunit, or both factors. In the form of the enzyme crystallized, one α-β pair contains ATP in the catalytic site (called α_{TP} and β_{TP} by Abrahams et al., 1994). β_{TP} is the site of binding of the short α helix of the γ subunit via hydrogen bonding of the backbone amide of residue 89, and the amino group of residue 96, with the so-called "DELSEED" region (β380-386) of this subunit. The catalytic site of the second α-β pair contains ADP ($\alpha_{DP}\beta_{DP}$). β_{DP} interacts with the γ subunit through its C-terminal domain. The third α-β pair has no catalytic site nucleotide ($\alpha_E\beta_E$). In β_E the catalytic site is disrupted, with the lower part of the nucleotide binding domain along with the C-terminal α-helical domain rotated by 30° with respect to their positions in β_{TP} and β_{DP}. This conformational difference requires movements of the C-terminal domain relative to the catalytic site domain of up to 20 Å. β_E is the site of the major specific interactions between the two large α helices of the γ subunit and α-β subunits, an interaction involving hydrogen bonding between Asp302, Thr304 and Asp305 with γ Arg268 and Gln269. Close by, α_E Asp336 makes a salt bridge with γ residue 266. α_{TP} also differs from α_E or α_{DP} in that the noncatalytic site is more open due to the hinging out of the C-terminal domain of β_E. As a result, in the interface between β_E and α_{TP}, there is no interaction between β Tyr354 and the adenine ring of bound noncatalytic nucleotide.

The conformational differences of the three α-β pairs in and around the catalytic sites result in one site being tightly closed (on β_{DP}), a second more open (on β_{TP}), and the third fully open (on β_E).

In the five subunit F_1, and in F_1F_0, the asymmetry of the enzyme must extend to the interaction of the three α-β pairs with other single copy subunits besides the γ subunit. We have been studying this asymmetry in ECF_1 by cross-linking experiments. When ECF_1 isolated from the mutant βE381C:ϵS108C is reacted with low concentrations of $CuCl_2$, there is essentially complete (>95%) disulfide bond formation between one β subunit and the ϵ subunit, and a second β subunit and the γ subunit (Aggeler et al., 1995a). These results indicate that the ϵ subunit interacts with a different β subunit than does the short α helix in the γ subunit. As the short α helix of γ is linked to β_{TP} in the nomenclature of Abrahams et al. (1994), the ϵ subunit must be bound to β_{DP} or β_E.

We have recently completed a study with the mutant αS411C:βE381C:ϵS108C (Aggeler, R. and Capaldi, R.A., unpublished data). In this mutant, there is a Cys residue both in the DELSEED region of β and at the equivalent site in the C-terminal region of the α subunit. In this mutant, there is high-yield cross-linking between an α and β subunit. The closest approach of an α Ser411 to a β Glu381 is 6.8 Å between the α carbons of α_E and β_{DP} (A. Leslie, personal communication). In the presence of the α-β cross-link, there can also be cross-linking of ε to a β in near 100% yield. If the α-β cross-link involves β_{DP}, then the ε subunit is bound to, and may define β_E.

X. ARRANGEMENT OF SUBUNITS IN THE F_0 PART

As described already, the simplest F_0 compositionally is that from bacteria with only three subunits, a, b, and c. The folding of the a subunit within the bilayer has been studied by several groups without a consensus on the topology. Models in which there are 5, 6, 7, or 8 transmembrane α helices have been proposed (reviewed in Fillingame, 1990; Deckers-Heberstreit and Altendorf, 1992). The arrangement of the b subunits is better understood. This polypeptide has an N-terminal hydrophobic stretch which spans the membrane, with the remainder of the subunit in the stalk and linked to the F_1 part (Cox et al., 1984; Schneider and Altendorf, 1987). Dunn (1992) has genetically constructed a water-soluble derivative of subunit b missing the transmembrane region, which forms a dimer in solution and appears to be predominantly α-helical in conformation.

The best characterized of the F_0 subunits is the c subunit. Girvin and Fillingame (1993, 1994, 1995) are determining the structure of this subunit by nuclear magnetic resonance. They are examining the polypeptide dissolved in chloroform methanol water mixture in which it appears to retain the same conformation, for the most part, as in the lipid milieu. The polypeptide is an α-helix-loop-α-helix with the loop region on the same side of the membrane as F_1. The two α helices appear gently curved, crossing at an angle of around 30° at a point close to Ala21 and Met65. Multiple c subunits—probably 10, as discussed already—interact in the F_0. Recent electron microscopy studies support a model in which the a and b subunits are outside the ring of c subunits (Birkenhäger et al., 1995).

XI. UNISITE CATALYSIS, ATP HYDROLYSIS, AND ATP SYNTHESIS IN A SINGLE CATALYTIC SITE

A single catalytic site can, by itself, hydrolyze ATP as indicated by studies of so-called "unisite" catalysis, observed when enzyme is in excess of substrate ATP. Kinetic constants for this unisite ATPase activity in MF_1, CF_1 and ECF_1 indicate that the reaction ATP , ADP + P_i is close to equilibrium (Grubmeyer et al., 1982; Al-Shawi and Senior, 1988). Also, binding of ATP is very strong while P_i-binding is relatively weak. Therefore, it has been argued, the energy derived from proton translocation must be used in ATP synthesis, primarily to release tightly bound product ATP (see Senior, 1990; Boyer, 1993) (although some must be utilized to favor P_i binding in catalytic sites). It is important to point out that unisite catalysis is not an artifact of F_1 preparations: For example, it can be followed in mitochondrial membranes as well as in purified CF_1F_0 reconstituted into liposomes, where the reaction can be coupled to proton translocation (Fromme and Gräber, 1990a, b).

In single turnover experiments in the presence of DMSO (up to 50%), it has proved possible to reverse ATP hydrolysis by F_1, that is, to synthesize ATP on the enzyme (Feldman and Sigman, 1982; Sakamoto and Tonomura, 1983; Yoshida, 1983; Beharry and Bragg, 1992). This finding is additional evidence that the equilibrium between ATP and ADP + P_i on the enzyme is close to unity, and has been taken to imply that a more hydrophobic, that is, closed, catalytic site is needed for ATP synthesis to favor binding of P_i.

XII. MULTISITE OR COOPERATIVE FUNCTIONING OF THE F_1 AND F_1F_0 COMPLEXES

Both ATP hydrolysis and ATP synthesis depend physiologically on cooperativity between catalytic sites. The evidence that binding of substrate in one catalytic site dramatically increases the rate of catalysis in a second site, first provided by Penefsky and colleagues (Cross et al., 1982), has been discussed by Boyer (1993), and will not be covered here. In MF_1, the rate increase of ATP hydrolysis under such cooperative or multisite conditions is 10^6 greater than in unisite catalysis (see Cross et al., 1982). For ECF_1, the rate enhancement is probably less, that is, 10^4-10^5 (see Senior, 1990).

The model of catalytic site cooperativity that has received the most attention is the alternating-site mechanism of Boyer (described in detail in Boyer, 1987, 1993; Cross, 1992). In this model, each catalytic site cycles through three states, an open state in which the site has released product (ADP+P_i in the direction of ATP hydrolysis) and is ready to accept substrate (ATP), a closed state in which substrate (ATP) has been tightly bound but no catalysis has occurred, and a partially open state in which bond cleavage (in the direction of ATP hydrolysis) or bond formation (in the direction of ATP synthesis) has occurred, but product release has not yet taken place. In such an alternating catalytic site mechanism, the three catalytic sites should be in different conformations at any one time, as is the case in the high-resolution structure of MF_1. However, the crystal form of Abrahams et al. (1994) has ADP in the closed site, AMP·PNP in the partially open site and contains azide, a potent inhibitor of the enzyme. This overall structure, then, is at best an inactive state, and may not be a physiologically relevant form.

A contentious issue at present is whether two catalytic sites can undergo rapid multisite catalysis, or if all three sites must participate for rates to approach V_{max}. It has been argued by Berden and colleagues that a two-site mechanism is the only way in which the enzyme could work (Edel et al., 1992, 1993). In contrast, Boyer has proposed that the enzyme can turn over relatively rapidly by either bi-site or tri-site catalysis (Boyer, 1993). Recent studies by Weber et al. (1993) using a mutant ECF_1 in which β Tyr331 in the catalytic site is replaced by a Trp are convincing evidence that all three sites must participate to obtain V_{max} rates of ATP hydrolysis. In this mutant, the binding of nucleotide ATP or ADP into catalytic sites can be monitored by changes in fluorescence of the introduced Trp and, thereby, differentiated from nucleotide binding into noncatalytic sites. The rate of bi-site catalysis measured in this way is less than 20% of that obtained when all three sites are participating.

XIII.   STUDIES OF THE DYNAMICS OF THE F_1F_0 COMPLEX DURING FUNCTIONING

Catalytic site cooperativity and coupling between catalytic sites and the proton channel, a distance of close to 100 Å, should involve the enzyme cycling through a number of conformations. This dynamic characteristic of ECF_1F_0 has been explored in our laboratory by both electron microscopy and by cross-linking studies. When ECF_1 is observed in the hexagonal projection, a central mass is seen within the ring of α and β subunits that is the projection down the γ subunit (Gogol et al., 1990; Wilkens and Capaldi, 1994). Several locations of the central mass are seen under all nucleotide conditions, even in the absence of nucleotides in catalytic sites, indicating a variety of positions of the γ subunit relative to the three α-β pairs.

Using novel bifunctional reagents, the TFPAMs, that contain a maleimide at one end (for reaction with introduced Cys residues) and a photo-activatable tetrafluorophenylazide group at the other end, which inserts in high yield into several chemical bonds in the protein (Aggeler et al., 1992), we have generated specific cross-links from residue 8 of γ to a β subunit (in the glycine rich loop) (Aggeler et al., 1993), from a Cys at position 286 of γ to an α subunit (Aggeler and Capaldi, 1992), and from a Cys at residue 108 of ϵ to the α subunit(s) (Aggeler et al., 1992). In each case, there was a loss of ATPase activity in proportion to the yield of cross-link obtained, confirming the importance of motion of the γ subunit and indicating that there are movements of the ϵ subunit in relation to α and β subunits during enzyme functioning.

More recently, we have created mutations in which Cys residues were placed to allow disulfide bond formation between different subunits in the enzyme. Among the mutants we have studied are: (1) β E381C, in which $CuCl_2$ treatment induces disulfide bond formation between β Cys381 and the intrinsic γ Cys87; (2) βE381C:ϵS108C, in which S-S bridges can be formed selectively between a β Cys381 and ϵ Cys108, or in which two disulfides can be obtained, that is, β Cys381-γ Cys87 and β Cys381- ϵ Cys108; (3) αS411C, in which a disulfide bond forms between α Cys411 and γ Cys87; (4) αS411C:ϵS108C, in which there is disul-

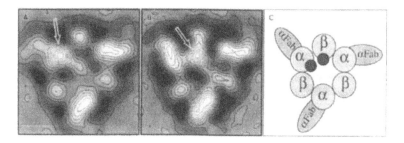

Figure 4. Movements of the ε subunit indicated by positions of this subunit resolved in cryoelectron microscopy using ε labeled with a 14Å gold particle. (See Wilkens and Capaldi, 1994 for details).

fide bond formation between an α Cys411 and γ Cys87, and a second S-S bridge between an α Cys411 and ε Cys108; and (5) αS411C:βE381C:εS108C, in which all of the above disulfide bonds can be formed along with an S-S bridge between adjacent α and β subunits (Aggeler et al., 1995a; Capaldi et al., 1995; Aggeler, R. and Capaldi, R.A., unpublished data). Formation of any of these disulfides blocks activity.

In one or more of these mutants, we have obtained, with $ATP + Mg^{2+}$ in catalytic sites, high-yield cross-linking between α and β subunits as discussed already, high-yield cross-linking between α and ε, as well as low-yield cross-linking of an α to the γ subunit. These three different interactions appear to involve different α subunits. There is also a high yield of β-γ cross-link product under these conditions. In $ADP + P_i + Mg^{2+}$, the α-β and a β-γ cross-link are formed in low yield, and there is high-yield cross-linking of β to ε. Thus, the γ and ε subunits appear to be able to move between α and β subunits. Movement of the ε subunit from α to β, a distance of close to 20 Å, is also indicated by the cryoelectron microscopy studies, shown in Figure 4, in which the ε subunit has been tagged with a gold particle (Wilkens and Capaldi 1994).

If the γ (and ε) subunit(s) determine catalytic site conformation and, thereby nucleotide occupancy, then movements of the small subunits between the three α-β subunit pairs could alternate catalytic site function, as suggested by the alternating site model. The data obtained so far are not direct proof that the γ and/or ε subunit switching involves all three α-β subunit pairs, and convincing evidence for rotational catalysis as proposed by Boyer (1993) and others is still lacking.

XIV. CONFORMATIONAL COUPLING BETWEEN CATALYTIC SITES AND THE PROTON CHANNEL

There is general acceptance that catalytic site events are coupled to proton translocation by conformational changes in the stalk subunits and, in particular, in the γ subunit (discussed in Capaldi et al., 1994). The first evidence that structural changes in the F_0 were transmitted to catalytic sites came from the work of Penefsky (1985) and Matsuno-Yagi et al. (1985) who showed that chemical modifications of the F_0 part affected the catalytic site binding of nucleotides as well as aurovertin fluorescence in the F_1 part. Also, Mendel-Hartvig and Capaldi (1991b) showed that DCCD binding to F_0 locks the F_1 part in specific conformations determined by which nucleotides were present in catalytic sites during modification by the inhibitor. In none of these experiments were the subunits transducing the conformational change identified. The structural data reviewed in this chapter point to the γ and ε subunits as involved and, to test this, we have been incorporating reporter groups, such as cross-linkers and fluorophores, into these subunits via cysteine residues introduced by site-directed mutagenesis, as described already.

In one set of experiments, we have incorporated the fluorescent probe coumarin maleimide at positions 8 and 106 of the γ subunit, and at position 108 of the ε subunit (in the mutants γS8C, γT106C, and εS108C, respectively) (Turina and Capaldi, 1994a). A fluorescence change in the probe was observed at each of these sites during unisite catalysis, involving a fluorescence enhancement on binding ATP (but not ADP), followed by fluorescence quenching as the ATP was converted to ADP. Careful kinetic experiments with ATPγS established that the fluorescence quenching followed the bond cleavage reaction and not P_i release from the catalytic site (Turina and Capaldi, 1994b). Importantly, these structural changes in the γ subunit were not seen in enzyme missing the ε subunit, although such preparations show high, cooperative ATPase activity. The fluorescence changes, therefore, are not monitoring cooperativity but, instead, we have argued, are monitoring conformational changes that are essential to the coupling mechanism. We have also found nucleotide-dependent changes in cross-linking between γ (TFPAM bound at a Cys at position 8) and β subunits that are lost when the ε subunit is removed (Aggeler and Capaldi, 1993; Aggeler et al., 1993). We conclude that the γ and ε subunits work in unison in the coupling process. As described earlier, both are linked to the α and β subunits at one end and to the c subunits at the other, as required for a conduit of conformational changes.

Additional evidence of the involvement of the γ subunit in the coupling process has been obtained in genetic studies. Futai and colleagues have obtained a mutant of the γ subunit in ECF$_1$ (γM23K) which has normal ATPase activity, but this ATPase activity is not coupled to proton translocation (Shin et al., 1992). Revertants of this mutation were obtained that had amino acid changes in the C-terminal α helix, both at the site where M23 contacts the C-terminal helix, and at a region where the C-terminal helix interacts with a loop of the β subunit (Nakamoto et al., 1993).

The above results suggest that movements of the γ subunit, possibly of the N-terminal α helix relative to the C-terminal α helix, are an essential part of the coupling process. They indicate that the ε subunit which binds at, or close to, these helices in the stalk part acts directly and/or controls these conformational changes, which are then linked to structural changes in the F_0 subunits. All of the above evidence of the function of γ and ε subunits in conformational coupling has been obtained in the direction of ATP hydrolysis. The important studies of Richter and McCarty (1987) on CF_1F_0 are evidence of conformational changes in the ε subunit during ATP synthesis.

XV. MECHANISM OF PROTON TRANSLOCATION IN THE F_0 PART

The F_1F_0 complexes of certain bacteria can use ATP hydrolysis to translocate Na^+ ions through the F_0, and can use a gradient of Na^+ to drive ATP synthesis (Dimroth, 1990). *Propionigenium modestum* can interchangeably translocate either H^+ or Na^+ when the total Na^+ concentration is low (Laubinger and Dimroth, 1989), with the H^+ and Na^+ competing for the same binding site, probably the buried carboxyl in the c subunit (Glu65, which is the equivalent residue to Asp61 in the *E. coli* c subunit) (Kluge and Dimroth, 1993). Attempts to convert ECF_1F_0 into an Na^+-translocating ATPase are in progress. By replacing hydrophobic residues around Asp61 with the more polar amino acids found in the c subunit of *P. modestum*, Zhang and Fillingame (1995b) have already generated an enzyme in which H^+ transport is inhibited by Li^+ in a way that appears to be competitive with H^+ binding.

The finding that F_1F_0-ATPases can couple Na^+ as well as H^+ translocation to ATP synthesis is important, in that it rules out a direct proton coupling mechanism (see Mitchell, 1985) and instead favors a mechanism in which the translocation of ions (H^+ or Na^+) induces a conformational change in the F_0 subunits that is transmitted through the stalk to the catalytic sites.

Mechanisms for proton translocation through the F_0 are mostly based on the results of mutagenesis experiments using the enzyme from *E. coli*. Such studies show the critical importance of Asp61 in the second, more C-terminal α helix of the c subunit in proton movements (reviewed in Fillingame, 1990, 1992). Mutation of Asp61 to Glu profoundly reduces activity and shows a specific structural requirement at this site. In the mutant cD61N, proton translocation is abolished, implying that a group capable of protonation and deprotonation is essential. Nevertheless, the position of the carboxyl group is not absolute. Miller et al. (1990) showed that a mutant in which the carboxyl had been relocated from position 61 to 24 (A24D, D61G) retained proton pumping function. The position of the Ala at 24 and Asp at 61 are close in the NMR-derived structure (closer than 5 Å) where the two α helices of the c subunit come together. Asp61 is the site of reaction of DCCD, one mole of which per 10 ± 1 c subunits is sufficient to block both passive and active proton translocation (Fillingame, 1990).

Mutagenesis studies show that the c subunits alone are not sufficient for proton translocation, but that the a subunit is also involved (Cain and Simoni, 1986, 1988, 1989; Lightowlers et al., 1987, 1988). Residue Arg210 of the a subunit is absolutely required as even the most conservative substitution to a Lys abolishes proton translocation (Lightowlers et al., 1987; Howitt and Cox, 1992). Function is also lost with some mutations of residues E219 (for example, E219L) and H245 (for example, H245Y), although the F_0 complex is assembled (Cain and Simoni, 1986, 1988). More recently, Vik and Antonio (1994) have relocated the essential glutamic acid from position 219 to 252, with a retention of proton-translocating activity.

The key to understanding how the a and c subunits might both be involved in proton translocation comes from the recent studies of Fraga et al. (1994), who sought third-site mutations that would optimize function in the D24G61 mutant. These optimizing mutants map to two general regions, the C-terminal helix of subunit c and the region of the a subunit around Glu219. These results have led Fillingame et al. (1995) to propose a model in which there is interaction between a bi-helical c subunit and an α helix of subunit a that includes Glu219. Possibly, the interface between the c subunit and the a subunit provides the proton channel as suggested by Fillingame et al. (1995) and by Vik and Antonio (1994). In order to be protonated and deprotonated, Asp61 must be in an aqueous region within the bilayer and an α helix of the a subunit including Glu219 and Arg210 would provide a hydrophilic face for a proton channel, the gating of which would then control proton translocation.

If it takes translocation of 3H+ to make an ATP (Senior, 1988), and if ATP hydrolysis moves 3H+ per ATP cleaved through the proton channel, then the overall process must be cooperative. One way in which this cooperativity could be achieved is if the different c subunits each become protonated and deprotonated in turn at Asp61, with rotation of the ring of c subunits bringing each sequentially into interaction with the a subunit. A model of this type, extending the original ideas of Cox et al. (1984), has been described in detail by Vik and Antonio (1994).

In order to drive ATP synthesis, the proton translocation step must alter the conformation/position of the γ and ε subunits. This may occur via structural changes in the loop region of the c subunits, which are seen to alter conformation when Asp61 is protonated and deprotonated (Fillingame, 1990), and which are the sites of binding of the γ and ε subunits as described earlier (see Watts et al., 1995).

XVI.   SUMMARY

The synthesis of ATP in all cells is almost exclusively accomplished by a single enzyme, an F_1F_0 type ATPase, found in the plasma membrane of prokaryotes, the mitochondrial inner membrane of animal cells, and in both the chloroplast thylakoid membrane and mitochondrial inner membrane of plants. There has been considerable recent progress in structure determination of this large multi-subunit enzyme by electron microscopy, X-ray crystallography, and various biophysical methods, and these recent data are discussed. They show the extrinsic F_1 part organized with 3 α and 3 β subunits alternating in a hexagonal ring around the γ subunit. The γ subunit and the δ and ε subunits of the F_1 part (in the enzyme from *Escherichia coli*) form a 45Å long thin stalk, which links into the F_0. In the *E. coli* enzyme, the membrane embedded F_0 part is composed of three different subunits a, b, and c in the molar ratio 1:2:10-12. The b subunits of the F_0 part also contribute to the stalk. Coupling between the three catalytic sites on β subunits in the F_1 part, and the proton channel in the ac_{10} complex of the F_0 part, appears to be via conformational changes in the γ and ε subunits of the stalk region. Movements of the γ and ε subunits in response to nucleotide binding changes in catalytic sites are reviewed, and models of the proton-pumping mechanism discussed.

REFERENCES

Abrahams, J.P., Leslie, A.G.W., Lutter, R., & Walker, J.E. (1994). Structure at 2.8 Å resolution of F_1-ATPase from bovine heart mitochondria. Nature 370, 621-628.

Aggeler, R., & Capaldi, R.A. (1992). Cross-linking of the γ subunit of *Escherichia coli* ATPase via cysteines introduced by site-directed mutagenesis. J. Biol. Chem. 267, 21355-21359.

Aggeler, R., & Capaldi, R.A. (1993). ATP hydrolysis-linked structural changes in the N-terminal part of the γ subunit of *Escherichia coli* F_1-ATPase examined by cross-linking studies. J. Biol. Chem. 268, 14576-14578.

Aggeler, R., Chicas-Cruz, K., Cai, S.-X., Keana, J.F.W., & Capaldi, R.A. (1992). Introduction of reactive cysteine residues in the ε subunit of *Escherichia coli* F_1-ATPase, modification of these sites with tetrafluorophenyl azide-maleimides, and examination of changes in the binding of the ε subunit when different nucleotides are in catalytic sites. Biochemistry 31, 2956-2961.

Aggeler, R., Cai, S.-X., Keana, J.F.W., Koike, T., & Capaldi, R.A. (1993). The γ subunit of the *Escherichia coli* F_1-ATPase can be cross-linked near the glycine-rich loop region of a β subunit when $ADP + Mg^{2+}$ occupies catalytic sites but not when $ATP + Mg^{2+}$ is bound. J. Biol. Chem. 268, 20831-20837.

Aggeler, R., Haughton, M.A., & Capaldi, R.A. (1995a) Disulfide bond formation between the COOH terminal domain of the β subunits and the γ and ε subunits of the *E. coli* F_1 ATPase. J. Biol. Chem. 270, 9185-9191.

Aggeler, R., Weinreich, F., & Capaldi, R.A. (1995b). Arrangement of the ε subunit in the *E. coli* ATP synthase from the reactivity of cysteine residues introduced at different positions in this subunit. Biochim. Biophys. Acta 1230, 62-68.

Al-Shawi, M.K., & Senior, A.E. (1988). Complete kinetic and thermodynamic characterization of the unisite catalytic pathway of *Escherichia coli* F_1-ATPase. Comparison with mitochondrial F_1-ATPase and application to the study of mutant enzymes. J. Biol. Chem. 263, 19640-19648.

Allison, W.S., Jault, J.-M., Zhuo, S., & Paik, S.R. (1992). Functional sites in F_1-ATPases: location and interactions. J. Bioenerg. Biomembr. 24, 469-477.

Amzel, L.M., McKinney, M., Narayanan, P., & Pedersen, P.L. (1982). Structure of the mitochondrial F_1 ATPase at 9 Å resolution. Proc. Natl. Acad. Sci. USA 79, 5852-5856.

Beckers, G., Berzborn, R.J., & Strotmann, H. (1992). Zero length cross-linking between subunits δ and I of the H^+ translocating ATPase of chloroplasts. Biochim. Biophys. Acta 1101, 97-104.

Beharry, S., & Bragg, P.D. (1992). Changes in the adenine nucleotide and inorganic phosphate content of *Escherichia coli* F_1-ATPase during ATP synthesis in dimethyl sulphoxide. Biochem. J. 286, 603-606.

Belogrudov, G.I., Tomich, J.M., & Hatefi, Y. (1995). ATP synthase complex: proximities of subunits in bovine submitochondrial particles. J. Biol. Chem. 270, 2053-2060.

Berzborn, R.J., Klein-Hitpass, L., Otto, J., Schünemann, S., Oworah-Nkruma, R., & Meyer, H.E. (1990). The additional subunit CF_0II of the photosynthetic ATP synthase and the thylakoid polypeptide, binding ferredoxin NADP reductase: are they different? Z. Naturforsch. 45c, 772-784.

Birkenhäger, R., Hoppert, M., Deckers-Hebestreit, G., Mayer, F., & Altendorf, K. (1995). The F_0 complex of the *Escherichia coli* ATP synthase: Investigation by electron spectroscopic imaging and immunoelectron microscopy. Eur. J. Biochem. 230, 58-67.

Boekema, E.J., Xiao, J., & McCarty, R.E. (1990). Structure of the ATP synthase from chloroplasts studied by electron microscopy. Localization of the small subunits. Biochim. Biophys. Acta 1020, 49-56.

Boyer, P.D. (1987). The unusual enzymology of ATP synthase. Biochemistry 26, 8503-8507.

Boyer, P.D. (1989). A perspective of the binding change mechanism for ATP synthesis. FASEB J. 3, 2164-2178.

Boyer, P.D. (1993). The binding change mechanism for ATP synthase—some probabilities and possibilities. Biochim. Biophys. Acta 1140, 215-250.

Cain, B.D., & Simoni, R.D. (1986). Impaired proton conductivity resulting from mutations in the a subunit of F_1F_0 ATPase in *Escherichia coli*. J. Biol. Chem. 261, 10043-10050.

Cain, B.D., & Simoni, R.D. (1988). Interaction between Glu219 and His245 within the a subunit of F_1F_0 ATPase in *Escherichia coli*. J. Biol. Chem. 263, 6606-6612.

Cain, B.D., & Simoni, R.D. (1989). Proton translocation by the F_1F_0 ATPase of *Escherichia coli*. Mutagenic analysis of the a subunit. J. Biol. Chem. 264, 3292-3300.

Capaldi, R.A., Aggeler, R., Turina, P., & Wilkens, S. (1994). Coupling between catalytic sites and the proton channel in F_1F_0-type ATPases. Trends Biochem. Sci. 19, 284-289.

Capaldi, R.A., Aggeler, R., & Wilkens, S. (1995). Conformational changes in the γ and ε subunits are integral to the functioning of the *E. coli* H$^+$ pumping ATPase (ECF$_1F_0$). Trans Biochem. Soc. 23, 767-770.

Collinson, I.R., Runswick, M.J., Buchanan, S.K., Fearnley, I.M., Skehel, J.M., van Raaij, M.J., Griffiths, D.E., & Walker, J.E. (1994a). F_0 membrane domain of ATP synthase from bovine heart mitochondria: purification, subunit composition, and reconstitution with F_1-ATPase. Biochemistry 33, 7971-7978.

Collinson , I.R., van Raaij, M.J., Runswick, M.J., Fearnley, I.M., Skehel, J.M., Orriss, G.L., Miroux, B., & Walker, J.E. (1994b). ATP synthase from bovine heart mitochondria. *In vitro* assembly of a stalk complex in the presence of F_1-ATPase and in its absence. J. Mol. Biol. 242, 408-421.

Cox, G.B., Jans, D.A., Fimmel, A.L., Gibson, F. & Hatch, L. (1984). The mechanism of ATP synthase: conformational change by rotation of the b subunit. Biochim. Biophys. Acta 768, 201-208.

Cozens, A.L., Walker, J.E., Phillips, A.L., Huttly, A.K., & Gray, J.C. (1986). A sixth subunit of ATP synthase, an F_0 component, is encoded in the pea chloroplast genome. EMBO J. 5, 217-222.

Cross, R.L. (1992). The reaction mechanism of F_0F_1 ATP syntheses in molecular mechanisms. In: Bioenergetics (L. Ernster, Ed),. pp 317-330. Elsevier, Amsterdam.

Cross, R.L., & Nalin, C.M. (1982). Adenine nucleotide binding sites on beef heart F_1 ATPase. Evidence for three exchangeable sites that are distinct from three noncatalytic sites. J. Biol. Chem. 257, 2874-2881.

Cross, R.L., Grubmeyer, C., & Penefsky, H.S. (1982). Mechanism of ATP hydrolysis by beef heart mitochondrial ATPase. Rate enhancements resulting from cooperative interactions between multiple catalytic sites. J. Biol. Chem. 257, 12101-12105.

Dallmann, H.G., Flynn, T.G., & Dunn, S.D. (1992). Determination of the 1-ethyl-3-[(3-dimethylamino)propyl]-carbodiimide-induced cross-link between the β and ε subunits of *Escherichia coli* F_1 ATPase. J. Biol. Chem. 267, 18953-18960.

Deckers-Hebestreit, G., & Altendorf, K. (1992). The F_0 complex of the proton-translocating F-type ATPase of *Escherichia coli*. J. Exp. Biol. 172, 451-459.

Dimroth, P. (1990). Bacterial energy transductions coupled to sodium ions. Res. Microbiol. 141, 332-336.

Dunn, S.D. (1992). The polar domain of the b subunit of *Escherichia coli* F_1F_0-ATPase forms an elongated dimer that interacts with the F_1 sector. J. Biol. Chem. 267, 7630-7636.

Edel, C.M., Hartog, A.F., & Berden, J.A. (1992). Inhibition of mitochondrial F_1 ATPase activity by binding of 2-azido ADP to a slowly exchangeable non-catalytic nucleotide binding site. Biochim. Biophys. Acta 1101, 329-338.

Edel, C.M., Hartog, A.F., & Berden, J.A. (1993). Identification of an exchangeable non-catalytic site on mitochondrial F_1 ATPase which is involved in the negative cooperativity of ATP hydrolysis. Biochim. Biophys. Acta 1142, 327-335.

Engelbrecht, S., Reed, J., Penin, F., Gautheron, D.C. & Junge, W. (1991). Subunit δ of chloroplast F_1F_0 ATPase and OSCP of mitochondrial F_1F_0 ATPase: a comparison by CD-spectroscopy. Z. Naturforsch. 46c, 759-764.

Falk, G., Hampe, A., & Walker, J.E. (1985). Nucleotide sequence of the *Rhodospirillum rubrum* ATP operon. Biochem. J. 228, 391-407.

Feldman, R.I., & Sigman, D.S. (1982). The synthesis of enzyme bound ATP by soluble chloroplast coupling factor 1. J. Biol. Chem. 257, 1676-1683.

Fernandez-Moran, H. (1962). Cell-membrane ultra-structure. Low-temperature electron microscopy and X-ray diffraction studies of lipoprotein components in lamellar systems. Circulation 26, 1039-1065.

Fillingame, R.H. (1990). Molecular mechanics of ATP synthesis by F_1F_0-type H+-transporting ATPases. The Bacteria 12, 345-391.

Fillingame, R.H. (1992). H+ transport and coupling by the F_0 sector of the ATP synthase: insights into the molecular mechanism of function. J. Bioenerg. Biomembr. 24, 485-491.

Fillingame, R.H., Girvin, M.E., & Zhang, Y. (1995). Correlations of structure and function in subunit c of *E. coli* F_1F_0 ATPase synthase. Trans. Biochem. Soc. 23, 760-766.

Foster, D.L., & Fillingame, R.H. (1982). Stoichiometry of subunits in the H+-ATPase complex of *Escherichia coli*. J. Biol. Chem. 257, 2009-2015.

Fraga, D., Hermolin, J., & Fillingame, R.H. (1994). Transmembrane helix-helix interactions in F_0 suggested by suppressor mutations to Ala24→Asp/Asp61→Gly mutant of ATP synthase subunit c. J. Biol. Chem. 269, 2562-2567.

Fromme, P., & Gräber, P. (1990a). ATP hydrolysis in chloroplasts: unisite catalysis and evidence for heterogeneity of catalytic sites. Biochim. Biophys. Acta 1020, 187-194.

Fromme, P., & Gräber, P. (1990b). Activation/inactivation and unisite catalysis by the reconstituted ATP synthase from chloroplasts. Biochim. Biophys. Acta 1016, 29-42.

Fromme, P., Gräber, P., & Salnikow, J. (1987). Isolation and identification of a fourth subunit in the membrane part of the chloroplast ATP synthase. FEBS Lett. 218, 27-30.

Futai, M., Noumi, T., & Maeda, M. (1989). ATP synthase (H+-ATPase): results by combined biochemical and molecular biological approaches. Ann. Rev. Biochem. 58, 111-136.

Futai, M., Iwamoto, A., Omote, H., & Maeda, M. (1992). A glycine-rich sequence in the catalytic site of F-type ATPase. J. Bioenerg. Biomembr. 24, 463-467.

Giraud, M.F., & Velours, J. (1994). ATP synthase of yeast mitochondria—isolation of the $F_1\delta$ subunit, sequence and disruption of the structural gene. Eur. J. Biochem. 222, 851-859.

Girvin, M.E., & Fillingame, R.H. (1993). Helical structure and folding of subunit c of F_1F_0 ATP synthase: 1H NMR resonance assignments and NOE analysis. Biochemistry 32, 12167-12177.

Girvin, M.E., & Fillingame, R.H. (1994). Hairpin folding of subunit c of F_1F_0 ATP synthase: 1H distance measurements to nitroxide-derivatized aspartyl-61. Biochemistry 33, 665-674.

Girvin, M.E., & Fillingame, R.H. (1995). Determination of local protein structure by spin label difference 2D NMR: the region neighboring Asp61 of subunit c of the F_1F_0 ATP synthase. Biochemistry 34, 1635-1645.

Gogol, E.P., Lücken, U., & Capaldi, R.A. (1987). The stalk connecting the F_1 and F_0 domains of ATP synthase visualized by electron microscopy of unstained specimens. FEBS Lett. 219, 274-278.

Gogol, E.P., Aggeler, R., Sagermann, M., & Capaldi, R.A. (1989a). Cryoelectron microscopy of *Escherichia coli* F_1 adenosinetriphosphatase decorated with monoclonal antibodies to individual subunits of the complex. Biochemistry 28, 4717-4724.

Gogol, E.P., Lücken, U., Bork, T., & Capaldi, R.A. (1989b). Molecular architecture of *Escherichia coli* F_1 adenosinetriphosphatase. Biochemistry 28, 4709-4716.

Gogol, E.P., Johnston, E., Aggeler, R., & Capaldi, R.A. (1990). Ligand-dependent structural variations in *Escherichia coli* F_1 ATPase revealed by cryoelectron microscopy. Proc. Natl. Acad. Sci. USA 87, 9585-9589.

Gräber, P., Böttcher, B., & Boekema, E.J. (1990). The structure of the ATP-synthase from chloroplasts. In: Bioelectrochemistry III Milazzo, G. and Blank, M. Eds.), pp 247-278. Plenum Press, New York.

Grubmeyer, C., Cross, R.L., & Penefsky, H.S. (1982). Mechanism of ATP hydrolysis by beef heart mitochondrial ATPase. Rate constants for elementary steps in catalysis at a single site. J. Biol. Chem. 257, 12092-12100.

Guelin, E., Chevallier, J., Rigoulet, M., Guelin, B., & Velours, J. (1993). ATP synthase of yeast mitochondria. Isolation and disruption of the ATP ε gene. J. Biol. Chem. 268, 161-167.

Hamasur, B., & Glaser, E. (1992). Plant mitochondrial F_1F_0 ATP synthase. Identification of the individual subunits and properties of the purified spinach leaf mitochondrial ATP synthase. Eur. J. Biochem. 205, 409-416.

Hatefi, Y. (1993). ATP synthesis in mitochondria. Eur. J. Biochem. 218, 759-767.

Hekman, C., Tomich, J.M., & Hatefi, Y. (1991). Mitochondrial ATP synthase complex. Membrane topography and stoichiometry of the F_0 subunits. J. Biol. Chem. 266, 13564-13571.

Howitt, S.M., & Cox, G.B. (1992). Second-site revertants of an arginine-210 to lysine mutation in the a subunit of the F_0F_1 ATPase from *Escherichia coli*: implications for structure. Proc. Natl. Acad. Sci. USA 89, 9799-9803.

Issartel, J.P., Lunardi, J., & Vignais, P.V. (1986). Characterization of exchangeable and non-exchangeable bound adenine nucleotides in F_1 ATPase from *Escherichia coli*. J. Biol. Chem. 261, 895-901.

Ketcham, S.R., Davenport, J.W., Warncke, K., & McCarty, R.E. (1984). Role of the γ subunit of chloroplast coupling factor 1 in the light dependent activation of photophosphorylation and ATPase activity by dithiothreitol. J. Biol. Chem. 259, 7286-7293.

Kluge, C., & Dimroth, P. (1993). Kinetics of inactivation of the F_1F_0 ATPase of *Propionigenium modestum* by DCCD in relation to H^+ and Na^+ concentration; probing the binding site for coupling ions. Biochemistry 32, 10378-10386.

Laubinger, W., & Dimroth, P. (1989). The sodium ion translocating ATPase of *Propionigenium modestum* pumps protons at low sodium ion concentrations. Biochemistry 28, 7194-7198.

Lien, S., & Racker, E. (1971). Partial resolution of the enzymes catalyzing photophosphorylation. Properties of silicotungstate-treated subchloroplast particles. J. Biol. Chem. 246, 4298-4307.

Lightowlers, R.N., Howitt, S.M., Hatch, L., Gibson, F., & Cox, G.B. (1987). The proton pore in the *Escherichia coli* F_0F_1-ATPase: a requirement for arginine at position 210 of the a subunit. Biochim. Biophys. Acta 894, 399-406.

Lightowlers, R.N., Howitt, S.M., Hatch, L., Gibson, F., & Cox, G. (1988). The proton pore in the *Escherichia coli* F_0F_1-ATPase: substitution of glutamate by glutamine at position 219 of the a subunit prevents F_0-mediated proton permeability. Biochim. Biophys. Acta 933, 241-248.

Lücken, U., Gogol, E.P., & Capaldi, R.A. (1990). Structure of the ATP synthase complex (ECF_1F_0) of *Escherichia coli* from cryoelectron microscopy. Biochemistry 29, 5339-5343.

Ludwig, B., Prochaska, L., & Capaldi, R.A. (1980). Arrangement of oligomycin-sensitive ATPase in the mitochondrial inner membrane. Biochemistry 19, 1516-1523.

Matsuno-Yagi, A., Yagi, T., & Hatefi, Y. (1985). Studies on the mechanism of oxidative phosphorylation: Effects of specific F_0 modifiers on ligand-induced conformational changes of F_1. Proc. Natl. Acad. Sci. USA 82, 7550-7554.

Mendel-Hartvig, J., & Capaldi, R.A. (1991a). Structure-function relationships of domains of the δ subunit in *Escherichia coli* adenosine triphosphatase. Biochim. Biophys. Acta 1060, 115-124.

Mendel-Hartvig, J., & Capaldi, R.A. (1991b). Nucleotide-dependent and dicyclohexylcarbodiimide-sensitive conformational changes in the ε subunit of *Escherichia coli* ATP synthase. Biochemistry 30, 10987-10991.

Miller, M.J., Oldenburg, M., & Fillingame, R.H. (1990). The essential carboxyl group in subunit c of the F_1F_0 ATP synthase can be moved and $H^{(+)}$-translocating function retained. Proc. Natl. Acad. Sci. USA 87, 4900-4904.

Mitchell, P. (1985). Molecular mechanics of proton motive F_0F_1 ATPases. Rolling well and turnstile hypothesis. FEBS Lett. 182, 1-7.

Nakamoto, R.K., Maeda, M., & Futai, M. (1993). The γ subunit of the *Escherichia coli* ATP synthase. Mutations in the carboxyl-terminal region restore energy coupling to the amino-terminal mutant γ Met-23→Lys. J. Biol. Chem. 268, 867-872.

Nalin, C.M., & McCarty, R.E. (1984). Role of a disulfide bond in the γ subunit in activation of the ATPase of chloroplast coupling factor 1. J. Biol. Chem. 259, 7275-7280.

Pedersen, P.L., Hullihen, J., Bianchet, M., Amzel, L.A., & Lebowitz, M.S. (1995). Rat liver ATP synthase: relationship of the unique substructure of the F_1 moiety to its nucleotide binding properties, enzymatic states and crystalline form. J. Biol. Chem. 270, 1775-1784.

Penefsky, H.S. (1985). Mechanism of inhibition of mitochondrial ATPase by DCCD and oligomycin. Relationship to ATP synthesis. Proc. Natl. Acad. Sci. USA 82, 1589-1593.

Penefsky, H.S., & Cross, R.L. (1991). Structure and mechanism of F_0F_1-type ATP synthases and ATPases. Adv. Enzymol. Relat. Areas. Mol. Biol. 64, 173-214.

Perlin, D.S., Latchney, L.R., Wise, J.G., & Senior, A.E. (1984). Specificity of the H^+ ATPase of *Escherichia coli* for adenine, guanine and inosine nucleotides in catalysis and binding. Biochemistry 23, 4998-5003.

Richter, M.L., & McCarty, R.E. (1987). Energy dependent changes in the conformation of the ε subunit of the chloroplast ATP synthase. J. Biol. Chem. 262, 15037-5040.

Sakamoto, J., & Tonomura, Y. (1983). Synthesis of enzyme bound ATP by mitochondrial soluble F_1 ATPase in the presence of dimethylsulfoxide. J. Biochem. 93, 1601-1614.

Schneider, E., & Altendorf, K. (1987). Bacterial ATP synthase (F_1F_0): purification and reconstitution of F_0 complexes and biochemical and functional characterization of their subunits. Microbiol. Rev. 51, 477-497.

Senior, A.E. (1988). ATP synthesis by oxidative phosphorylation. Physiol. Rev. 68, 177-231.

Senior, A.E. (1990). The proton-translocating ATPase of *Escherichia coli*. Ann. Rev. Biophys. Biophys. Chem. 19, 7-41.

Senior, A.E., Weber, J., & Al-Shawi, M. (1995). Catalytic mechanism of *E. coli* F_1 ATPase. Biochem. Soc. Transac. 23, 747-752.

Shin, K., Nakamoto, R.K., Maeda, M., & Futai, M. (1992). F_1F_0 ATPase γ subunit mutations perturb the coupling between catalysis and transport. J. Biol. Chem. 267, 20835-20839.

Story, R.M., Weber, I.T., & Steitz, T.A. (1992). Structure of the recA protein-ADP complex. Nature 355, 374-376.

Tiedge, H., Schäfer, G., & Mayer, F. (1983). An electron microscopic approach to the quaternary structure of mitochondrial F_1 ATPase. Eur. J. Biochem. 132, 37-45.

Tsuprun, V.L., Mesyanzhinova, I.V., Koslov, I.A., & Orlova, E.V. (1984). Electron microscopy of beef heart mitochondrial F_1 ATPase. FEBS Lett. 167, 285-290.

Tsuprun, V.L., Orlova, E.V., & Mesyanzhinova, I.V. (1989). Structure of the ATP synthase studied by electron microscopy and image processing. FEBS Lett. 244, 279-282.

Turina, P., & Capaldi, R.A. (1994a). ATP hydrolysis-driven structural changes in the γ subunit of *Escherichia coli* ATPase monitored by fluorescence from probes bound at introduced cysteine residues. J. Biol. Chem. 269, 13465-13471.

Turina, P., & Capaldi, R.A. (1994b). ATP binding causes a conformational change in the γ subunit of the *Escherichia coli* F_1 ATPase which is reversed on bond cleavage. Biochemistry 33, 14275-14280.

Vik, S.B., & Antonio, B.J. (1994). A mechanism of proton translocation by F_1F_0 ATP synthases suggested by double mutants of the a subunit. J. Biol. Chem. 269, 30364-30369.

Walker, J.E., Runswick, M.J., & Saraste, M. (1982a). Subunit equivalence in *Escherichia coli* and bovine heart mitochondrial F_1F_0 ATPases. FEBS Lett. 146, 393-396.

Walker, J.E., Saraste, M., & Gay, N.J. (1982b). *E. coli* F_1-ATPase interacts with a membrane protein component of a proton channel. Nature (London) 298, 867-869.

Walker, J.E., Fearnley, I.M., Gay, N.J., Gibson, B.W., Northrop, F.D., Powell, S.J., Runswick, M.J., Saraste, M., & Tybulewicz, V.L.J. (1985). Primary structure and subunit stoichiometry of F_1-ATPase from bovine mitochondria. J. Mol. Biol. 184, 677-701.

Walker, J.E., Lutter, R., Dupuis, A., & Runswick, M.J. (1991). Identification of the subunits of F_1F_0 ATPase from bovine heart mitochondria. Biochemistry 30, 5369-5378.

Watts, S.D., Zhang, Y., Fillingame, R.H., & Capaldi, R.A. (1995). The γ subunit in the *Escherichia coli* ATP synthase complex (ECF_1F_0) extends through the stalk and contacts the c subunits of the F_0 part. FEBS Lett. 368, 235-238.

Weber, J., Wilke-Mounts, S., Lee, R.S.-F., Grell, E., & Senior, A.E. (1993). Specific placement of tryptophan in the catalytic sites of *Escherichia coli* F_1-ATPase provides a direct probe of nucleotide binding: maximal ATP hydrolysis occurs with three sites occupied. J. Biol. Chem. 268, 20126-20133.

Wilkens, S., & Capaldi, R.A. (1994). Asymmetry and structural changes in ECF_1 examined by cryoelectron microscopy. Biol. Chem. Hoppe Seyler 375, 43-51.

Wilkens, S., Dahlquist, F.W., McIntosh, L.P., Donaldson, L.W., & Capaldi, R.A. (1995). Structural features of the ε subunit of the *Escherichia coli* ATP synthase (ECF_1F_0) from NMR-spectroscopy. Nature Struct. Biol. 2, 961-967.

Yoshida, M. (1983). The synthesis of enzyme bound ATP by the F_1 ATPase from the thermophilic bacterium PS3 in 50% dimethylsulfoxide. Biochem. Biophys. Res. Commun. 114, 907-912.

Zhang, Y., & Fillingame, R.H. (1995a). Subunits coupling H^+ transport and ATP synthesis in the *E. coli* ATP synthase: Cys-Cys crosslinking of F_1 subunit ε to the polar loop of F_0 subunit c. J. Biol. Chem. 270, 24609-24614..

Zhang, Y., & Fillingame, R.H. (1995b). Changing the ion binding specificity of the *E. coli* H^+ transporting ATP synthase by directed mutagenesis of subunit c. J. Biol. Chem. 270, 87-93.

Zhang, Y., Oldenburg, M., & Fillingame, R.H. (1994). Suppressor mutations in F_1 subunit ε recouple ATP driven H^+ translocation in uncoupled Q42E subunit c mutant of *E. coli* F_1F_0 ATP synthase. J. Biol. Chem. 269, 10221-10224.

STRUCTURE AND MECHANISM OF F-TYPE ATPases

Ursula Gerike and Peter Dimroth

I. INTRODUCTION

The physiological importance of F-type ATPases in contrast to P-type and V-type ATPases is to synthesize ATP by expending the free energy of an electrochemical

Advances in Molecular and Cell Biology
Volume 23B, pages 381-402.
Copyright © 1998 by JAI Press Inc.
All right of reproduction in any form reserved.
ISBN: 0-7623-0287-9

proton gradient ($\Delta\mu H^+$) (Pedersen and Amzel, 1993). These enzymes supply the cells with the ATP necessary for transport, biosyntheses, muscle contraction, and other mobility functions and are therefore key enzymes of energy conservation in all living cells. Under conditions of low $\Delta\mu H^+$, for example, in bacteria growing anaerobically, F-type ATPases also hydrolyze ATP to generate a $\Delta\mu H^+$. Therefore, these enzymes are often designated as ATPases. The F-type ATPases are found in the plasma membranes of eubacteria and the inner membranes of chloroplasts and mitochondria. Two reactions are involved in the synthesis of ATP: the chemical synthesis of ATP from ADP and inorganic phosphate and the coupled proton movement across the membrane. An intriguing, unresolved question is how the two partial reactions are coupled to each other.

The overall structure of F-type ATPases is a membrane-associated F_1 moiety, which catalyzes the hydrolysis/synthesis of ATP, and a membrane-integrated F_0 portion, which is responsible for ion translocation across the membrane. The two parts are connected by a stalk region (Hazard and Senior, 1994a,b; Collinson et al., 1994) (Figure 1). There is no variability in the subunit composition of F_1. In all organisms investigated so far it consists of five different subunits viz, α, β, γ, δ and ε with a stoichiometry of 3:3:1:1:1. Depending on the organism, the F_0 consists of three (for example, *E. coli*, subunits a, b, and c; stoichiometry 1:2:9-11) up to nine (bovine mitochondria; subunits a, b, c, d, e, F6, A6L, f, and g) (Collinson et al., 1994) different subunits. Thus at least 21 subunits are used to build up the ATPase molecule.

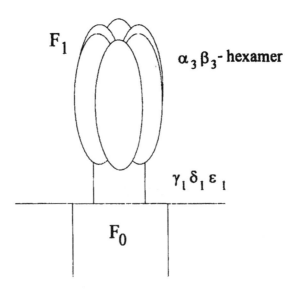

Figure 1. Schematic representation of F_1F_0-type ATPases. The number of different F_0 subunits varies from 3 to 9 depending on the organism.

II. THE THREE-DIMENSIONAL STRUCTURE OF F_1 FROM BOVINE HEART MITCHONDRIA

Crystals of the F_1 moiety of bovine heart mitochondria diffracting at 0.28 nm resolution have been obtained, providing details of the arrangements of the α, β, and γ subunits. From this structure, a model of the ATP synthesis/hydrolysis mechanism has been deduced (Abrahams et al., 1994). It was shown that the $\alpha_3\beta_3$ hexagon is arranged, like the segments of an orange, around the C-terminal part of the γ subunit. The γ subunit passes through a large internal cavity connected to the outside by crevices between neighboring α and β subunits. The structures of the α and β subunits are almost identical and consist of three domains: the N-terminal β-barrel domain, the central nucleotide-binding domain, containing the conserved P-loop (nucleotide-binding motif GXXXXGKT/S), and a C-terminal bundle domain (α-helices). The nucleotide-binding sites are located at the interfaces between α and β subunits. The three catalytic sites are predominantly on the β subunits with some contribution from side chains of the α subunits, and the three noncatalytic sites are predominantly on the α subunits. The most important finding is the asymmetry of the 3 β subunits, particularly in the region of the active sites leading to an empty nucleotide-binding site (β_E), one with bound ATP (β_{TP}) and one with bound ADP (β_{DP}). This asymmetry is associated with the asymmetric positioning of the γ subunit relative to the $\alpha_3\beta_3$ hexagon.

The three dimensional structure of bovine mitochondrial F_1-ATPase coincides with the binding change mechanism of ATP synthesis (Figure 2) (Boyer, 1989). This model is based on the assumption that each catalytic site (located on the β subunits) exists in three different conformational states and that they move, phase shifted with respect to each other, through the same cycle of conformational states. The three conformational states are: L(loose), T(tight), and O(open). In state L, ADP and inorganic phosphate (P_i) are loosely bound and binding of ADP is preferred over binding of ATP. In state T, ADP and P_i or ATP are firmly bound. In state O, ATP is loosely bound and binding of ATP is preferred over binding of ADP. In the first step of the cycle, ADP and P_i bind to the L-site of an enzyme which already contains firmly bound ATP at the T-site (1). Energy input by proton flux converts the T-site into an O-site, the L-site into a T-site and the O-site into a L-site (2). This conformational change is accompanied by binding changes: ATP and P_i become firmly bound in the T-site, leading to the freely reversible synthesis of tightly bound ATP (3). The conversion of the T-site occupied by ATP into an O-site (2) reduces the ATP binding affinity at this site and therefore leads to its dissociation from the enzyme (4). A new cycle of ATP synthesis is initiated by the binding of ADP and P_i to the empty L-site (1).

The asymmetry of the three β subunits in the three-dimensional structure of bovine mitochondrial F_1 may represent the three conformational states in the ATP synthesis model of Boyer (1989), with subunits β_{TP}, β_{DP}, and β_E reflecting the T, L, and O states, respectively.

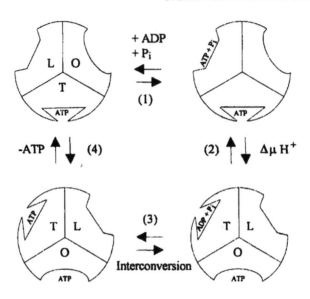

Figure 2. Proposed model for ATP synthesis at the F_1 moiety according to the "binding change mechanism." The catalytic sites on β subunits interact and interconvert between three forms: O, open site with very low affinity for ligands and catalytically inactive; L, loose site, loosely binding ligands and catalytically inactive; T, tight site, tightly binding ligands and catalytically active. The proton-induced conformational changes convert a T-site with bound ATP to an O-site, releasing the bound nucleotide. Concomitantly, an L-site with loosely bound ADP and phosphate is converted to a T-site, where the substrates are tightly bound and ATP forms. Fresh substrates bind to an O-site, converting it to an L-site, and so on. The energy-requiring steps are substrate binding and product release. One third of a catalytic cycle is shown (Boyer, 1989; Cross, 1994).

III.   THE *E. COLI* ENZYME

The *E. coli* F_1F_0-ATPase is biochemically and genetically the most well characterized F_1F_0-ATPase. By purification (Foster and Fillingame, 1979; Friedl et al., 1979) and sequencing of its genes (Gay and Walker, 1981; Nielsen et al., 1981; Kanazawa et al., 1981) it was shown that the *E. coli* ATPase consists of eight different subunits, whose genes are organized in the *unc* (uncoupled) operon. The order of the genes is *unc*I, B, E, F, H, A, G, D, C coding for protein i which is not part of the ATPase complex and subunits a, b, c, δ, α, γ, β and ε. The function of the i protein is unknown, but deletion mutants have clearly shown that it is not essential for enzyme function (Schneppe at al., 1991).

   The F_1 moiety can easily be dissociated from bacterial membranes by low ionic strength EDTA-containing buffer of high pH. The lowest F_1 subassembly, which still possesses ATP hydrolyzing activity, consists of the $\alpha_3\beta_3$ hexamer and subunit γ (Senior, 1990). Addition of subunit ϵ to this subassembly leads to a reduced ATP hydrolyzing activity indicating a regulator role for subunit ϵ within F_1 (Tang et al., 1994). Apart from its function as regulator, subunit ϵ is thought to play a role in the binding and energy coupling of F_1 to F_0 (Zhang et al., 1994). The main functions of subunit δ are the binding of F_1 to F_0 and the coupling of ATP hydrolysis on F_1 to H^+-transport across F_0. It is a highly helical protein, with two proposed long helices which are thought to interact with the four hydrophilic helices of the two copies of subunit b (Hazard and Senior, 1994a, b). Mutagenesis studies in the conserved C-terminal region of subunit δ (K145-R167) revealed an uncoupled phenotype, where ATP-dependent H^+-transport is impaired although ATPase activity and H^+-transport through the isolated F_0 moiety are unaffected (Hazard and Senior, 1994a,b).

The F_0 moiety of *E. coli* is able to translocate protons, if a K^+-diffusion potential is applied to reconstituted proteoliposomes (Negrin et al., 1980; Schneider and Altendorf, 1982). This activity is not inhibited by adding Na^+ or Li^+ ions, showing that protons are the only ion species transported by the *E. coli* enzyme. Adding purified F_1 to F_0-containing proteoliposomes inhibited the $\Delta\psi$-driven proton translocation and restored ATP-dependent H^+-transport. The F_1 can therefore be dissociated reversibly from F_0 and reconstituted to give a functional complex. All three *E. coli* F_0 subunits a, b, and c were purified, and reconstitution experiments were performed to incorporate all possible combinations of these subunits into liposomes. It was shown that all three subunits are required for a functional F_0 (Schneider and Altendorf, 1985). A first step towards understanding which of the three F_0 subunits are involved in ion binding and transport is to solve the structure of the individual subunits. From DNA sequencing and computer modeling of the derived amino acid sequence, structures were proposed for subunits a, b, and c (Figure 3).

Subunit a is a highly hydrophobic protein which is deeply embedded into the membrane. The topology is not solved and five to eight transmembrane spanning-

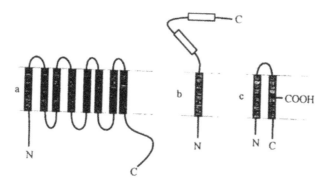

Figure 3. Subunit a, b and c from *Escherichia coli* F_0 and their proposed orientation in the membrane (Modified from Läuger, 1991).

helices have been predicted (Lewis et al., 1990). Subunit b consists of a membrane anchor, which is formed by approximately 30 N-terminal amino acids and a cytoplasmatic part, which protrudes with its two helical domains into the F_1 moiety, possibly interacting with subunit δ. Two proline residues link the membrane part to the cytoplasmatic part of the protein.

Subunit c, which is also called proteolipid due to its high hydrophobicity, is like subunit a fully integrated into the membrane. Its hairpin-like structure with two hydrophobic stretches linked by a polar cytoplasmatic loop was proposed in 1981 (Sebald and Hoppe, 1981) and confirmed by two-dimensional NMR spectroscopy of subunit c of *E. coli* dissolved in chloroform/methanol/H_2O (Girvin and Fillingame, 1993,1994). The gross structure of subunit c seems to be conserved among proteolipids from all known species. Involvement in ion transport was demonstrated for subunits a and c only. Subunit b plays a major role in binding of F_1 to F_0 (Hazard and Senior, 1994a,b) and in assembly of F_0 (Steffens et al., 1987), while no essential residues for H^+-transport have been identified. Intensive mutational studies have been carried out with subunit a to identify residues involved in proton translocation. Essential roles were demonstrated for amino acids R210, E219, and H245. These are located in the conserved C-terminal part of subunit a (regions D190-L220, F244-Y263) (Cain and Simoni, 1988, 1989).

More is known about subunit c (Figure 4). A prominent feature is the harboring of the binding site for the proton translocation inhibitor dicyclohexylcarbodiimide (DCCD). DCCD reacts specifically with an acidic residue (D61 for *E. coli*) in the C-terminal helix (helix 2) of subunit c thereby inhibiting not only proton transport but also ATP hydrolysis/synthesis of F_1F_0. The acidic residue in helix 2 is conserved in all proteolipids characterized so far. It is represented either by an aspartate or a glutamate residue. Other conserved structural elements are the glycine-rich region in the center of the N-terminal transmembrane helix (helix 1), with the consensus GXG/AXGXG/A (G23-G29) and a region in the polar cytoplasmatic loop (K34-L45). Within this region, residues 41R-Q-P-43 are highly conserved with no variation reported so far for arginine and proline.

The results from mutational studies within the glycine-rich region are very interesting. Mutations involving the conserved alanine residue at position 24 resulted in a reduced reactivity towards DCCD (Fillingame et al., 1991). Similar results were obtained with mutants I28V and I28T (Hoppe et al., 1980) indicating that the amino acids of position 24 and 28, although located on helix 1, are important for the proper binding of DCCD. They may therefore be close to D61. Moving the acidic residue D61 from helix 2 to position 24 in helix 1 (mutant A24D/D61G) restored in part the ATP synthesis activity (Miller et al., 1990). These results indicate that the two helices are located opposite of each other and that the functionally essential carboxylic acid residue is positioned toward the center of the hairpin. The proximity of D61 to A24 and A25 on helix 1 (1.2 nm) was confirmed by NMR studies after modification of D61 with N-(2,2,6,6-tetramethylpiperidine-1-oxy)-N'-cyclohexylcarbodiimide (NCCD), a nitroxide analogue of DCCD. The reduced reactivity of the I28T mutant

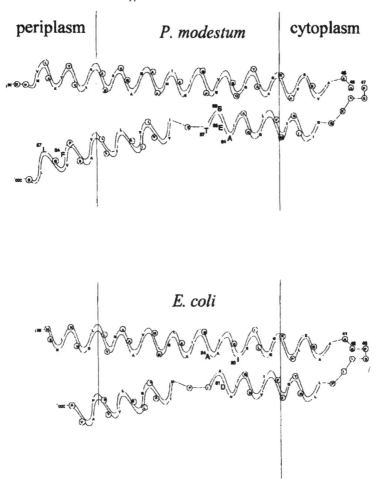

Figure 4. Model for folding of subunit c of *Escherichia coli* in the membrane. Mutational, labeling and NMR studies are supporting this model (From Miller et al., 1990). *P.modestum* subunit c was designed according to the model of *E. coli* subunit c.

toward DCCD was shown to correlate with a conformational change around D61 (Girvin and Fillingame 1993, 1994). The combined function of subunit a and subunit c in H$^+$-translocation was shown by suppressor mutations of mutant A24D/D61G in which the ATP synthase function is partially impaired. The suppressor mutations, leading to a more functional ATP synthase, were found close to the essential amino acids R210 and E219 on subunit a (Fraga et al., 1994a). Mutagenesis of the polar loop region revealed that residues R41 and Q42 are involved in energy transduction from F_1 to F_0 (Fraga et al., 1994b). Second site suppressor mutations of mutant Q42E all resided in subunit e of F_1 (Zhang et al., 1994). These results provide strong evidence that subunit ε and the polar loop region of subunit c take part in energy coupling between F_1 to F_0.

IV.   THE F_1F_0 ATPase OF *P. MODESTUM* SHEDS NEW LIGHT ON THE ATPase CATALYTIC MECHANISM

The discovery of the sodium ion-dependent F_1F_0-ATPase of *P. modestum* has greatly advanced our understanding of the ion-translocating mechanism across the membrane and its coupling to ATP synthesis. The fact that it uses sodium ions instead of protons (the only known coupling ion for F_1F_0-ATPases so far) excludes any mechanism which is restricted to protons as coupling ions. Mitchell (1976) proposed that the coupling protons move through the F_0 proton channel into the catalytic center of ATP synthesis at the F_1 moiety, where they remove an oxygen atom of a bound phosphate molecule, forming water, and making the rest of the phosphate molecule highly active for condensation with ADP to form ATP. As Na^+ ions are unable to replace protons in the removal of the oxygen from the phosphate molecule, this model is no longer valid. The existence of a Na^+-translocating F_1F_0-ATPase is, however, consistent with the conformational coupling mechanism proposed by Boyer (1975). Here, the electrochemical gradient induces conformational changes within F_0 which are transmitted to the catalytic sites on F_1 leading to the endergonic synthesis of ATP from ADP and inorganic phosphate. This mechanism is equally compatible with H^+ and Na^+ ions.

A.   Origin and Structure of the *P. modestum* ATPase

The *P. modestum* ATPase derives from the strictly anaerobic bacterium *Propionigenium modestum* which is able to grow on succinate as sole source of carbon and energy (Schink and Pfennig, 1982). The medium has to contain 100-150 mM NaCl, reflecting the marine origin of the organsim. Energy conservation is maintained by a sodium ion cycle. A membrane-bound methylmalonyl-CoA decarboxylase uses the free energy of decarboxylation of methylmalonyl-CoA to propionyl-CoA and CO_2 to generate an electrochemical sodium ion gradient (Hilpert et al., 1984). This is subsequently used by the Na^+-dependent F_1F_0-ATPase to synthesize ATP. No other energy-conserving mechanisms, like substrate-level phosphorylation or electron-transport phosphorylation, are present in this organism. Hence, ATP derives only from decarboxylation phosphorylation (for review see Dimroth 1987; 1990a, b; 1991; 1997).

The ATPase of *P. modestum* was purified (Laubinger and Dimroth, 1987,1988) and shown to belong to the F-type ATPases by subunit composition and inhibitor studies (for example, insensitivity to vanadate, sensitivity to DCCD). Cloning and sequencing of the ATPase operon of *P. modestum* revealed the same number and order of the genes as the *E. coli unc* operon (Amann et al., 1988; Ludwig et al., 1990; Kaim et al., 1992; Krumholz et al., 1992) (Figure 5).

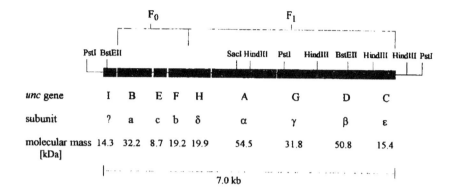

Figure 5. Organization of the *unc* operon of the *P. modestum* ATPase and the molecular masses of the deduced proteins. The function of the gene product of *uncI* is not yet verified and is indicated by a question mark.

A comparison of the amino acid sequences with the corresponding subunits of the *E. coli* ATPase revealed that the percent identity of the F_0 subunits a, b, and c is low (25.2%, 30.1%, and 25%). This is also the case for subunits δ, γ, and ε of the F_1 moiety (28.3%, 38.4%, and 27.4%), whereas subunits α and β show high homology (54.8% and 70.5% identity, respectively) (Gerike and Dimroth, 1993). The relationships between corresponding subunits of the *P. modestum* and *E. coli* ATPase are similar to those between other members of the F_1F_0 family, where a high overall homology is only found for subunits α and β (Walker et al., 1984, 1985). The similarities in the primary structure provide further evidence that the *P. modestum* ATPase belongs to the F-type family of ATPases.

B. Features of Sodium Ion Transport Across *P. modestum* F_0

As mentioned above, the unique feature of the *P. modestum* ATPase is its use of Na+ ions for coupling. The *P. modestum* ATPase uses Li+ as alternate coupling ion, albeit at an affinity about 10 times lower than that for Na+ ions. At low sodium concentrations (< 1 mM) the *P. modestum* ATPase switches to protons as coupling ions (Laubinger and Dimroth, 1989). Increasing the Na+ concentration from 0 to 1 mM leads to a progressive decrease in ATP-dependent proton transport. This behavior indicates competition between Na+ and H+ for the same binding site. This conclusion was confirmed by measuring ATPase activity with the purified enzyme in a Na+-and pH-dependent manner (Kluge and Dimroth 1993a). There was virtually no ATPase activity detectable at pH 9.0 in the absence of sodium ions. With increasing sodium ion concentration, ATPase activity increased with strong positive cooperativity (n_H=2.6) and reached a maximum at about 1 mM NaCl. Thus, at pH 9.0 at least three Na+-binding sites need to be occupied for maximal ATPase activity. In contrast, at pH 6.5 a certain amount of ATPase activity was found even in the absence of Na^+ ions. Upon NaCl addition, the activity increased reaching its maximum at about 12 mM NaCl. At more acidic pH values (pH 6.5), all three binding sites may be occupied by protons which leads to only partial activation of the enzyme. Activation of ATPase activity by sodium ions was only observed with the F_1F_0 complex, but not with the isolated F_1 moiety, showing that the ion binding site is located on F_0. Therefore, the binding of Na^+ ions to the F_0 part seems to be essential for initiating ATP hydrolysis on F_1, indicating long-ranging conformational changes from the ion binding sites on F_0 to the catalytic sites on F_1.

The location of the the binding site for the coupling ions on F_0 was further demonstrated by ion transport studies with F_0-containing liposomes (Kluge and Dimroth, 1992). To measure H^+ or Na^+ transport into the F_0-containing liposomes, it was essential to apply an inside negative membrane potential (potassium diffusion potential). The F_0 liposomes were unable to accumulate Na^+ if a sodium concentration gradient instead of a membrane potential was applied. Thus, a membrane potential-dependent step is obligatory for Na^+ accumulation by the F_0 liposomes.

At low (<2 mM) Na^+ concentrations the F_0-liposomes were also able to translocate protons. The proton transport into the inside negative compartment of the F_0 liposomes was inhibited by the addition of NaCl to the outside but was not significantly affected by adding NaCl to the inside. These results demonstrate that the accessibility of the Na^+-binding site in F_0 is strictly oriented with respect to the applied membrane potential. With an inside negative membrane potential, Na^+-ions have access to the externally oriented binding site and are transported to the inside. The reversal of this reaction is, however, apparently kinetically inhibited. Therefore, internal Na^+ ions do not compete with protons for binding and translocation from the outside to the inside.

$^{22}Na^+$ movement in the direction of a positive potential was also not found, if unlabeled Na^+ ions were present on the positive side of the membrane (exchange). In the absence of a potential, however, $^{22}Na^+$ moved to the opposite side of the membrane, where unlabeled Na^+ was present, even if the transport of $^{22}Na^+$ was against the concentration gradient (counterflow). These results indicate that the Na^+ translocation is kinetically controlled by the membrane potential and follows typical features of a transporter type mechanism (Figure 6).

In the absence of a potential, the binary complex of F_0 and Na^+ can move freely over the membrane, and Na^+ can bind and dissociate on both sides (steps 1-3). Under these conditions, reorientation of the binding site of the unloaded F_0 is kinetically prevented, and unidirectional Na^+ transport does not occur. In the presence of a membrane potential, the unloaded negatively charged carrier is forced to move from the negative to the positive charged membrane side (step 4), leading to unidirectional Na^+ transport from the positive to the negative membrane side.

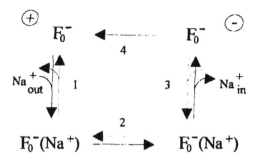

Figure 6. Schematic representation of reactions involved in Na$^+$ influx and counterflow as catalyzed by the F$_0$ moiety of the *P. modestum* ATPase. Na$^+$ counterflow proceeds by the reversible reactions 1, 2, and 3 (*solid and dashed lines*) in the absence of a membrane potential. Na$^+$ uptake (*solid lines*) in addition requires the membrane potential-dependent reorientation of the unloaded carrier (step 4). At membrane potentials of about -100 mV, the velocity of step 4 must be so high that counterflow does not occur. F$_0$ catalyzes Na$^+$ efflux, if the membrane potential is reversed (outside negative) (From Kluge and Dimroth, 1992).

C. Glutamate-65 of Subunit c is Part of the Sodium Ion Binding Site in the *P. modestum* Enzyme

The experiments with F$_0$-containing liposomes described above demonstrated that the Na$^+$ binding site(s) is restricted to the F$_0$ moiety of the enzyme. In addition, Na$^+$-activating profiles of ATP hydrolysis at pH 9.0 showed that about 3 Na$^+$ binding sites are present on F$_0$. As subunit c is the only F$_0$ subunit with a high enough copy number to account for a simultaneous binding of 3 Na$^+$ ions, the Na$^+$ ion binding site should be located on this subunit (Kluge and Dimroth, 1993a). Further evidence for the location of the sodium ion binding site on subunit c was obtained by measuring the kinetics of ATPase inactivation by DCCD at different pH values and Na$^+$-concentrations (Kluge and Dimroth, 1993a, b). DCCD specifically reacts with the conserved carboxylic acid residue in the C-terminal helix of subunit c, thereby abolishing ion transport and ATP hydrolysis. It has been proposed that the protonated form of this carboxylic acid is the reactive species (Kluge and Dimroth, 1993a). The mechanism could involve proton transfer to DCCD, followed by nucleophilic attack of the carboxylate at the protonated DCCD molecule resulting in the O-acyl isourea derivative which is then rearranged to the more stable N-acyl urea component (Azzi et al., 1984). E65 is the proposed carboxylic acid residue in the *P. modestum* c subunit reacting with DCCD. It is at the analogue position to D 61 of the *E. coli* c subunit (Kaim et al., 1992).

Inactivation of the ATPase by DCCD occurred with a high reaction rate, which was about 10^7 times faster than the reaction of DCCD with acetic acid under similar conditions. This high rate of inactivation is in accord with the remarkable specificity for modification of a single amino acid residue in the protein (Kluge and Dimroth, 1993a, b). As anticipated from the proposed reaction mechanism, the inhibition by DCCD was highly pH-dependent. The rate constant of inactivation increased by an order of magnitude between pH 8.5 and pH 5.5 following a titration curve with a pK of 7.0 if no Na^+ was present. It was therefore concluded that the DCCD-reactive glutamate residue dissociates with a pK of 7.0 in the absence of sodium ions and that this residue must be protonated in order to get modified. It was further shown that the addition of sodium ions protected E65 of subunit c from DCCD modification at all pH values tested. The pK of E65 shifted to 6.2 and 5.8 in the presence of 0.5 mM and 1 mM NaCl, respectively. These results support a model for the competition of H^+ and Na^+ in binding to the glutamate residue at the active site (Figure 7).

DCCD only reacts with the undissociated E65, whereas dissociated glutamate (glu⁻) is the Na^+-binding species. Binding of sodium ions to E65 shifts the equilibrium to the side of the dissociated glutamate residue. The concomitant decrease of the concentration of undissociated glutamate residues decreases the rate constant for inactivation of the ATPase by DCCD. For H^+-translocation, the bound species may be H_3O^+ (Kluge and Dimroth, 1993a, b, 1994). Li^+ ions also protect the enzyme from modification with DCCD but ten times higher Li^+ than Na^+ concentrations were required to achieve similar protection. The reactivity of E65 for modification with DCCD was retained in the c subunit purified by extraction with chloro-

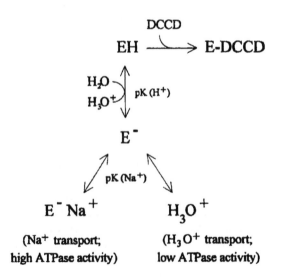

Figure 7. Model for the interaction of Na^+, H_3O^+ and DCCD with the active-site glutamate residue (E⁻) of subunit c of the *P. modestum* ATPase (From Kluge and Dimroth, 1994b).

form/methanol and transferred to an aqueous buffer containing dodecylmaltoside. Subunit c became labeled after incubation with [^{14}C]DCCD, and the presence of Na$^+$ ions and high pH values protected the protein from this modification (Kluge and Dimroth, 1994).

The very high rate and specificity of the reaction of DCCD with E65 of subunit c could be of some significance in relation to the ion translocation mechanism. Conceivably the reactivity toward DCCD reflects recognition of part of the DCCD structure by the protein that resembles the structure of an amino acid coming close to this site during natural catalysis. The guanidino group like structure of the protonated DCCD which is the presumed transition state, resembles the side chain of arginine. An arginine residue might therefore come into close contact to E65 during the catalytic cycle. EIPA (ethylisopropylamiloride), another guanidino group-containing compound, protected the enzyme competitively from modification with DCCD (Kluge and Dimroth, 1993a). Thus, EIPA like DCCD seems to bind close to E65 on subunit c. A hypothetical mechanism of Na$^+$(or H$_3$O$^+$) translocation by the F$_1$F$_0$-ATPase involves binding of one of these coupling ions to a binding site on F$_0$ including the conserved acidic residue of subunit c on one side of the membrane. A conformational change reorients the occupied ion binding site to the other side of the membrane and at the same time brings an arginine residue close to the conserved acidic residue on subunit c leading to the release of the bound coupling ion by an ion exchange type of mechanism. Conserved arginine residues, that are functionally important, are found in the C-terminal helix of subunit a (R210 *E. coli*; R226 *P. modestum*) (Cain and Simoni, 1989; Kaim et al., 1992) and in the polar loop region of subunit c (R41 *E. coli*; R45 *P. modestum*) (Ludwig et al., 1990; Fraga et al., 1994b) and may represent such candidates.

Another sodium ion-dependent ATPase has been found in *Acetobacterium woodii*. The enzyme was proposed to belong to the F-type ATPase family by some inhibitor studies (Reidlinger and Müller, 1994), and by sequencing part of its genes (ΔuncA, uncG, uncD, and uncC) (Forster et al., 1995). However, only six subunits have been identified corresponding to subunits α, β, γ, ϵ, c, and another subunit of about 19 kDa that may represent subunit δ, a, or b. Whether the ATPase of *A. woodii* consists in fact of only six different subunits has to await the sequencing of the upstream region of uncA of *A. woodii*. The ATPase of *A. woodii* differs from the *P. modestum* enzyme in its sensitivity toward nitrate, a specific inhibitor of V-type ATPases. Furthermore, it was not possible to dissociate the F$_1$ moiety from the membrane by conventional means. Sodium ions stimulate ATP hydrolysis of the *A. woodii* ATPase with an apparent K$_m$ of 0.4mM at pH 7.5 and protect the enzyme from modification by DCCD. Li$^+$ ions had a similar protective effect as Na$^+$ ions although about six times higher Li$^+$ concentrations were needed (Spruth et al., 1995). The enzyme was inhibited by amiloride derivatives with hydrophobic side chains [EIPA, hexamethylene-amiloride (HMA), benzamil and phenamil] but not by amiloride itself. Similarly *A. woodii* ATPase was protected from DCCD modification by substituted amiloride derivatives. However protection was not found when benzamil and phenamil were used (Spruth et al., 1995). Thus, some features of the *A. woodii* enzyme are shared with the *P. modestum* enzyme while others are clearly distinct. Further experiments are therefore required to decide whether the *A. woodi* ATPase belongs to the family of F-type ATPases.

D.   Getting Closer to the Ion Binding Site by Genetical Means

Much information has been obtained by investigating the *P. modestum* ATPase biochemically. A clear role in binding of the coupling ions has been assigned to E65 of subunit c of the *P. modestum* enzyme. Mutational studies are a good tool to determine additional residues that may contribute to ion binding. The *P. modestum* F_1F_0 ATPase genes have been cloned and sequenced (see Figure 5). The primary structure did not reveal striking differences in comparison to H^+-translocating F_1F_0-ATPases that could be connected to the change in coupling ion specificity. Residues essential for H^+-translocating activity, which are conserved in H^+-translocating ATPases, are also present in the *P. modestum* enzyme.

In a first attempt to get closer to the ion binding site, hybrid ATPases between the *E. coli* and the *P. modestum* F_1F_0-ATPase were constructed *in vivo*. The close relationship between the *E. coli* and *P. modestum* F_1F_0 ATPases is shown by several features: the same number and order of the genes, the high homology of subunits α and β (54.8% and 70.5% identity, respectively), and conservation of functional essential residues (for example, the acidic residue in the C-terminal helix of subunit c). The most important indication, however, is the construction of a hybrid enzyme *in vitro* consisting of the F_0 part of *P. modestum* and the F_1 part of *E. coli* (Laubinger et al., 1990). This hybrid ATPase was able to transport Na^+ ions as the result of ATP hydrolysis, although with a poor coupling ratio (0.007 Na^+ per ATP hydrolyzed). This observation further supports previous data that the F_0 part is entirely responsible for the coupling ion specificity (see above).

E. coli unc mutants which are deficient in either subunit a or c were complemented with the respective genes of the *P. modestum* ATPase (Gerike and Dimroth, 1994). Due probably to the low conservation of structure, no functional ATPase hybrids were obtained although the respective genes of the *P. modestum* ATPase were expressed. Surprisingly, even the whole *P. modestum* ATPase was not functional in *E. coli* strain (DK8 pUGP2) (Figure 8). Again, the expression of all subunits of the *P. modestum* ATPase in *E. coli* was demonstrated by labeling plasmid encoded proteins specifically with [^{35}S]L-methionine. However, ATP hydrolyzing activity was not detectable. A reason for the expression of an inactive enzyme could be the absence of an assembly factor in the *E. coli* host cells which is specifically required for the F_1F_0 ATPase of *P. modestum*. Involvement of assembly factors for the correct assembly of F_1 and F_0 were demonstrated for the F_1F_0.ATPase of *Saccharomyces cerevisiae* (Ackerman and Tzagoloff, 1990; White and Ackermann, 1995).

A hybrid ATPase consisting of F_0 of *P. modestum* (PF_0) and F_1 of *E. coli* (EF_1) was constructed on plasmid (DK8 pBWF018) (Figure 8). Expression in an *E. coli*

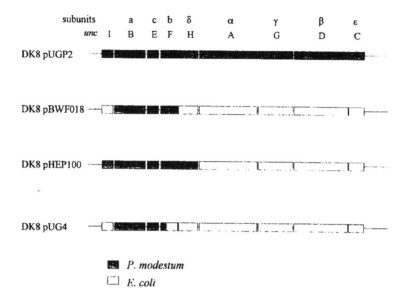

Figure 8. Survey of hybrid *P. modestum/E. coli unc* operons encoded on plasmids and expressed in *E. coli unc* mutant strain DK8 (ΔuncB-C).

unc deletion strain showed that this ATPase is not able to synthesize ATP, either. This finding was not surprising, however, because the analogue hybrid ATPase constructed *in vitro* (Laubinger et al., 1990) showed a poor coupling ratio as described above. Characterization of the EF_1PF_0 hybrid ATPase synthesized in *E. coli* demonstrated the lack of coupling of the F_1 of *E. coli* to the F_0 of *P. modestum*. The ATP hydrolyzing activity could be demonstrated but no functionality of the PF_0 moiety of the EF_1PF_0 hybrid ATPase was detectable (Gerike et al., 1995).

Recent expression studies of F_0 in *E. coli* mutants missing the whole *unc* operon demonstrated that the F_0 moiety is in a low state of conductivity, if the F_1 subunits are not coexpressed. These findings were independent of whether the F_0 genes were expressed from plasmids or whether they were chromosomally integrated (Brusilow and Monticello, 1994). After adding F_1 to the membranes of the F_0 expression clone, significant ATP-dependent proton pumping activity could be demonstrated. These results seem to indicate that the F_1 moiety is necessary to convert F_0 from an inactive into its mature catalytically active form. The PF_0 moiety of DK8pBWFO18 synthesizing an EF_1PF_0 hybrid may not be functional because a specific interaction with PF_1 is required to convert it into its mature catalytically active form.

It was, however, possible to construct a functional *P. modestum/E. coli* hybrid ATPase in *E. coli* consisting of the F_0 moiety and subunit δ of *P. modestum* and the remaining F_1 subunits of *E. coli* (Kaim and Dimroth, 1993,1994). This mutant was denoted DK8 pHEP100 (Figure 8). The growth of DK8 pHEP100 on succinate minimal medium was strictly dependent on Na+ ions. This sodium ion dependence coincided with the sodium ion dependence of the purified hybrid ATPase. ATP-dependent proton transport decreased and ATP hydrolyzing activity increased with increasing NaCl concentrations. This behavior is typical of the *P. modestum* enzyme and again demonstrates that the ion specificity derives from F_0 and not from F_1.

According to these data, subunit δ has to derive from the same *P. modestum* origin as the F_0 moiety to obtain a functional sodium ion-dependent *P. modestum/E. coli* hybrid ATPase in *E. coli*. It is therefore possible that the *P. modestum* δ subunit is sufficient to induce the proper maturation of the PF_0 moiety. It was proposed previously that the two copies of subunit b together with subunit δ are responsible for a proper binding of F_1 to F_0 (Hazard and Senior 1994a, b). The same origin of subunit β and subunit δ may therefore be important for tight binding between the F_1 moiety and the F_0 part in hybrid ATPases. To investigate this hypothesis, a hybrid ATPase was constructed consisting of subunits a and c of *P. modestum*, the F_1 subunits of *E. coli* and a hybrid b subunit with the membrane anchor deriving from *P. modestum* and the hydrophilic cytoplasmic region (see also Figure 3) deriving from *E. coli*. This hybrid enzyme (DK8 pUG4) (Figure 8) is an analogue of DK8 pHEP100 but with subunits δ and the main part of subunit b deriving from *E. coli* and not from *P. modestum*. The hybrid b subunit was constructed because it cannot be ruled out that the membrane part of subunit b has to be of the same origin as subunits a and c to provide a functional F_0. The hybrid ATPase of DK8pUG4, however, was not functional. Immunoblot studies with *E. coli* anti-b antibody showed that the hybrid subunit b is degraded (Gerike et al., 1995). The degradation of hybrid subunit b might be due to a weak binding of F_1 to F_0 caused by an incorrect secondary structure of hybrid subunit b. The weak binding of F_1 to F_0 in *E. coli* mutant strain DK8 pUG4 was demonstrated by washing the membranes once. Seventeen percent of the original activity remained bound to the membrane of DK8 pUG4 compared to 80% of the wildtype *E. coli* DH5. It has already been shown that subunit b of *E. coli* is degraded if a defective assembly of the *E. coli* F_1F_0 ATPase occurs (Hazard and Senior, 1994b). Furthermore, it was shown that the C-terminal part of subunit b is needed for correct assembly of F_0 (Steffens et al., 1987). As the C-terminal part of hybrid subunit b derives from the *E. coli* b subunit, the proper assembly of *P. modestum* F_0 may be impaired.

The *E. coli* enzyme uses only H+ as coupling ion and does not use either Na+ or Li+. The *P. modestum* enzyme, however, has the capacity to transport Na+, Li+ and H+ ions, where Na+ is the superior coupling ion (Kluge and Dimroth, 1993a, b). A mutational analysis aiming to convert an H+-coupled ATPase into a Na+-coupled one was performed with the *E. coli* enzyme by replacing the amino acids around residue D61 of subunit c by the corresponding amino acids of the *P. modestum* c subunit (Zhang and Fillingame, 1995). With this approach a mutant (E239) was obtained which contained a Li+-binding site. This was shown by the inhibition of ATP-dependent H+ transport in the presence of 50 mM LiCl. The newly created metal ion binding site was apparently specific for Li+, because Na+ ions were without effect on the proton pumping activity. Li^+ ions seem to bind to the mutant ATPase, but are probably not transported, because ATP hydrolysis is not activated but inhibited, by Li^+ addition. There are apparently structural constraints in this *E. coli* mutant that prevent the proper conformational changes for release of the Li^+ ion from its binding pocket and delivery to the other surface. The amino acid exchanges in the mutant E239 were shown to be V60AD61EA62SI63T and thus reflect the region surrounding residue E65 of subunit c of *P. modestum* (A64E65S66T67) (see also Figure 4). It is proposed that the newly designed Li^+-binding site could be formed, at least in part, by the E61 and S62 side chain oxygens, with perhaps the E61 peptide carbonyl oxygen acting as an additional liganding group. The lack of Na^+ ion binding may be due to the larger ionic diameter of Na^+ (about 0.19 nm) compared to Li^+ (about 0.13 nm). Na^+-binding may require smaller or more flexible residues around position 61 in *E. coli* mutant E239. Another reason could be the requirement of an extra liganding group in the form of a hydroxyl or carbonyl group. Residue Q32 of the *P. modestum* c subunit provides the additional liganding groups for Na^+-binding (Kain et al., 1977).

The availability of a functional sodium ion-dependent hybrid ATPase in *E. coli* represents an excellent tool to investigate the ion-binding site by site-specific or random mutagenesis. Changing the ion specificity from Na^+ to H^+ should result in mutants which are able to grow on succinate minimal medium without Na^+ addition. Recent mutational studies with the sodium ion-dependent *P. modestum/E. coli* hybrid ATPase (DK8 pHEP100) led to the desired change in ion specificity (Kaim and Dimroth, 1995). Here, the mutant (MPC8487) lost its ability to transport Na^+ ions, but retained its capacity for Li^+ and H^+ transport. The proton transporting activity was in fact improved over that of the parent hybrid enzyme such that the *E. coli* clone expressing the mutant ATPase could use H^+ as coupling ions for ATP synthesis, as evidenced by its Na^+-independent growth on succinate. The new phenotype was created by a double mutation in the c subunit. The two amino acid exchanges (F84LL87V) were found unexpectedly within the last six amino acids of subunit c of *P. modestum*, a region that does not exist in the *E. coli* c subunit as judged from amino acid alignments (see also Figure 4). This region might therefore play an important role in proper functioning of the *P. modestum* ATPase as an Na^+-coupled enzyme. The altered amino acid residues might interact differently with residues in their environment within the same or the opposite helix. This could create a conformational change in which the size of the metal binding pocket is narrowed so that the smaller Li^+ ions may still be complexed and translocated, whereas the larger Na^+ ions can no longer bind.

V. A MODEL FOR THE INTERACTION OF A AND C SUBUNITS

The mechanism of linking ion translocation across the F_0 portion of the ATPase to the chemical reaction of ATP synthesis/hydrolysis at the F_1 portion is still poorly understood. Based on the rotational model that emerged from the F_1 structure, the γ-subunit rotates relative to the $\alpha_3\beta_3$ hexagon (Abrahams et al., 1994). A logical extension of this model assumes the rotation of additional subunits in the stalk region and in the F_0 moiety. The movement of certain F_0 subunits relative to other F_0 subunits should be directly connected to ion translocation across the membrane. Evidence from mutational analyses of the *E. coli* ATPase revealed that besides the c subunit the a subunit is intimately involved in the proton translocation mechanism (Cain and Simoni, 1988; Vik and Antonio, 1994). The c subunits (about 10 copies) are supposed to form a ring to which the single a subunit is attached from the outside (Schneider and Altendorf, 1987).

A model linking ion translocation to a rotation of the c subunits relative to the a subunit can be envisaged with an access and a release channel for the coupling ions. The proton binding site on D61 of the c subunits of the *E. coli* enzyme may have access to the periplasm, except for one c subunit that is in contact with the a subunit, where the proton has access to the channel leading to the cytoplasm. For a complete crossing of the membrane, the proton must therefore bind to D61 from the periplasm, the ring of c subunits must rotate to make the proper contact to the a subunit, and the proton must be released to the cytoplasm. It is conceivable that proton transfer and rotation are driven by electrical forces through the attraction of positively and negatively charged groups in the membrane (on c and a subunits) or through repulsion of groups carrying the same charge. It is interesting in this context that the three amino acids of the *E. coli* a subunit K210, E219, and H245 may exist in a neutral and in a charged form. This mechanism of ion translocation can easily be adapted to Na^+ tansport in case of the *P. modestum* ATPase. All that is needed is to modify the cation-binding site such that Na^+ ions fit in and are properly complexed by liganding groups in the vicinity of the active site carboxylate. It is interesting that the R210 residue (*E. coli* numbering) of the a subunit is found in all ATPase sequences including that of *P. modestum*. Position 219 contains either aspartate or histidine. ATPases from mitochondria which have histidine at position 219 have glutamate at position 245, while those with glutamate at position 219 may have histidine, glycine, proline or serine at position 245.

Results from mutagenesis studies indicated a severe defect in the F_0-mediated proton translocation by an a H245 $\rightarrow$ E mutation, but interestingly, the double mutant (aE219 $\rightarrow$ H, H245 $\rightarrow$ E) yielded an ATPase complex with improved proton translocation as compared to the single mutants. These results may indicate a close interaction of residues 219 and 245 of the a subunit and a requirement for an acidic residue in either of these positions in order to catalyze proton tanslocation (Cain and Simoni, 1988). The *P. modestum* ATPase has methionine at position 219 and aspartate at position 245 of the a subunit (*E. coli* numbering) (Kaim et al., 1992). The enzyme is therefore related to the other ATPases by retainment of an acidic amino acid in one of these positions. A methionine at position 219, however, has never been found before in an ATPase sequence. Whether this substitution is essential for the change in ion specificity is unknown.

VI. CONCLUSION AND OUTLOOK

Data discussed in this chapter have helped to establish the theory of Boyer (1975) that conformational changes couple ion translocation on F_0 to ATP synthesis on F_1 within the F_1F_0 ATP synthase. But the modes of conformational changes are still unknown. The structure of F_1 of bovine heart mitochondria implies that rotational movements are involved as the $\alpha_3\beta_3$ hexamer rotates relative to subunit γ. A similar rotation could take place within F_0, where a ring of c subunits might rotate relative to subunit a. The rotational movements within the two moieties have to be connected by movements within the stalk region. From mutational studies of the *E. coli* enzyme subunits β, δ, and ϵ might be contributing to this function.

The crystallization and structure determination of F_0 will be a major aim in the future. Knowing the F_0 structure would solve the orientation of the individual subunits within F_0 and would therefore shed new light on the understanding of the ion translocation mechanism. Mutational studies with the hybrid *P. modestum/E. coli* ATPase will be a powerful tool to determine the amino acids contributing ligands for metal ion (Na^+ or Li^+) binding. With the F_1 structure available, future mutational studies within F_1 will help to understand the movements of the $\alpha_3\beta_3$ hexamer relative to subunit γ. Attention should also be given to the stalk region which catalyzes a major step in energy transduction from F_0 to F_1.

REFERENCES

Abrahams, J. P., Leslie, A. G. W., & Walker, J. E. (1994). Structure at 2.8Å resolution of F_1-ATPase from bovine heart mitochondria. Nature 370, 621-628.

Ackerman, S. H., & Tzagoloff, A. (1990). *ATP10*, a yeast nuclear gene required for the assembly of the mitochondrial F_1-F_0 complex. J. Biol. Chem. 265, 9952-9959.

Amann, R., Ludwig, W., Laubinger, W., Dimroth, P., & Schleifer, K.H. (1988). Cloning and sequencing of the gene encoding the beta subunit of the sodium ion translocating ATP synthase of *Propionigenium modestum*. FEMS Microbiol. Lett. 56, 253-260.

Azzi, A., Casey, R. P., & Nalecz, M. J. (1984). The effect of *N,N'*-dicyclohexylcarbodiimide on enzymes of bioenergetic relevance. Biochim. Biophys.Acta 768, 209-226.

Boyer, P. D. (1975). A model for conformational coupling of membrane potential and proton translocation to ATP synthesis and active transport. FEBS Lett. 58, 1-6.

Boyer, P. D. (1989). A perspective of the binding change mechanism for ATP synthesis. FASEB J. 3, 2164-2178.

Brusilow, W. S. A., & Monticello, R. A. (1994). Synthesis and assembly of the F_0 proton channel from F_0 genes cloned into bacteriophage λ and integrated into the *Escherichia coli* chromosome. J. Biol. Chem. 269, 7285-7289.

Cain, B. D., & Simoni, R.D. (1988). Interaction between glu-219 and his-245 within the *a* subunit of F_1F_0-ATPase in *Escherichia coli*. J. Biol. Chem. 263, 6606-6612.

Cain, B. D., & Simoni, R.D. (1989). Proton translocation by the F_1F_0-ATPase of *Escherichia coli*. J. Biol. Chem. 264, 3292-3300.

Collinson, I. R., van Raaij, M. J., Runswick, M. J., Fearnley, I. M., Skehel, J. M., Orriss, G. L., Miroux, B., & Walker, J. E. (1994). ATP synthase from bovine heart mitochondria: *In vitro* assembly of a stalk complex in the presence of F_1-ATPase and in its absence. J. Mol. Biol. 242, 408-421.

Cross, R.L. (1994) Our primary source of ATP. Nature 370, 594-595.

Dimroth, P. (1987). Sodium ion transport decarboxylases and other aspects of sodium ion cycling in bacteria. Microbiol. Rev. 51, 320-340.

Dimroth, P. (1990a). Mechanism of sodium transport in bacteria. Phil. Trans. R. Soc. Lond. B 326, 465-477.

Dimroth, P. (1990b). Energy transduction by an electrochemical gradient of sodium ions. In: 41.Colloquium Mosbach: The Molecular Basis of Bacterial Metabolism (Hauska, G., & Thauer, R., Eds.), pp. 114-127, Springer-Verlag, Heidelberg.

Dimroth, P. (1991). Na^+-coupled alternative to H^+-coupled primary transport systems in bacteria. BioEssays 13, 463-468.

Dimroth, P. (1997). Primary sodium ion translocating enzymes. Biochim. Biophys. Acta 1318, 11-51.

Fillingame, R. H., Oldenburg, M., & Fraga, D. (1991). Mutation of alanine 24 to serine in subunit c of the *Escherichia coli* F_1F_0-ATP synthase reduces reactivity of aspartyl 61 with dicyclohexylcarbodiimide. J. Biol. Chem. 266, 20934-20939.

Forster, A., Daniel, R., & Müller, V. (1995). The Na^+-translocating ATPase of *Acetobacterium woodii* is a F_1F_0-type enzyme as deduced from the primary structure of its β, γ, and ε subunits. Biochim. Biophys. Acta 1229, 393-397.

Foster, D. L., & Fillingame, R. H. (1979). Energy-transducing H+-ATPase of *Escherichia coli*. J. Biol. Chem. 254, 8230-8236.

Fraga, D., Hermolin, J., & Fillingame, R. H. (1994a). Transmembrane helix-helix interactions in F_0 suggested by suppressor mutations to ala^{24}→ asp/asp^{61} → gly mutant of ATP synthase subunit c. J. Biol. Chem. 269, 2562-2567.

Fraga, D., Hermolin, J., Oldenburg, M., Miller, M. J., & Fillingame, R. H. (1994b). Arginine 41 of subunit c of *Escherichia coli* H^+-ATP synthase is essential in binding and coupling of F_1 to F_0. J. Biol. Chem. 269, 7532-7537.

Friedl, P., Friedl, C., & Schairer, H. U. (1979). The ATP synthetase of *Escherichia coli* K12: purification of the enzyme and reconstitution of energy-transducing activities. Eur. J. Biochem. 100, 175-180.

Gay, N. J., & Walker, J. E. (1981). The *atp* operon: Nucleotide sequence of the promoter and the genes for the membrane proteins and the δ-subunit of *Escherichia coli* ATP-synthase. Nucleic Acids Res. 9, 3919-3926.

Gerike, U., & Dimroth, P. (1993). N-terminal amino acid sequences of the subunits of the Na^+-translocating F_1F_0 ATPase from *Propionigenium modestum*. FEBS Lett. 316, 89-92.

Gerike, U., & Dimroth, P. (1994). Expression of subunits a and c of the sodium-dependent ATPase of *Propionigenium modestum* in *Escherichia coli*. Arch. Microbiol. 161, 495-500.

Gerike, U., Kaim, G.,& Dimroth, P. (1995). *In vivo* synthesis of ATPase complexes of *Propionigenium modestum* and *Escherichia coli* and analysis of their function. Eur. J. Biochem. 232, 596-602.

Girvin, M. E., & Fillingame, R. H. (1993). Helical structure and folding of subunit c of F_1F_0 ATP synthase: ^{1}H NMR resonance assignments and NOE analysis. Biochemistry 32, 12167-12177.

Girvin, M. E., & Fillingame, R. H. (1994). Hairpin folding of subunit c of F_1F_0 ATP synthase: ^{1}H distance measurements to nitroxide-derivatized aspartyl-61. Biochemistry 33, 665-674.

Hazard, A. L., & Senior, A. E. (1994a). Mutagenesis of subunit δ from *Escherichia coli* F_1F_0 ATP synthase. J. Biol. Chem. 269, 418-426.

Hazard, A. L., & Senior, A. E. (1994b). Defective energy coupling in δ subunit mutants of *Escherichia coli* F_1F_0 ATP synthase. J. Biol. Chem. 269, 427-432.

Hilpert, W., Schink, B., & Dimroth, P. (1984). Life by a new decarboxylation-dependent energy conservation mechanism with Na^+ as coupling ion. EMBO J. 3, 1665-1670.

Hoppe, J., Schairer, H. U., & Sebald, W. (1980). Identification of amino-acid substitutions in the proteolipid subunit of the ATP synthase from dicyclohexylcarbodiimide-resistant mutants of *Escherichia coli*. Eur. J. Biochem. 112, 17-24.

Kanazawa, H., Mabuchi, K., Kayano, T., Tamuar, F., & Futai, M. (1981). Nucleotide sequence of genes coding for dicyclohexylcarbodiimide-binding protein and the alpha subunit of proton-translocating ATPase of *Escherichia coli*. Biochem. Biophys. Res. Commun. 100, 219-225.

Kaim, G., & Dimroth, P. (1993). Formation of a functionally active sodium-translocating hybrid F_1F_0 ATPase in *Escherichia coli*. by homologous recombination. Eur. J. Biochem. 218, 937-944.

Kaim, G., & Dimroth, P. (1994). Construction, expression and characterization of a plasmid-encoded Na$^+$-specific ATPase hybrid consisting of *Propionigenium modestum* F_0-ATPase and *Escherichia coli* F_1-ATPase. Eur. J. Biochem. 222, 615-623.

Kaim, G., Ludwig, W., Dimroth, P., & Schleifer, K. H. (1992). Cloning, sequencing and *in vivo* expression of genes encoding the F_0 part of the sodium-ion-dependent ATP synthase of *Propionigenium modestum* in *Escherichia coli*. Eur. J. Biochem. 207, 463-470.

Kaim, G., & Dimroth, P. (1995). A double-mutation in subunit c of the Na$^+$-specific F_1F_0-ATPase of *Propionigenium modestum* results in a switch from Na$^+$- to H$^+$-coupled ATP synthesis in the *Escherichia coli* host cells. J. Mol. Biol. 253, 726-738.

Kaim, G., Wehrle, F., Gerike, U., & Dimroth, P. (1997). Molecular basis for the coupling ion selectivity of F_1F_0 ATP synthases: Probing the liganding groups for NA$^+$ and Li$^+$ in the c subunit of the ATP synthase from *Propionigenium modestum*. Biochemistry 36, 9185-9194.

Kluge, C., & Dimroth, P. (1992). Studies on Na$^+$ and H$^+$ translocation through the F_0 part of the Na$^+$ translocating F_1F_0 ATPase from *Propionigenium modestum*: Discovery of a membrane potential-dependent step. Biochemistry 31, 12665-12672.

Kluge, C., & Dimroth, P. (1993a). Kinetics of inactivation of the F_1F_0 ATPase of *Propionigenium modestum* by dicyclohexylcarbodiimide in relationship to H$^+$ and Na$^+$ concentration: probing the binding site for the coupling ions. Biochemistry 32, 10378-10386.

Kluge, C., & Dimroth, P. (1993b). Specific protection by Na$^+$ or Li$^+$ of the F_1F_0 ATPase of *Propionigenium modestum* from the reaction with dicyclohexylcarbodiimide. J. Biol. Chem. 268, 14557-14560.

Kluge, C., & Dimroth, P. (1994). Modification of isolated subunit c of the F_1F_0 ATPase from *Propionigenium modestum* by dicyclohexylcarbodiimide. FEBS Lett. 340, 245-248.

Krumholz, L. R., Esser, U., & Simoni, R. D. (1992) Characterization of the genes coding for the F_1F_0 subunits of the sodium dependent ATPase of *Propionigenium modestum*. FEMS Microbiol. Lett. 91, 37-42.

Läuger, P. (1991). Electrogenic ion pumps. Sinauer Associates, Inc., Sunderland.

Laubinger, W., & Dimroth, P. (1987). Characterization of the Na$^+$-stimulated ATPase of *Propionigenium modestum* as an enzyme of the F_1F_0 type. Eur. J. Biochem. 168, 475-480.

Laubinger, W., & Dimroth, P. (1988). Characterization of the ATP synthase of *Propionigenium modestum* as a primary sodium pump. Biochemistry 27, 7531-7537.

Laubinger, W., & Dimroth, P. (1989). The sodium translocating adenosinetriphosphatase of *Propionigenium modestum* pumps protons at low sodium ion concentrations. Biochemistry 28, 7194-7198.

Laubinger, W., Deckers-Hebestreit, G., Altendorf, K., & Dimroth, P. (1990). A hybrid adenosinetriphosphatase composed of F_1 of *Escherichia coli*. and F_0 of *Propionigenium modestum* is a functional sodium ion pump. Biochemistry 29, 5458-5463.

Lewis, M. J., Chang, J. A., & Simoni, R. D. (1990). A topological analysis of subunit a from *Escherichia coli* F_1F_0-ATP synthase predicts eight transmembrane segments. J. Biol. Chem. 265, 10541-10550.

Ludwig, W., Kaim, G., Laubinger, W., Dimroth, P., Hoppe, J., & Schleifer, K. H. (1990). Sequence of subunit c of the sodium ion translocating adenosine triphosphate synthase of *Propionigenium modestum*. Eur. J. Biochem. 193, 395-399.

Miller, M. J., Oldenburg, M., & Fillingame, R. H. (1990). The essential carboxyl group in subunit c of the F_1F_0 ATP synthase can be moved and H$^+$-translocating function retained. Proc. Natl. Acad. Sci. USA 87, 4900-4904.

Mitchell, P. (1976). Vectorial chemistry and the molecular mechanism of chemiosmotic couplings: Power transmission by proticity. Biochem. Soc. Trans. 4, 399-430.

Negrin, R. S., Foster, D. L., & Fillingame R. H. (1980). Energy-transducing H^+-ATPase of *Escherichia coli*. J. Biol. Chem. 255, 5643-5648.

Nielsen, J., Hansen, F. G., Hoppe, J., Friedl, P., & von Meyenburg, K. (1981). The nucleotide sequence of the *atp* genes coding for the F_0 subunits a, b, c and the F_1 subunit δ of the membrane bound ATP synthase of *Escherichia coli*. Mol. Gen. Genet. 184, 33-39.

Pedersen, P.L. & Amzel, L.M. (1993). ATP synthases. Structure, reaction center, mechanism, and regulation of one of nature's most unique machines. J. Biol. Chem. 268, 9937-9940.

Reidlinger, J., & Müller, V. (1994). Purification of ATP synthase from *Acetobacterium woodii* and identification as a Na^+-translocating F_1F_0-type enzyme. Eur. J. Biochem. 223, 275-283.

Schink, B., & Pfennig, N. (1982). *Propionigenium modestum* gen. nov. sp. nov., a new strictly anaerobic, nonsporing bacterium growing on succinate. Arch. Microbiol. 133, 209-216.

Schneider, E., & Altendorf, K. (1982). ATP synthase (F_1F_0) of *Escherichia coli*. K12. Eur. J. Biochem. 126, 149-153.

Schneider, E., & Altendorf, K. (1985). All three subunits are required for the reconstitution of an active proton channel (F_0) of *Escherichia coli*. ATP synthase (F_1F_0). EMBO J. 4, 515-518.

Schneider, E. & Altendorf, K. (1987). Bacterial adenosine 5'-triphosphate synthase (F_1F_0): purification and reconstitution of F_0 complexes and biochemical and functional characterization of their subunits. Microbiol. Rev. 51, 477-497.

Schneppe, B., Deckers-Hebestreit, G., & Altendorf, K. (1991). Detection and localization of the *i* protein in *Escherichia coli*. cells using antibodies. FEBS Lett. 292, 145-147.

Sebald, W., & Hoppe, J. (1981). On the structure and genetics of the proteolipid subunit of the ATP synthase complex. Curr. Top. Bioenerg. 12, 1-64.

Senior, A.E. (1990). The proton-translocating ATPase of *Escherichia coli*. Ann. Rev. Biophys. and Biophys. Chem. 19, 7-41.

Spruth, M., Reidlinger, J., & Müller, V. (1995). Sodium ion dependence of inhibition of the Na^+-translocating F_1F_0-ATPase from *Acetobacterium woodii*. Probing the site(s) involved in ion transport. Biochim. Biophys. Acta 1229, 96-102.

Steffens, K., Schneider, E., Deckers-Hebestreit, G., & Altendorf, K. (1987) F_0 portion of *Escherichia coli* ATP synthase. J. Biol. Chem. 262, 5866-5869.

Tang, C., Wilkens, S., & Capaldi, R. A. (1994). Structure of the γ subunit of *Escherichia coli* F_1 ATPase probed in trypsin digestion and biotin-avidin binding studies. J. Biol. Chem. 269, 4467-4472.

Vik, S. B., & Antonio, B. J. (1994). A mechanism of proton translocation by F_1F_0 ATP synthases suggested by double mutants of the a subunit. J. Biol. Chem. 269, 30364-30369.

Walker, J. E., Saraste, M., & Gay, N. J. (1984). The *unc* operon. Nucleotide sequence, regulation and structure of ATP-synthase. Biochim. Biophys. Acta 768, 164-200.

Walker, J. E., Fearnley, I. M., Gay, N. J., Gibson, B. W., Northrop, F. D., Powell, S. J., Runswick, M. J., Saraste, M., & Tybulewicz, V. L. J. (1985). Primary structure and subunit stoichiometry of F_1-ATPase from bovine mitochondria. J. Mol. Biol. 184, 677-701.

White, M., & Ackerman, S. H. (1995). Bacterial production and characterization of ATP11, a yeast protein required for mitochondrial F_1-ATPase assembly. Arch. Biochem. Biophys. 319, 299-304.

Zhang, Y., Oldenburg, M., & Fillingame, R. H. (1994). Suppressor mutations in F_1 subunit ε recouple ATP-driven H^+ translocation in uncoupled Q42E subunit *c* mutant of *Escherichia coli* F_1F_0 ATP synthase. J. Biol. Chem. 269, 10221-10224.

Zhang, Y., & Fillingame, R. H. (1995). Changing the ion binding specificity of the *Escherichia coli* H^+-transporting ATP synthase by directed mutagenesis of subunit *c*. J. Biol. Chem. 270, 87-93.

STRUCTURE, FUNCTION, AND REGULATION OF THE VACUOLAR (H$^+$)-ATPASES

Michael Forgac

Advances in Molecular and Cell Biology
Volume 23B, pages 403-453.
Copyright © 1998 by JAI Press Inc.
All right of reproduction in any form reserved.
ISBN: 0-7623-0287-9

I. INTRODUCTION

The vacuolar (H^+)-ATPases (or V-ATPases) function to acidify vacuolar compartments in eukaryotic cells (for reviews, see Forgac, 1989; Forgac, 1992; Bowman et al., 1992; Kane and Stevens, 1992; Gluck, 1992; Sze et al., 1992; Anraku et al., 1992; Kibak et al., 1992; Nelson, 1992). V-ATPases have been identified in a wide variety of intracellular compartments, including clathrin-coated vesicles, endosomes, lysosomes, Golgi-derived vesicles, chromaffin granules, synaptic vesicles, multivesicular bodies, and the central vacuoles of yeast, *Neurospora* and plants.

Acidification of vacuolar compartments plays an important role in a number of intracellular processes (Forgac, 1989), including receptor recycling and receptor-mediated endocytosis, intracellular targeting of newly synthesized lysosomal enzymes from the Golgi to lysosomes, macromolecular processing and degradation in secretory and digestive organelles, and coupled transport of small molecules in various vacuolar compartments. V-ATPases are also present in the plasma membrane of certain specialized cells, including renal intercalated cells (Brown et al., 1987), macrophages, and neutrophils (Swallow et al., 1990), osteoclasts (Chatterjee et al., 1992), and tumor cells (Martinez-Zaguilan et al., 1993), where they function in renal acidification, cytoplasmic neutralization, bone resorption, and metastasis, respectively. V-ATPases in the intestinal epithelium of insects function to drive K^+ transport into the insect midgut (Wieczorek et al., 1991).

To accomplish this broad range of biological functions, the V-ATPases have evolved to both conserve their basic mechanism of ATP-dependent proton transport and to facilitate their selective targeting and regulation in different cellular compartments. Thus, while there appears to be a core of subunits that are common to all V-ATPases, there also exist unique polypeptides as well as isoforms of several of the subunits which allow for the differential targeting and/or regulation of V-ATPases in different cellular membranes. This review is focused on our current knowledge of the structure, function and mechanisms of regulation of this important class of (H^+)-ATPases.

II. FUNCTION OF THE V-ATPases

A. Receptor-Mediated Endocytosis

Ligand-Receptor Dissociation

Receptor-mediated endocytosis is the process by which eukaryotic cells selectively take up specific macromolecules from their extracellular environment (for reviews, see Forgac, 1988; Brown and Goldstein, 1986; Trowbridge et al., 1993). Among the macromolecules internalized by this process are the cholesterol carrier low-density lipoprotein (LDL), the iron-carrying protein transferrin, asialoglyco-proteins, and many hormones and growth factors, including insulin and epidermal growth factor (EGF).

This process begins with the binding of the macromolecule to be internalized to a specific receptor expressed on the cell surface. Ligand-receptor complexes become concentrated in specialized regions of the plasma membrane termed clathrin-coated pits (Pearse and Robinson, 1990). Clathrin is a cytoskeletal complex composed of a 180 kDa heavy chain and two 30-35 kDa light chains which is capable of assembling into a polyhedral matrix at the cell surface (the coated pit) which then pinches off from the plasma membrane in a complex maturation process to give a clathrin-coated vesicle.

For some receptors, such as the LDL receptor, clustering happens even in the absence of bound ligand, while for other receptors, such as the EGF receptor, internalization is activated upon ligand-binding (Forgac, 1988). Clustering of ligand-receptor complexes in clathrin-coated pits occurs as a result of the binding of specific tyrosine-containing internalization signals in the cytoplasmic tail of the internalized receptor to a class of clathrin-associated proteins called adaptors (Pearse and Robinson, 1990; see pages 428-429).

Following the formation of clathrin-coated vesicles, these vesicles rapidly lose their clathrin coat to form uncoated endosomes. Uncoating is catalyzed by a member of the heat shock-70 family of proteins termed the uncoating ATPase (HSC70) (Chappell et al., 1986). Following removal of the clathrin coat, the uncoated endosome then undergoes a series of membrane fusion events which result in the transfer of ligand-receptor complexes to larger and more complex endosomal structures. These fusion events have been extensively studied (Diaz et al., 1988) and appear to share many properties with the fusion between synaptic vesicles and the presynaptic membrane (Sollner et al., 1993).

Ligand-receptor complexes are eventually delivered via these fusion events to an early endosomal compartment termed CURL (for the Compartment of Uncoupling Receptor and Ligand) (Geuze et al., 1983). CURL is a compartment possessing a significant lumenal space surrounded by a membrane containing many tubular projections. As a result of the acidic environment within CURL, most internalized ligands dissociate from their receptors. In the case of transferrin, acidification causes release of iron from transferrin but the apotransferrin remains bound to its receptor (Dautry-Varsat et al., 1983). The ligands released into the lumen of CURL then move on through later endosomal compartments for ultimate delivery to the lysosome. For receptors which undergo recycling, unoccupied receptors cluster in the tubular projections of CURL which pinch off to give recycling vesicles. These recycling vesicles then carry unoccupied receptors back to the plasma membrane. Acidification of CURL by V-ATPases is thus essential for both ligand-receptor dissociation and receptor recycling (Figure 1).

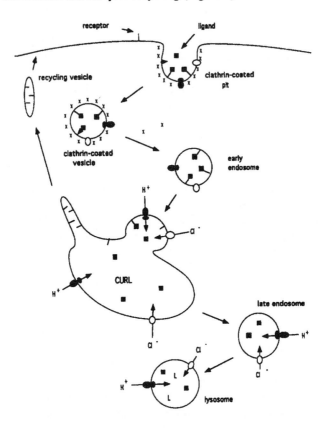

Figure 1. Function of vacuolar acidification in receptor-mediated endocytosis. Following internalization via clathrin-coated pits and coated vesicles, ligand-receptor complexes are delivered to the acidic compartment CURL (for Compartment of Uncoupling Receptor and Ligand) (Geuze et al., 1983) where the low pH activates dissociation of these complexes, thus facilitating recycling of receptors to the cell surface and targeting of ligands to lysosomes. A low pH within endosomes is also required for formation of carrier vesicles that mediate transfer of ligands from endosomes to lysosomes. Vacuolar acidification is carried out by the V-ATPases (?) and requires the activity of a parallel chloride channel (O) to dissipate the membrane potential generated.

Vesicle Budding

That acidification plays additional roles in endocytic membrane traffic is clear from early results that indicated that treatment of cells with weak bases, such as chloroquine or ammonia, which neutralized internally acidic compartments, prevented recycling of receptors, such as the asialoglycoprotein receptor, even in the complete absence of ligand (Schwartz et al., 1984). Internalized receptors thus become trapped within endosomal compartments on treatment with these agents. In addition, weak bases also block the movement of internalized ligands from endosomes to lysosomes (Harford et al., 1983).

Insight into the role that endosomal acidification is playing in these cases has come from *in vitro* studies of endosomal budding and fusion. Thus, although fusion of endosomal compartments with each other *in vitro* occurs independently of acidification (Diaz et al. 1988), formation of carrier vesicles which carry the luminal contents from early to late endosomes is blocked by bafilomycin (Clague et al., 1994), a specific inhibitor of the V-ATPases (Bowman, E.J. et al., 1988a). It is thus possible that the budding of recycling vesicles from CURL which carry unoccupied receptors back to the cell surface is also dependent upon an acidic lumenal pH.

Viral and Toxin Entry

Endosomal acidification also provides a pathway for the entry of certain envelope viruses and bacterial toxins. Envelope viruses such as influenza and Semliki forest virus enter the cell by receptor-mediated endocytosis. Upon delivery to an acidic intracellular compartment, proteins in the coats of these viruses mediate fusion between the viral membrane and the endosomal membrane, thus releasing the encapsulated nucleic acid into the cytoplasm of the target cell (Marsh et al., 1983). The influenza hemagglutinin accomplishes this by undergoing a major, acid-dependent conformational change which exposes hydrophobic regions of the hemagglutinin molecule that insert in the endosomal membrane (White, 1992). Infection by these viruses can thus be blocked by weak bases which prevent release of nucleic acid from the viral capsid.

Certain toxins also use the acidic endosomal environment to gain access to the cytoplasm (Sandvig and Olsnes, 1980). Diphtheria toxin is an A-B toxin in which the A chain blocks protein synthesis through ADP-ribosylation. Entry of the A chain requires the presence of the B chain which, on exposure to low pH, forms a channel through which the A chain can pass. Toxin entry is blocked by incubation of cells with weak bases. Thus, both envelope viruses and toxins utilize endosomal acidification to gain entry to the cell.

B.   Intracellular Membrane Traffic

Mannose-6-phosphate Receptor System

The targeting of newly synthesized lysosomal enzymes from the Golgi to lysosomes is also dependent upon vacuolar acidification. Lysosomal enzymes bind to a receptor in the trans-Golgi which recognizes an exposed mannose-6-phosphate (Man-6-P) recognition marker (Kornfeld, 1992; Sahagian and Novikoff, 1994). The Man-6-P receptor:lysosomal enzyme complexes then cluster in clathrin-coated pits in the trans-Golgi through interaction with the AP-1 adaptor complex (Pearse and Robinson, 1990). Following pinching off of the coated pits to form coated vesicles and removal of the clathrin coat, the Man-6-P receptor:lysosomal enzyme complexes are delivered to an acidic prelysosomal compartment. The low pH within this compartment (which appears to correspond to a late endosomal compartment) activates dissociation of lysosomal enzymes from Man-6-P receptors (Creek and Sly, 1984). The lysosomal enzymes which are released into the lumen are then targeted to lysosomes, while the Man-6-P receptors are recycled to the trans-Golgi for additional rounds of lysosomal enzyme targeting.

Neutralization of this acidic uncoupling compartment with weak bases prevents dissociation of lysosomal enzymes from Man-6-P receptors, resulting in saturation of the available receptors and secretion of unbound lysosomal enzymes via a constitutive secretory pathway (Creek and Sly, 1984). Thus in both receptor-mediated endocytosis and intracellular targeting of lysosomal enzymes, a low vacuolar pH is used as a signal to activate ligand-receptor dissociation and facilitate receptor recycling (Figure 2).

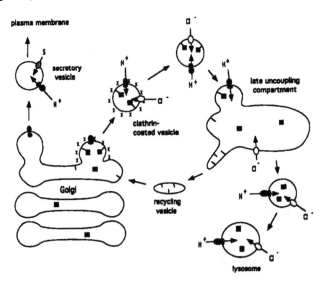

Figure 2. Function of vacuolar acidification in intracellular membrane traffic. Newly synthesized lysosomal enzymes (■) bind to the Man-6-P receptor (I) in the trans-Golgi. Following their delivery to a late uncoupling compartment via clathrin-coated pits and coated vesicles, the low pH within the uncoupling compartment causes dissociation of the complex, thus facilitating recycling of Man-6-P receptors to the trans-Golgi and delivery of lysosomal enzymes to lysosomes.

Vacuolar Targeting in Yeast

Yeast also selectively target newly synthesized, soluble hydrolases from the Golgi to the central vacuole, but do not employ the Man-6-P receptor system. Instead, recognition occurs by an as yet unidentified receptor in the trans-Golgi. Nevertheless, yeast depend upon an acidic pH within a post-Golgi, prevacuolar compartment for efficient delivery of soluble vacuolar hydrolases to the vacuole. Thus, in yeast mutants which are defective in genes encoding subunits of the V-ATPase, a significant mistargeting of soluble vacuolar hydrolases is observed (Rothman et al., 1989; Klionsky et al., 1992). Missorting of membrane bound vacuolar proteins is also observed but appears to require more prolonged neutralization of vacuolar compartments (Morano and Klionsky, 1994).

Other Functions in the Golgi

Vacuolar acidification also plays a role in targeting of secreted proteins to secretory vesicles. Thus, treatment with weak bases blocks sorting of prohormones to vesicles involved in regulated secretion (Moore et al., 1983), most likely in a late or post-Golgi compartment. Acidification of the trans-Golgi is also essential for the correct processing of glycoproteins. Thus, bafilomycin blocks maturation of viral glycoproteins in the trans-Golgi (Palokangas et al. 1994), and cells expressing the cystic fibrosis phenotype show defective acidification of the trans-Golgi and abberant sialylation of proteins (Barasch et al., 1991). Movement of glycoproteins through the constitutive secretory pathway in HepG2 cells is also blocked in the trans-Golgi by concanamycin, a bafilomycin analog (Yilla et al., 1993). Interestingly, bafilomycin has been shown to increase secretion of the amyloid beta protein in kidney cells transfected with the beta-amyloid precursor protein (Haass et al., 1995), although the significance of this result for the study of Alzheimer's disease remains to be determined.

Although the cis and medial Golgi are not acidic by the criteria of labeling with DAMP (Anderson and Orci, 1988), it was found that interaction of the KDEL receptor with resident ER proteins in the cis Golgi (which is necessary for return of these proteins to the ER) is strongly pH-dependent in the range 5.5-7.0 (Wilson et al., 1993). It is thus possible that subtle differences in the pH of the ER and cis-Golgi might allow this receptor to bind ligands in the cis-Golgi and release them in the ER. It should also be noted that even in compartments that are not acidic, the V-ATPases may play a role by virtue of the fact that they are electrogenic and able to establish a membrane potential in the absence of a compensating anion conductance.

C. Functions in Lysosomes and Central Vacuoles

Both secondary lysosomes in animal cells and the central vacuoles of plants, yeast, and *Neurospora* contain a variety of digestive enzymes which are responsible for the breakdown of macromolecules. These digestive enzymes, including many proteases, such as the cathepsins, have a low pH optima, and therefore require V-ATPases to acidify the vacuolar interior in order to function. Treatment of cells with weak bases or bafilomycin therefore results in defective breakdown and accumulation of undegraded macromolecules within these organelles.

Lysosomes also contain a variety of coupled transport systems that are responsible for the release of macromolecular degradation products. Thus, there are several systems which couple efflux of amino acids to the proton gradient or the membrane potential (interior positive) generated by the V-ATPase (Pisoni et al., 1985). The absence of particular transporters results in intralysosomal accumulation of the solute and lysosomal storage diseases, such as cystinosis, which occurs in patients lacking the lysosomal transporter for cystine (Pisoni et al., 1985).

In addition to lysosomes, phagosomes (Lukacs et al., 1990) and multivesicular bodies (van Dyke et al., 1985) also contain V-ATPases which acidify their interior and activate hydrolases within their lumen. In the case of phagosomes in *Dictyostelium discoideum*, acidification appears to be the consequence of fusion with organelles, termed acidisomes, which contain a high density of V-ATPases (Nolta et al., 1991).

The central vacuoles of plants, yeast and *Neurospora* also function as storage compartments and contain transport systems that couple the uptake of various solutes to the proton motive force generated by the V-ATPase (Sze et al., 1992; Anraku et al., 1992; Bowman, B.J. et al., 1992). Thus a Ca^{2+}/H^+ antiporter in the vacuolar membrane of plants and yeast couples Ca^{2+} uptake to the vacuolar proton gradient, thereby maintaining a low cytoplasmic Ca^{2+} concentration. Yeast mutants lacking a functional V-ATPase become hypersensitive to extracellular Ca^{2+} because of their inability to maintain a low cytosolic Ca^{2+} concentration by employing this pathway (Anraku et al., 1992). Coupled vacuolar transporters also exist for a variety of other solutes, including phosphate, malate, amino acids, and sugars (Sze et al., 1992). This ability to store large quantities of low molecular weight solutes also imparts to the central vacuole an important role in osmotic regulation.

D. Functions in Secretory Organelles

Secretory organelles such as chromaffin granules and synaptic vesicles must store high concentrations of low-molecular-weight solutes which are released upon fusion with the plasma membrane. Uptake of these solutes is coupled via specific transport proteins to the proton gradient or membrane potential generated by the V-ATPases. In chromaffin granules, an electrogenic catecholamine/H^+ antiporter couples the uptake of catecholamines to both the pH gradient and the membrane potential (Johnson et al., 1979), while the interior positive membrane potential serves to drive the influx of negatively charged nucleotides (Aberer et al., 1978).

In synaptic vesicles, accumulation of some neurotransmitters, such as glutamate, is driven solely by the membrane potential (Moriyama et al., 1990), while uptake of other transmitters, such as norepinephrine, is coupled to both the membrane potential and the pH gradient via electrogenic amine/H^+ antiporters which have a solute:proton stoichiometry of 2 (Rudnick and Clark, 1993). Agents which dissipate the protonmotive force across the synaptic vesicle membrane, such as the uncoupler CCCP, thus block uptake of neurotransmitters. Certain neuron blockers, such as chlorpromazine and propranonol, also block transmitter uptake by accumulation within synaptic vesicles and dissipation of the pH gradient (Moriyama et al., 1993).

Secretory vesicles involved in storage and secretion of peptide hormones, such as insulin, contain proteases responsible for prohormone processing. Because these proteases have a low pH optima, processing requires acidification of the secretory vesicles by V-ATPases (Rhodes et al., 1987). Thus vacuolar acidification plays multiple roles in the biogenesis of secretory vesicles.

E. Functions in the Plasma Membrane

Renal Intercalated Cells

The V-ATPases play an important role in renal acidification (Figure 3). In the collecting duct of the mammalian kidney, intercalated cells contain a high density of V-ATPases in their apical membrane oriented so as to carry out proton transport into the extracellular space (Brown et al., 1987). The V-ATPases in this membrane are present at such high density that they virtually form a two-dimensional lattice. V-ATPases have also been identified in the plasma membrane of other cells in mammalian kidney, including brush border microvilli in the proximal tubule, and in the thick ascending limb and the distal convoluted tubule (Gluck, 1992).

In the mitochondria-rich cells of the turtle bladder, which are closely related to renal intercalated cells, the density of V-ATPases is controlled by exocytic fusion of intracellular vesicles with the apical plasma membrane (Gluck et al., 1982). Thus, an increase in the cytoplasmic CO_2 concentration induces insertion of intracellular V-ATPases into the apical membrane, resulting in increased acid secretion from the cell. Cytoplasmic alkalinization results in rapid reversal of this process by reinternalization of the apical V-ATPases.

Macrophages and Neutrophils

Macrophages also contain constitutively active V-ATPases at the plasma membrane which pump protons out of the cytoplasm (Swallow et al., 1990)(Figure 3). This was demonstrated by measurement of a bafilomycin-sensitive proton efflux which occurred in acid-loaded cells incubated in the absence of Na^+ or bicarbonate. The function of the V-ATPase in the macrophage plasma membrane is uncertain, but it may assist in neutralizing the cytoplasm under conditions of significant acid load, as in the local environment of an infection or tumor. V-ATPases have also

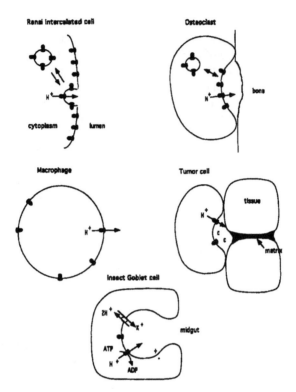

Figure 3. Function of V-ATPases in the plasma membrane of specialized cells. V-ATPases function in renal acidification by intercalated cells, in bone resorption by osteoclasts, in cytosolic alkalinization in macrophages and neutrophils, in metastasis by tumor cells and in K^+ transport in the insect midgut. In renal intercalated cells, the density of V-ATPases in the apical membrane is rapidly controlled by reversible exocytosis/endocytosis of pumps. In both osteoclasts and tumor cells, V-ATPases acidify a localized, sealed region of the extracellular environment that is required for the activity of degradative enzymes. In insect cells, V-ATPases create a membrane potential that drives K^+ transport into the midgut via an electrogenic K^+/H^+ antiporter

been demonstrated in the plasma membrane of neutrophils (Nanda et al., 1992), although the constitutive level of V-ATPase activity in the neutrophil plasma membrane is much lower than in macrophages. Activation of neutrophils with phorbol esters, however, causes a significant increase in V-ATPase-dependent proton transport across the plasma membrane, an effect dependent upon the activity of protein kinase C. Because neutrophils undergo a large respiratory burst and subsequent cytoplasmic acidification when activated in response to an infection, expression of V-ATPases at the plasma membrane may be necessary to maintain a neutral cytosolic pH.

Osteoclasts

Osteoclasts are responsible for bone resorption. These cells adhere to the surface of bone and seal off a portion of the extracellular space into which they secrete enzymes that degrade the bone matrix. Bone resorption requires an acidic pH, both for the activity of the secreted hydrolases and to assist in dissolving the hydroxyapatite matrix. Acidification of this extracellular space is accomplished through the activity of V-ATPases that are targeted to the domain of the plasma membrane surrounding the extracellular compartment (Blair et al., 1989)(Figure 3). Inhibition of the osteoclast V-ATPase with bafilomycin would therefore be predicted to inhibit bone resorption.. The possibility of selectively inhibiting the plasma membrane V-ATPase is of considerable interest in the treatment of osteoporosis.

In addition to its plasma membrane localization, the osteoclast V-ATPase is also distinguished from other V-ATPases by its sensitivity to vanadate (Chatterjee et al., 1992), an inhibitor of the P-type ATPases (Pedersen and Carafoli, 1987). The osteoclast V-ATPase may also express a unique isoform of the catalytic A subunit that could account for its novel inhibitor sensitivity (Hernando et al., 1995).

Tumor Cells

Certain tumor cells also appear to express V-ATPases at the plasma membrane (Martinez-Zaguilan et al., 1993), where they may serve a number of possible functions. One likely function is in tumor invasion (Figure 3). Thus, metastatic cells are able to invade tissues by virtue of their ability to secrete digestive enzymes into their environment which break down extracellular matrix proteins, thus facilitating their entry into the target tissue. A number of transformed cell lines secrete lysosomal proteases, including cathepsin L (Dong et al., 1989), which require a low pH for optimal activity. Gillies and coworkers have observed a correlation between metastatic potential and the level of bafilomycin sensitive proton transport across the plasma membrane (Martinez-Zaguilan et al., 1993). V-ATPases targeted to the plasma membrane may thus facilitate metastasis by acidification of the extracellular environment.

A second possible function of plasma membrane V-ATPases in transformed cells is in alkalinization of the cytoplasm. Transformation of cells is often accompanied by an increase in the cytoplasmic pH. Although this increase is most often due to increased activity of the Na^+/H^+ antiporter (Schuldiner and Rozengurt, 1982), transformation of cells has also been achieved by expression of a P-type proton ATPase at the plasma membrane (Perona and Serano, 1988), suggesting the possibility that expression of functional V-type proton pumps at the cell surface may contribute to transformation. How V-ATPases might be expressed at the surface of tumor cells is unknown, but the observation that ras transformed cells show decreased lysosomal acidification (Jiang et al., 1990), suggests the possibility that intracellular V-ATPases may be retargeted to the plasma membrane, leaving the vacuolar compartments less acidic.

Insect Midgut

The V-ATPases play a unique role in the insect midgut (Figure 3). Located in the apical membrane of the goblet cells which line the midgut, the V-ATPases are oriented to pump protons into the midgut lumen (Wieczorek et al., 1991). The lumenal pH, however, is alkaline rather than acidic. This is the result of the presence of an electrogenic K^+/H^+ antiporter and the creation, by the V-ATPase, of a large (250 mV) membrane potential, lumenal positive. This membrane potential results from the unidirectionl flux of protons into the midgut lumen driven by the V-ATPase and the absence of anion channels to dissipate this potential. Because the K^+/H^+ antiporter has a H^+/K^+ stoichiometry of greater than one, this potential drives protons out of the midgut lumen and K^+ in, resulting in net potassium secretion into the midgut.

III. TRANSPORT AND INHIBITOR PROPERTIES OF THE V-ATPases

The V-ATPases catalyze unidirectional proton transport from the cytoplasmic to the lumenal side of the membrane uncoupled to the movement of other cations (Forgac and Cantley, 1984). As a result, the V-ATPases are electrogenic, creating a membrane potential which is lumenal positive (Forgac et al., 1983). In order for significant acidification to occur, this membrane potential must be dissipated by the movement of other ions. *In vivo*, this is most often accomplished through the activity of a chloride channel, which allows for chloride flux in response to the positive interior membrane potential (Glickman et al., 1983; Xie et al., 1983; Arai et al., 1989). Because of the dependence of acidification on chloride conductance, regulation of this chloride channel represents an important potential mechanism for controlling vacuolar acidification (see page 442).

Although the V-ATPases resemble the F-ATPases in being unidirectional ATP-driven proton pumps, the V-ATPases have not been shown to carry out the reverse reaction, namely proton gradient-driven ATP synthesis, which is the primary funcion of the F-ATPases in mitochondria, chloroplasts and bacteria (Senior, 1990; Penefsky and Cross, 1991; Pedersen and Amzel, 1993). This may in part be related to the lower H^+/ATP stoichiometry of the V-ATPases. Thus, while the F-ATPases have a H^+/ATP stoichiometry of 3-4 (Berry and Hinkle, 1983), measurements on the V-ATPase in chromaffin granules suggest a H^+/ATP stoichiometry of 2 (Johnson et al., 1982). This lower stoichiometry favors proton pumping since more energy is available from ATP hydrolysis to pump each proton. Other properties of the V-ATPases may also help to insure that this enzyme operates only in the direction of ATP-driven proton transport.

Early studies on the V-ATPases revealed a distinct inhibitor profile which distinguished the V-ATPases from both the F-ATPases and the P-ATPases (Pedersen and

Carafoli, 1987; Forgac, 1989). Unlike the P-ATPases, such as the Na^+-K^+-ATPase and the Ca^{2+}-ATPase, the V-ATPases are not sensitive to vanadate, a transition state analog for these enzymes. This is consistent with the absence of a phosphorylated intermediate during the catalytic cycle of the V-ATPases (Forgac and Cantley, 1984). As noted above, however, the osteoclast V-ATPase has been reported to be vanadate sensitive (Chatterjee et al., 1992). The V-ATPases can also be distinguished from the F-ATPases by the insensitivity of the former to oligomycin and aurovertin and their sensitivity to sulfhydryl reagents, such as N-ethylmaleimide (NEM) (Forgac, 1989). The V-ATPases are also more sensitive to NBD-Cl, a reagent which reacts with a key tyrosine residue in the F-ATPases (Senior, 1990; Penefsky and Cross, 1991) but which inhibits the V-ATPases by reaction with the same cysteine residue that is responsible for sensitivity to NEM (Arai et al., 1987b).

By far the most specific inhibitor of the V-ATPases is bafilomycin, a macrolide antibiotic isolated from *Streptomyces griseus* (Bowman, E.J. et al., 1988a). Bafilomycin A_1 inhibits V-ATPases with a K_i of 1-10 nM, three to four orders of magnitude more potently than it inhibits the P-ATPases. Concanamycin, a structurally related macrolide, is an even more potent inhibitor of the V-ATPases, with a K_i of approximately 0.1 nM (Drose et al., 1993). While bafilomycin has been used to inhibit V-ATPase activity in intact cells (Yoshimori et al., 1991), the concentrations required are somewhat higher, generally in the range of 0.1-1.0 μM, presumably due to the lipophilic nature of bafilomycin that results in its adsorption onto cellular membranes. Information concerning the location of the bafilomycin binding site on the V-ATPase complex has begun to emerge (see pages 431-432).

IV. STRUCTURE AND SUBUNIT FUNCTION OF THE V-ATPases

A. Structural Model

Our current model for the structure of the V-ATPase complex, which incorporates only those subunits observed for virtually all V-ATPases, is shown in Figure 4. The V-ATPase is a 750 kDa macromolecular complex composed of 10 subunits organized into two structurally and functionally distinct domains. The peripheral V_1 domain, which is located on the cytoplasmic side of the membrane and has a calculated molecular weight of 500 kDa, has the structure $A_3B_3C_1D_1E_1F_1$, and is responsible for nucleotide-binding and hydrolysis. The catalytic nucleotide-binding sites are located on the 70 kDa A subunits while additional nucleotide binding sites are located on the 60 kDa B subunits. The accessory subunits of molecular mass 40(C), 34(D), 33(E), and 14(F) kDa function to attach V_1 to V_0 and to couple ATP hydrolysis to proton translocation. The integral V_0 domain, which has a calculated molecular weight of 250 kDa, has the structure $100_1 38_1 19_1 c_6$, and is responsible for proton translocation, with the 17 kDa c subunit playing an important role in the for-

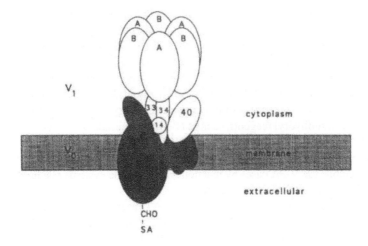

Figure 4. Structural model of the V-ATPase. The 500 kDa peripheral V_1 domain has the structure $A_3B_340_134_133_114_1$ (Arai et al., 1988) and is responsible for ATP hydrolysis, with the nucleotide binding sites located on the 70 kDa A subunits (catalytic) and the 60 kDa B subunits (noncatalytic). The 250 kDa integral V_0 domain is responsible for proton transport and has the structure $100_138_119_1c_6$, with the 17 kDa c subunit participating directly in proton translocation. Unlike the case with the homologous F-ATPases, the separate V_1 and V_0 domains are not independently active (Puopolo et al., 1990; Zhang, J. et al., 1992).

mation of a proton pore. Differences in subunit composition observed between V-ATPases from different sources will be discussed below.

As can be seen, the V-ATPases are remarkably similar in overall structure to the F-ATPases (Senior, 1990; Penefsky and Cross, 1991; Pedersen and Amzel, 1993). Thus, in both cases, a hexameric complex of nucleotide binding subunits together with several accessory polypeptides makes up the peripheral hydrolytic domain (V_1 or F_1) which is attached to an integral domain (V_0 or F_0) which contains multiple copies of a DCCD-reactive proteolipid (subunit c). As discussed below, the evolutionary relationship between the V and F-ATPases is supported by sequence homology between the nucleotide binding subunits (A and B of the V-ATPases and alpha and beta of the F-ATPases) as well as the DCCD-reactive c subunit.

The publication of the 2.8 Å crystal structure of the bovine heart mitochondrial F_1 (Abrahams et al., 1994) represents a major advance in our understanding of both the F-ATPases and the related V-ATPases. The F_1 structure, in which the alpha, beta, and most of the gamma subunits can be localized, clearly shows an alternating hexameric arrangement of alpha and beta subunits with the six nucleotide-binding sites located near the interface of these subunits. The gamma subunit, which is an extended alpha helix, runs through the center of this hexameric complex and presumably makes contact at its distal end with the remaining F_1 subunits or with the F_0

domain. Interestingly, the top portion of the gamma subunit comes into closer contact with one alpha-beta subunit pair than with the others, and the three beta subunits show different nucleotide occupancy and significant structural differences that indicate an asymmetry to the F1 complex. These observations have been interpreted as support for a binding change mechanism in which each catalytic nucleotide-binding site is sequentially occupied by ATP, ADP, or no nucleotide during the catalytic cycle of the F-ATPase (Boyer, 1993). The structure of the F-ATPase nucleotide binding sites is discussed in more detail below.

Further support for the similarity in overall structure of the F and V-ATPases comes from electron micoscopic examination of purified preparations of the V-ATPases which display a ball and stalk structure reminiscent of the F-ATPases (Dschida and Bowman, 1992). There are, however, important differences between these structures, particularly in the existence of projections which originate near the base of the stalk in the V-ATPases.

B. Subunit Composition

Shown in Table 1 is the subunit composition of V-ATPases purified from various sources together with the yeast genes encoding the corresponding polypeptides. As can be seen, the V-ATPases share a common overall structure with a number of polypeptides that may be unique to certain members of the V-ATPase class. Thus all V-ATPases thus far identified include the two nucleotide-binding subunits of approximate molecular weight 70 and 60 kDa, termed the A and B subunits, respectively, as well as the 17 kDa c subunit which forms part of the V_0 domain. In addition, all V-ATPases possess two peripheral subunits of approximately 40 kDa (subunit C) and 33 kDa (subunit E) as well as a V_0 subunit of approximately 38 kDa (except for the beet V-ATPase).

Subunits which are shared by nearly all V-ATPases include V_1 proteins around 34 kDa (subunit D) and 14 kDa (subunit F) and a V_0 subunit around 100 kDa. A powerful tool in identifying functionally important V-ATPase subunits has come from studies of the yeast V-ATPase. Thus, yeast mutants unable to assemble a functional V-ATPase show a conditional lethal phenotype in which the mutant cells are unable to grow at neutral pH but are able to grow at acid pH (5.5) (Nelson and Nelson, 1990). Interestingly, V-ATPase mutants that are unable to carry out endocytosis are not viable, even at acid pH (Munn and Riezman, 1994). V-ATPase mutants are also hypersensitive to Ca^{2+} in the medium (Anraku et al., 1992), due to the function of the vacuole in maintaining a low cytosolic Ca^{2+} concentration.

In addition to the common V-ATPase subunits, there are some polypeptides that may be unique to certain V-ATPases. Thus a 50 kDa polypeptide, which corresponds to the 50 kDa subunit of the AP-2 adaptor complex, has thus far been identified only in the V-ATPase from clathrin-coated vesicles (Myers and Forgac, 1993b), while a 54 kDa protein (not homologous to AP50) is essential for the yeast V-ATPase (Ho et al., 1993b) but has not been identified in animal cells. Similarly, a

Table 1. Subunit Composition of V-ATPases

Domain	Code	Bovine Coated Vesicles	Bovine Kidney Micro.	Bovine Chrom. Granule	Plant Vacuole (Oat)	Plant Vacuole (Beet)	Neuro. Vacuole	Yeast Vacuole	Yeast Gene
V1	A	73	70	72	70	67	67	69	vma1
	B	58	56	57	60	55	57	57	vma2
	54	-	53(?)	-	44	52	51	54	vma13
	AP-50	50	50(?)	-	-	-	-	-	-
	C	40	42	41	42	44	48	42	vma5
	D	34	33	34	36	-	-	32	vma8
	E	33	31	33	29	32	30	27	vma4
	F	14	12	-	12	-	16	14	vma7
V0		100	100	-	115	-	100	100	95	vph1/stv1
		45	45	45	45	-	-	-	-	-
		38	38	38	39	32	-	40	36	vma6
		19	19	15	20	-	-	-	-	-
	c	17	14	16	16	16	16	17	vma3

Note: * V-ATPase preparations were isolated from bovine brain coated vesicles (Arai, et al., 1987b), bovine kidney microsomes (Gluck and Caldwell, 1987), bovine chromaffin granules (Moriyama and Nelson, 1989), oat vacuoles (Ward and Sze, 1991), beet vacuoles (Parry et al., 1989), Neurospora vacuoles (Bowman, B.J. et al., 1989) and yeast vacuoles (Kane et al., 1989). Yeast vma genes were identified as described in the text.

45 kDa V_0 subunit has been observed only in V-ATPase preparations from animal cells (Supek et al., 1994a). It is also possible that in certain cases subunits of similar molecular weights may be encoded by unrelated genes. The information that is currently available concerning the structure and function of each of the V-ATPase subunits is discussed below.

C. Domain Structure and Function

The V_1 Domain

The peripheral V_1 domain contains five to eight subunits (depending upon the source) which form a soluble complex of molecular mass 500-600 kDa. For the V-ATPase from clathrin-coated vesicles (Arai et al., 1987b), quantitative amino acid analysis following SDS-PAGE and transfer to Immobilon indicate that the V_1 domain contains three copies of the 73 kDa A subunit, three copies of the 58 kDa B subunit and single copies of polypeptides of molecular mass 50, 40(C), 34(D) and

33(E) kDa (Arai et al., 1988). A 14 kDa polypeptide (subunit F), first identified in insects and yeast (Graf et al., 1994; Graham et al., 1994; Nelson, H. et al., 1994), and a 54 kDa polypeptide (the product of the vma13 gene in yeast), (Ho et al., 1993b) are also present in many V-ATPase preparations.

The V_1 subunits have been shown to be peripheral by their removal from the membrane in the absence of detergents using chaotropic agents such as KI and KNO_3 (Arai et al., 1989; Moriyama and Nelson, 1989; Adachi et al., 1990b). This dissociation is activated in the presence of ATP at a site which has an affinity of 200 nM (Arai et al., 1989). Because intracellular organelles, such as clathrin-coated vesicles, are oriented with the cytoplasmic surface exposed, these results indicate that the V_1 domain is present on the cytoplasmic surface of the membrane. Further data supporting this idea comes from studies of the coated vesicle V-ATPase in which it was demonstrated that all of the V1 subunits can be labeled in intact coated vesicles with membrane impermeant reagents, such as [^{125}I]-sulfo-SHPP (Arai et al., 1988). In addition, both the A and B subunits are cleaved in intact vesicles by trypsin, indicating their exposure on the surface of the vesicles (Adachi et al., 1990a).

Following removal of the V_1 subunits with chaotropic agents, it has been shown that they can be reassembled *in vitro* onto the V_0 sector to give a functional V-ATPase (Puopolo and Forgac, 1990; Ward et al., 1991; Puopolo et al., 1992b). In the absence of the V_0 domain, the V_1 subunits from the coated vesicle V-ATPase are able to assemble into a V_1 subcomplex which contains the A, B, 50, D, and E subunits but which lacks the C subunit (Puopolo et al., 1992b). Unlike the corresponding F_1 domain, which retains its ability to hydrolyze MgATP, the reassembled V_1 domain is unable to hydrolyze MgATP (Puopolo and Forgac, 1990; Puopolo et al., 1992b), although some Ca^{2+}-ATPase activity of the V_1 domain has been reported (Xie and Stone, 1988). Because CaATP is unable to support proton transport by the intact V-ATPase and because the CaATPase activity is not inhibited by bafilomycin, the significance of this activity remains uncertain.

Although unable to hydrolyze MgATP, the V_1 domain does contain all of the nucleotide binding sites of the V-ATPase, which are located on the A and B subunits (see below). Covalent crosslinking studies using the reversible cross-linking reagent DTSSP indicate that there is extensive contact between the A and B subunits (Adachi et al., 1990b), consistent with the crystal structure of F_1 showing a hexameric arrangement of alternating alpha and beta subunits (Abrahams et al., 1994). In addition, the C, D and E subunits of V_1 all appear to make contact with the V_0 c subunit (Adachi et al., 1990b), suggesting that these V_1 polypeptides serve to bridge the peripheral V_1 domain with the integral V_0 domain.

The V_0 Domain

The V_0 domain is a 250-300 kDa integral complex which contains three to five subunits and is responsible for proton translocation across the membrane. In clathrin-coated vesicles, the V_0 domain contains five subunits of molecular weight 100, 45, 38, 19, and 17 (c) kDa (Arai et al., 1987b), which are present at a stoichiometry of six copies of the c subunit and single copies of the remaining polypeptides (Arai et al., 1988). The 45 kDa polypeptide, recently cloned from adrenal medulla (Supek et al., 1994a), was not originally identified as part of the coated vesicle V-ATPase because of its extremely diffuse appearance and poor visualization by silver staining, properties that are likely attributable to its glycosylation.

Following removal of the V_1 domain and solubilization of the membrane, the V_0 domain remains together as a complex which sediments with an apparent molecular mass of approximately 250 kDa (Zhang, J. et al., 1992). Protease digestion in intact coated vesicles indicates that both the 100 and 38 kDa subunits are exposed on the cytoplasmic surface of the membrane (Adachi et al., 1990a) while labeling with membrane impermeant reagents indicate that 100, 19, and 17 kDa subunits all possess significant lumenal domains (Arai et al., 1988).

Although the arrangement of subunits in the V_0 domain is not clear, some information has emerged concerning the structure of the homologous F_0 domain. For the $E.coli$ F_0, which has the structure ab_2c_{10-12} (Schneider and Altendorf, 1987; Fillingame, 1992), the highly hydrophobic c subunits appear to be arranged in a ring in the plane of the membrane with the a and b subunits to one side (Birkenhager et al., 1995).

One important difference between F_0 and V_0 concerns their proton channel activities. Thus, while the free F_0 domain is known to form a functional proton pore (Schneider and Altendorf, 1987; Fillingame, 1992), the homologous V_0 domain does not form an open proton channel (Zhang, J. et al., 1992). This was demonstrated both for the native V_0 domain, which is present in approximately equimolar amounts with the intact V-ATPase complex in native clathrin-coated vesicles, and also for the isolated reconstituted V_0 domain. These studies demonstrated that V_0 is unable to passively conduct protons from the cytoplasmic to the lumenal side of the membrane, but studies of the V_0 domain in plant vacuoles have demonstrated that V_0 is also unable to conduct protons in the lumenal to cytoplasmic direction (Ward and Sze, 1991).

Although native V_0 is not an open proton channel, the V_0 domain does possess the information necessary to form a functional proton channel. Thus, dissociation and separation of the V_0 subunits by gel filtration followed by reassembly and reconstitution demonstrated that the reassembled V_0 domain is able to catalyze a DCCD-inhibitable proton conductance (Zhang, J. et al., 1994). This result has been confirmed in studies in which V_0 has been acid treated, resulting in a passive proton conductance (Crider et al., 1994). These studies suggest that there may be some factor removed by gel filtration or acid treatment which controls proton conductance through the V_0 domain. Identification of such a factor may have important implications for regulation of vacuolar acidification in $vivo$.

In addition to carrying out proton conductance, the V_0 domain has also been shown to function as the binding site for bafilomycin. This was first demonstrated using chromaffin granule membranes from which the V_1 domain had been removed which were shown to protect the intact V-ATPase from inhibition by bafilomycin (Hanada et al., 1990). More recent studies have demonstrated that isolated, reconstituted V_0 is also capable of binding bafilomycin and have identified the binding site for bafilomycin in the V_0 complex (Zhang, J. et al., 1994). Proton conductance through acid-treated V_0 has also been reported to be bafilomycin sensitive (Crider et al., 1994). A discussion of the role of individual V_0 subunits in proton conductance and bafilomycin binding is presented in later sections.

D. Structure and Function of Nucleotide Binding A and B Subunits

The A Subunit

Evidence that the subunit of approximate molecular mass 70 kDa (subunit A) was essential for activity initially came from studies demonstrating that this polypeptide was labeled by reagents, such as [^{3}H]NEM and [^{14}C]NBD-Cl, which inhibited V-ATPase activity (Bowman, E.J. et al., 1986; Mandala and Taiz, 1986; Arai et al., 1987b; Randall and Sze, 1987; Uchida et al., 1988). Moreover, it was shown that by inhibition of activity and labeling was prevented in the presence of ATP, suggesting that the A subunit possessed a nucleotide-binding site essential for activity. Additional evidence for the participation of the A subunit in nucleotide binding comes from proteolysis studies demonstrating that tryptic cleavage of the A subunit at a site 2 kDa from the amino terminus is inhibited in the presence of the nucleotide analog TNP-ATP (Adachi et al., 1990a) and by the ability of the A subunit to be labeled using 2-azido[^{32}P]ATP (Zhang, J. et al., 1995) and [^{32}P]ATP (Moriyama and Nelson, 1987) following UV irradiation.

Further evidence for the presence of a nucleotide-binding site on the A subunit was derived from the amino acid sequence of the A subunit as deduced from the cDNA sequence (Zimniak et al., 1988; Bowman, E.J. et al., 1988b; Hirata et al., 1990; Puopolo et al., 1991). These results indicated that the A subunit has significant sequence homology with other nucleotide binding proteins, including the alpha and beta subunits of the F-ATPases (Walker et al., 1985), suggesting a common evolutionary origin for these proteins. Sequence homology is highest in certain conserved regions, termed the Walker consensus sequences (Walker et al., 1985), which have been identified from labeling and mutagenesis studies of the F-ATPases (Senior, 1990; Penefsky and Cross, 1991; Futai et al., 1994) to be essential for activity.

The first studies to identify residues located at the catalytic nucleotide binding site of the V-ATPases were those aimed at localizing the cysteine residue(s) responsible for the sensitivity of the V-ATPases to sulfhydryl reagents, such as NEM. Feng and Forgac (1992a) demonstrated that like NEM, cystine also inhibited the V-ATPases in an ATP-protectable manner, but unlike NEM, this inhibition was reversible upon treatment with DTT. Moreover, prereaction with cystine prevented the irreversible inhibition of the V-ATPases by NEM, indicating that cystine and NEM were reacting with the same essential cysteine residue. Cystine was shown to react through thio-disulfide exchange with a single cysteine residue, Cys254, of the A subunit (Feng and Forgac, 1992a). Cys254 is located in the highly conserved glycine rich loop sequence GXGKTV which forms part of the Walker consensus A sequence (Walker et al., 1985). From the crystal structure of the F_1 beta subunit, the glycine rich loop is in close contact with the triphosphates of ATP bound at the catalytic site, and thus may participate in ATP hydrolysis (Abrahams et al., 1994). The importance of this region for catalysis is supported by extensive mutagenesis data on the beta subunit of the *E. coli* F-ATPase (Senior, 1990; Penefsky and Cross, 1991; Futai et al., 1994).

Further work has demonstrated that Cys254 is capable of forming a disulfide bond with a second cysteine residue, Cys532, located in the same A subunit (Feng and Forgac, 1994). Formation of this disulfide bond results in DTT-reversible inhibition of V-ATPase activity . The ability of Cys254 and Cys532 to form a disulfide bond indicates that they must be within 5-6 Å of each other in the tertiary structure of the protein. The homologous regions of the F-ATPase beta subunit do in fact appear to be separated by approximately this distance in the crystal structure of the mitochondrial F-ATPase (Abrahams et al., 1994). The relative positions of Cys254 and Cys532 predicted from the F_1 crystal structure are shown in Figure 5. Interestingly, formation of the disulfide bond between Cys254 and Cys532, although causing inhibition of V-ATPase activity, does not prevent binding of nucleotides to the A subunit, as indicated by labeling using the photoaffinity analog 2-azido[^{32}P]ATP (Feng and Forgac, 1994). Thus, disulfide bond formation appears to inhibit V-ATPase activity by distorting the position of residues required for catalysis rather than by blocking nucleotide binding. The unoccupied beta subunit in F_1 shows considerable movement of the C-terminal domain relative to the glycine-rich loop region, possibly induced by the movement of the gamma subunit (Abrahams et al., 1994). If a similar structural change must accompany the catalytic cycle of the V-ATPases, it is possible that disulfide bond formation between Cys254 and Cys532 inhibits activity by preventing this movement between the two domains of the A subunit. Evidence that disulfide bond formation may play an important role in regulation of V-ATPase activity *in vivo* is discussed in a subsequent section.

Recent studies using 2-azido[^{32}P]ATP have provided additional information concerning the nucleotide-binding site on the A subunit (Zhang, J. et al., 1995). The A subunit is labeled in an ATP-protectable manner by 2-azido[^{32}P]ATP and this labeling correlates well with inhibition of V-ATPase activity. As with the F-ATPases (Cross et al., 1987), both rapidly and slowly exchangeable nucleotide-binding sites were identified on the V-ATPases using 2-azido[^{32}P]ATP, although with significant quantitative differences in the rate of nucleotide exchange at the slowly exchangeable sites (Zhang, J. et al., 1995). Rapidly exchangeable sites corresponding to catalytic sites on the F-ATPases are located principally on the beta subunits, with some residues contributed by the alpha subunits (Abrahams et al., 1994). Con-

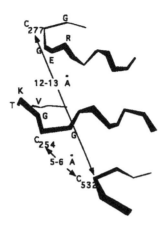

Figure 5. Structural model of the catalytic nucleotide-binding site on the V-ATPase A subunit. The structure shown is based on the X-ray structure of the F-ATPase beta subunit (Abrahams et al., 1994), the localization of Cys254 to the catalytic site on the V-ATPase A subunit (Feng and Forgac, 1992a) and the ability of Cys254 and Cys532 of the A subunit to form a disulfide bond (Feng and Forgac, 1994), indicating their separation by 5-6 A. ATP is bound such that the triphosphates are surrounded by the glycine rich loop (GCGKTV), with the adenine ring in close proximity to Cys532, as indicated from photolabeling studies using 2-azido[^{32}P]ATP (Zhang, J. et al., 1995).

versely, the slowly exchangeable sites correspond to noncatalytic nucleotide-binding sites which are located principally on the alpha subunits with some residues contributed by the beta subunits.

For the V-ATPases, inhibition of activity occurs upon modification of rapidly exchangeable sites, with minimal inhibition occurring upon modification of slowly exchangeable sites (Zhang, J. et al., 1995). Complete inhibition is achieved upon modification of a single A subunit per V-ATPase complex, consistent with the co-operativity observed between catalytic nucleotide binding sites on the F-ATPases (Senior, 1990; Penefsky and Cross, 1991; Pedersen and Amzel, 1993). Interestingly, both the A and B subunits are labeled upon modification of the rapidly exchangeable sites, suggesting that, as with the F-ATPases, the catalytic sites are located near the interface beteen the A and B subunits. On the other hand, only the A subunit is labeled upon modification of slowly exchangeable sites (Zhang, J. et al., 1995), a property again shared with the F-ATPases (Cross et al., 1987) The inability of the B subunit to be labeled upon modification of the slowly exchangeable sites may be due to the presence at these sites of residues which are not stably modified by 2-azido[^{32}P]ATP (Weber et al., 1993).

Sequencing of peptides labeled by 2-azido[^{32}P]ATP has revealed that modification of the A subunit at rapidly exchangeable sites results in labeling of residues in the vicinity of Cys532 (Zhang, J. et al., 1995), which is homologous to a region which contributes several aromatic residues to the adenine binding pocket of the F-ATPase beta subunit (Abrahams et al., 1994). Thus similar regions appear to contact the adenine ring at the catalytic sites of the F and V-ATPases. Modification of the V-ATPase at slowly exchangeable sites results in labeling of the A subunit at a site near the glycine-rich loop (Zhang, J. et al., 1995). While the glycine-rich loop region of the F-ATPase beta subunit is in close proximity to the triphosphates of ATP bound at the catalytic site, the region just upstream of this sequence is in close proximity to the noncatalytic site located at the intersubunit interface (Abrahams et al., 1994). A similar folding of the V-ATPase A subunit is thus suggested.

The A subunit in plants appears to be encoded by two genes which possess organelle specific targeting information (Gogarten et al., 1992) (see pages 438-439). The A subunit in animals, while encoded by a single gene (Puopolo et al., 1991), appears to exist in two isoforms that result from alternative splicing (Hernando et al., 1995). This is of particular interest because of the absence in one isoform of the glycine-rich loop containing Cys254 (see pages 434-437). The A subunit in yeast is unique in that it is encoded by a gene which gives rise to a 100 kDa protein which is post-translationally spliced to give the mature 70 kDa A subunit together with a 30 kDa internal fragment which encodes an endonuclease (Hirata et al., 1990; Kane et al., 1990). This remarkable example of protein splicing appears to be autocatalytic, although the reason why these two proteins are the result of a single transcript remains uncertain.

The B Subunit

Evidence that the 60 kDa B subunit participates in nucleotide binding initially came from studies of the plant V-ATPase, where it was shown that the B subunit is modified by the photoactivated analog BzATP (Manolson et al., 1985). Additional protein chemical evidence supporting this conclusion came from proteolysis studies which demonstrated that, like the A subunit, tryptic cleavage of the B subunit at a site 2 kDa from the amino terminus was inhibited in the presence of TNP-ATP (Adachi et al., 1990a).

As with the A subunit, compelling evidence for the presence on the B subunit of a nucleotide binding site came from sequence analysis of the corresponding cDNAs (Bowman, B.J. et al., 1988; Manolson et al., 1988; Nelson, H. et al., 1989; Sudhof et al., 1989; Bernasconi et al., . 1990; Puopolo et al., 1992a; Nelson, R. et al., 1992) which indicated significant sequence homology between the B subunit and other nucleotide binding proteins, including the A subunit and the alpha and beta subunits of F_1. Although the overall identity between these proteins is only 20-25%, regions shown to be important for nucleotide binding or hydrolysis from mutagenesis studies of the F-ATPase (Senior et al., 1990; Penefsky and Cross, 1991) are very highly conserved. One exception to this conservation is the glycine-rich loop or Walker consensus A sequence which is present in the V-ATPase A subunit and the F-ATPase alpha and beta subunits, but is absent from the B subunit. This, together with the data on the A subunit described above, supported a model in which the nucleotide binding sites on the B subunit were, like those on the F-ATPase alpha subunit, noncatalytic.

The function of the noncatalytic nucleotide-binding sites on the F-ATPase alpha subunit remains somewhat uncertain. Mutagenesis of these sites leads in many cases to the failure of the F-ATPase complex to assemble, suggesting the possibility that they play some role stabilizing the assembled complex (Maggio et al., 1987; Jounouchi et al., 1993). The extremely slow rate of nucleotide exchange observed at these sites for the F-ATPase makes it clear that they do not change occupancy during the normal turnover of the enzyme, although they may play some regulatory role. Recent studies in which a tryptophan residue has been introduced into the alpha to monitor ATP binding indicate that noncatalytic sites do not need to be occupied in order for catalysis to occur (Weber et al., 1994).

For the V-ATPases, slowly exchangeable nucleotide binding sites have also been observed using the photoaffinity analog 2-azido[^{32}P]ATP. but these sites exchange nucleotide at least 20-fold more rapidly than the slowly exchangeable sites on the F-ATPases (Zhang, J. et al, 1995). Moreover, modification of these sites leads to labeling of the A subunit rather than the B subunit, most likely because the B subunit residues in the vicinity of the adenine ring are not stably modified by 2-azido[^{32}P]ATP.

The B subunit has been observed to exist as multiple isoforms in animal tissues and to be expressed in a tissue-specific manner (Bernasconi et al., 1990; Puopolo et al., 1992a; Nelson, R. et al., 1992). Thus the isoform initially isolated from bovine brain is expressed in all bovine tissues tested (Puopolo et al., 1992a) whereas the isoform initially isolated from bovine kidney is expressed predominantly in kidney, with other tissues expressing only low levels of this isoform (Nelson, R. et al., 1992). The possible role of B subunit isoforms in targeting of V-ATPases to different cellular destinations is discussed in a later section (page 439).

E. Structure and Function of Accessory V_1 Subunits

Subunit C (40 kDa)

Subunit C, which has an approximate molecular weight of 40 kDa, was first cloned from bovine adrenal medulla (Nelson, H. et al., 1990) and subsequently from yeast, where it is encoded by the vma5 gene (Beltran et al., 1992; Ho et al., 1993a). The two proteins have predicted molecular weights of 44 and 42.3 kDa, respectively, and are 37% identical. The C subunit sequence reveals no putative transmembrane regions, consistent with its peripheral nature, and shows, unlike the A and B subunits, no sequence homology to any of the F-ATPase subunits. Disruption of the vma5 gene in yeast results in the typical vma phenotype, including loss of vacuolar acidification and V-ATPase activity and the inability of the mutant to grow at neutral pH (Beltran et al., 1992; Ho et al., 1993a).

Studies of the coated vesicle V-ATPase have revealed that the A, B, 34 (D), and 33 (E) kDa subunits can reassemble into a 500 kDa complex in the absence of the 40 kDa C subunit, indicating that subunit C is not essential for assembly of the remaining V_1 subunits with each other (Puopolo et al., 1992b). This is consistent with results obtained in yeast which demonstrate that in vma5 mutants, the A, B, and E subunits can still assemble with each other (Doherty and Kane, 1993). Nevertheless, it was demonstrated by coimmunoprecipitation of the C and E subunits using a monoclonal antibody specific for the E subunit that these two proteins are able to form a complex (Puopolo et al., 1992b).

In addition, it was observed that reassembly of the V_1(-40 K) subcomplex from the coated vesicle V-ATPase onto the V_0 domain gave a reassembled complex lacking the C subunit which possessed approximately 50% of the activity obtained with the full complement of subunits (Puopolo et al., 1992b). This reassembled complex lacking the subunit C, however, was unstable to detergent solubilization and immunoprecipitation, suggesting subunit C is essential for both maximal activity and stability of the V-ATPase complex. Moreover, subunit C appears to be essential for Ca^{2+}-ATPase activity by the peripheral domain (Peng et al., 1993). These results may explain why no functional V-ATPase activity is observed in vma5 mutants lacking subunit C (Beltran et al., 1992; Ho et al., 1993a).

Subunit D (34 kDa)

Subunit D has recently been cloned from both bovine adrenal medulla and yeast, where it is encoded by the vma8 gene (Nelson, H. et al., 1995). Although the bovine coated vesicle protein migrates with an apparent molecular mass of 34 kDa (Arai et al., 1987b), the mass predicted from the cDNA is 28.3 kDa. The yeast and bovine proteins display 55% identity, and although no sequence homology exists between subunit D and any of the F-ATPase subunits, similar structural motifs have been suggested to be present in subunit D and the gamma subunit of F_1 (Nelson, H. et al., 1995). The F-ATPase gamma subunit exists as an extended alpha helix which stretches approximately 90 Å and forms a central stalk around which the alpha$_3$beta$_3$ hexamer of F_1 is arranged (Abrahams et al., 1994). It is postulated that the gamma subunit may change orientation with respect to the three alpha-beta subunit pairs during the catalytic cycle of the enzyme. If subunit D is actually the V-ATPase counterpart to the gamma subunit, it presumably plays an essential role in the coupling of ATP hydrolysis to proton translocation.

Subunit E (33 kDa)

Initially cloned from bovine kidney (Hirsch et al., 1988) and subsequently from yeast (Foury, 1990), subunit E has a predicted molecular mass of 26.1 kDa, although the apparent molecular mass on SDS-PAGE of the bovine protein is 31-33 kDa (Arai et al., 1987b; Gluck and Caldwell, 1987). The yeast and bovine proteins are 34% identical at the amino acid level, although no sequence homology between subunit E and any of the F-ATPase subunits has been detected. As with the other vma genes, disruption of vma4 gene which encodes subunit E gives the typical vma phenotype in yeast (Foury, 1990; Ho et al., 1993a). Unlike subunit C, but as is the case with subunits A and B, the absence of subunit E prevents assembly of the V_1 domain in yeast (Doherty and Kane, 1993).

Cross linking studies indicate that subunit E, like subunits C and D, makes contact with the 17 kDa c subunit of the V_0 domain (Adachi et al., 1990b), suggesting that these subunits serve to bridge the periperal catalytic domain to the integral V_0 sector. Immunoprecipitation studies have also revealed an interaction between the C and E subunits (Puopolo et al., 1992b). Addition of recombinant subunit E to a depleted preparation of V_1 subunits was observed to give a 6-fold stimulation of Ca^{2+}-ATPase activity by the peripheral domain (Peng et al., 1994).

Although only one gene has been described to encode subunit E, heterogeneity has been observed in this subunit from bovine kidney. Greater heterogeneity was observed for the V-ATPase isolated from kidney brush border membranes than from kidney cortex microsomes (Wang and Gluck, 1990). In addition, antibodies have been isolated which react differently with different subpopulations of bands migrating at a molecular mass of approximately 31 kDa (Hemken et al., 1992). Moreover, these antibodies showed differential reactivity towards brush border membranes by immunofluorescence. Although the source of the microheterogeneity has not been identified (phosphorylation and glycosylation have been ruled out), these results suggest the possibility that different forms of the E subunit may have a role in targeting or regulation of V-ATPase activity in renal cells.

Subunit F (14 kDa)

Subunit F, with a molecular mass of 14 kDa, was first cloned from an insect cDNA library (Graf et al., 1994) and subsequently from yeast, where it is assigned the designation vma7 (Graham et al., 1994; Nelson, H. et al., 1994). As with the other V_1 subunits, subunit F can be stripped from the membrane in the absence of detergents. Moreover, antibodies against subunit F inhibited V-ATPase activity and proton transport (Graf et al., 1994). Unlike other V_1 subunits, however, its absence in yeast not only perturbs assembly of V_1 onto the V_0 domain, but also assembly of the V_0 domain iteself (Graham et al., 1994). Thus, the absence of the other V_1 subunits does not perturb the assembly and vacuolar targeting of the V_0 domain (Kane et al., 1992), whereas in the absence of subunit F, the 17 kDa c subunit was completely absent and the 100 kDa subunit was greatly reduced in vacuolar membranes (Graham et al., 1994). Despite its importance in assembly and/or targeting of V_0, no binding of subunit F to the V_0 domain was detected. Subunits of similar molecular weight have been detected in other V-ATPase preparations, including those from *Neurospora* (Bowman, B.J. et al., 1989), bovine renal microsomes (Gluck and Caldwell, 1987) and clathrin-coated vesicles (Hall and Forgac, unpublished results), although the identity of these proteins with respect to subunit F awaits primary sequence information.

54 kDa Subunit (vma13)

A 54 kDa V_1 subunit has been identified in yeast which has unique properties with respect to activity and assembly of the V-ATPase complex (Ho et al., 1993b). This subunit which is the product of the vma 13 gene, is essential for activity of the V-ATPase complex, as is the case for the other V-ATPase subunits. The vma 13 gene product is unique, however, in that the remainder of the V-ATPase complex can assemble in its absence (Ho et al., 1993b). The V-ATPase complex assembled in the absence of the 54 kDa subunit, however, is less stable than that assembled in its presence, suggesting that this protein plays a role in both activity and stability of the V-ATPase complex in yeast. Although proteins of similar molecular weight have been observed in purified V-ATPase preparations from *Neurospora* (Bowman, B.J. et al., 1989), plants (Parry et al., 1989) and bovine kidney microsomes (Gluck and Caldwell, 1987), the relation between these proteins and the vma 13 gene product remains uncertain.

50 kDa Subunit (AP50)

It has been demonstrated that the 50 kDa subunit of the AP-2 adaptor complex (AP50) is also a subunit of the V-ATPase from clathrin-coated vesicles (Myers and Forgac, 1993b). Thus AP50 co-purifies and co-immunoprecipitates with the V-ATPase complex and is present in a stoichiometry of one mole of AP50 per V-ATPase complex (Myers and Forgac, 1993b). Moreover, reassembly of the V_1 domain in the absence of the V_0 domain results in reassembly of AP50 with the A, B, D and E subunits into a 500 kDa complex (Puopolo et al., 1992b). AP50 was shown to bind to one or more of the subunits A, B and D, (Myers and Forgac, 1993b).

AP-2 is the plasma-membrane-specific adaptor complex which functions to bridge the cytoplasmic tails of internalized receptors with the 180 kDa heavy chain of clathrin (Pearse and Robinson, 1990). AP-2 is made up of two 100 kDa chains, termed alpha and beta adaptin, and two polypeptides of molecular mass 50 and 17 kDa. AP-1, the Golgi specific adaptor complex, is made up of homologous 100 kDa subunits, termed gamma and beta, together with polypeptides of 47 and 20 kDa. The adaptor complexes thus confer selectivity on the process of association of macromolecular receptors with clathrin-coated pits, both at the plasma membrane and in the Golgi. A function for the 50 kDa subunit (or the homologous 47 kDa polypeptide) has not yet been identified, since the 100 kDa chains appear to contain binding sites for both receptor tails and clathrin (Pearse and Robinson, 1990).

More recently it has been demonstrated that AP50 is essential for both activity and assembly of the coated vesicle V-ATPase *in vitro* (Liu et al., 1994). Thus, removal of AP50 by treatment with cystine results in a V-ATPase complex lacking both ATPase and proton transport activity, whereas reversal of the cystine treatment with reducing agents prior to purification results in restoration of AP50 together with proton transport and V-ATPase activity (Liu et al., 1994). The loss of activity was shown not to be due to blocking of the catalytic site on the A subunit (see above), since all purified V-ATPase preparations were treated with DTT prior to assay of activity. In addition, a purified V_1 fraction lacking AP50 was shown to be unable to assemble to give a functional V-ATPase complex unless supplemented with purified AP-2, from which the AP50 polypeptide was extracted during treatment with KI and ATP (Liu et al., 1994). These results indicate that AP50 is essential for both activity and *in vitro* assembly of the V-ATPase from clathrin-coated vesicles, and suggest that interaction beteen AP50 and the V-ATPase may play a role in targeting and/or assembly of the V-ATPase complex *in vivo*. It is uncertain whether V-ATPase complexes from other sources associate with a polypeptide similar to AP50 or whether this interaction is unique to the V-ATPase from clathrin-coated vesicles.

F. Structure and Function of the c Subunit

The 17 kDa c subunit of the V-ATPase was first characterized as the site of reaction with the carboxyl reagent dicyclohexylcarbodiimide (DCCD) (Bowman, E.J. 1983; Uchida et al., 1985; Manolson et al., 1985; Mandala and Taiz, 1986; Randall and Sze, 1986; Arai et al., 1987a). DCCD was shown to inhibit proton transport by the V-ATPases in the 10-100 μM concentration range, under which conditions [^{14}C]DCCD reacts selectively with the c subunit. Quantitative amino acid analysis reveals the presence of six copies of the c subunit per V-ATPase complex (Arai et al., 1988), and yet complete inhibition of proton transport activity occurs upon modification of only one c subunit per complex (Arai et al., 1987a). The c subunit was shown to be a highly hydrophobic protein, both by its high content of nonpolar amino acids (Arai et al., 1988) and by its extraction with organic solvents (Arai et al., 1987a). It is this latter property, rather than the presence of covalently bound lipid, that accounts for subunit c being termed a proteolipid.

Like the V-ATPase c subunit, the F-ATPase c subunit, which has a molecular mass of 8 kDa, is also a highly hydrophobic protein which is responsible for the sensitivity of the F-ATPases to DCCD (Fillingame, 1992). This protein is composed of two transmembrane helices, the C-terminal of which contains a buried carboxyl group that is the site of reaction with DCCD. This buried carboxylate thus plays a critical role in proton translocation through the F_0 domain. The F-ATPases contain 10-12 copies of the c subunit per complex, with complete inhibition of activity also observed upon modification of a single c subunit per F-ATPase complex (Hermolin and Fillingame, 1989). Extensive mutagenesis (Miller et al., 1990; Zhang and Fillingame, 1995) and NMR analysis (Girvin and Fillingame, 1995) have provided a detailed picture of the structure of the F-ATPase c subunit, which exists as a hairpin loop of two buried alpha helices connected by a polar loop which faces the cytoplsmic side of the membrane and plays a role in binding of F_1. Recent studies have identified substitutions in subunit c which alter the ion binding specificity of F_0 (Zhang and Fillingame, 1995).

The cDNA or gene encoding the 17 kDa c subunit of the V-ATPases has been cloned and sequenced from multiple sources, including bovine adrenal medulla (Mandel et al., 1988), yeast (Umemoto et al., 1990), mouse brain (Hanada et al., 1991), plant (Lai et al., 1991), and human (Hasebe et al., 1992). The primary sequence reveals the presence of four transmembrane helices, with a single buried carboxyl group in the fourth transmembrane helix that is the likely site of reaction with DCCD. Sequence homology between the two halves of the 17 kDa c subunit and between each half and the 8 kDa F-ATPase c subunit suggests that the V-ATPase c subunit arose by gene duplication of the F-ATPase gene (Mandel et al., 1988). Interestingly, the buried carboxyl group that would be predicted to exist in the middle of the second transmembrane helix of the V-ATPase protein has not been conserved. Thus, although the total number of transmembrane helices contributed by the c subunit to the integral domain (24) has been conserved between the two classes of ATPase, the number of buried carboxyl groups participating in proton translocation by the V-ATPases has been reduced by a factor of two. This has been suggested to account for the difference in H^+/ATP stoichiometry between the V and the F-ATPases (Cross and Taiz, 1990).

Mutational analysis of the V-ATPase c subunit in yeast has confirmed that the buried glutamic acid residue in the fourth transmembrane helix, as well as a number of other residues scattered throughout the protein, are essential for vacuolar acidification to occur (Noumi et al., 1991). This analysis is, however, complicated by the fact that only the growth phenotype of the mutants was analyzed. Thus it is unclear whether mutants that are unable to grow at neutral pH are truly defective in proton translocation or whether these mutations prevent stable expression of the c subunit or assembly of the V-ATPase complex. Suppressor analysis has confirmed the general folding pattern of four transmembrane helices (Supek et al., 1994b).

In yeast the c subunit is encoded by the vma 3 gene (Umemoto et al., 1990). Deletion of this gene not only prevents assembly of the V_1 domain onto the vacuolar membrane but also results in the absence of the 100 kDa V_0 subunit from the vacuole (Kane et al., 1992). Thus, the 100 kDa subunit appears to require assembly with the c subunit for stability and transport to the vacuolar membrane. In addition to vma3, yeast contain a second gene, vma11, which encodes a homologous polypeptide that is also essential for vacuolar acidification (Umemoto et al., 1991). The vma 3 and vma11 gene products are 57% identical at the amino acid level. Neither vma3 nor vma11 can substitute for each other, even on expression at high levels, indicating that they are both necessary for acidification of the vacuole. Whether the vma11 gene product functions in the vacuole or in some other intracellular compartment where acidification must occur in order to target the V-ATPase to the vacuole remains to be determined. Although multiple cDNAs for the proteolipid have also been detected in plants (Lai et al., 1991), these species are much more highly homologous (97-99%) than vma3 and vma11.

It has been reported that reconstitution of the c subunit from clathrin coated vesicles extracted with toluene gives a functional proton conductance which is inhibited by DCCD (Sun et al., 1987). More recently, dissociation of the V_0 complex with deoxycholate and trichloroacetate followed by separation of the V_0 subunits from each other by gel filtration gave a c subunit preparation which also displayed DCCD-inhibitable proton conductance activity (Zhang, J. et al., 1994). The rate of DCCD-inhibitable proton conductance, however, was greatly increased in the presence of the remaining V_0 subunits, which by themselves were unable to conduct protons. These results indicate that although the c subunit alone is capable of conducting protons at a low rate, formation of an efficient proton channel requires the presence of all the V_0 subunits. In the case of the F-ATPases, no proton conductance was observed using any of the isolated F_0 subunits alone, but only when all three subunits (a, b, and c) were coreconstituted together (Schneider and Altendorf, 1985). This is consistent with mutagenesis studies indicating that the F_0 a subunit plays a critical role in proton translocation (Cain and Simoni, 1986). Thus the F and V-ATPases appear to differ in the minimum subunit requirements necessary to form a functional proton pore.

It has been suggested from biochemical and immunocytochemical studies that the 17 kDa c subunit forms gap junction-like structures (Leitch and Finbow, 1990). The isolated structures appear to contain a hexameric arrangement of c subunits, with each monomer containing a bundle of four transmembrane alpha helices (Holzenberg et al., 1993). It is unclear, however, whether the polypeptides present in these preparations are identical to the V-ATPase c subunit or closely related members of a family of highly hydrophobic proteolipids. The c subunit has also been demonstrated to accumulate (along with the mitochondrial proteolipid) in Batten disease (a lysosomal storage disease) (Faust et al., 1994) and has been found in association with the E5 oncoprotein of bovine papillomavirus (Goldstein et al., 1991). E5 is a 44 amino acid, highly hydrophobic peptide which is able to transfom cells. Antibodies against E5 are able to coimmunoprecipitate the c subunit from transfected cells. In addition, substitution of a glycine for glutamine at position 17 of E5 disrupts both its ability to transform cells and its association with the c subunit (Goldstein et al., 1992). These results suggest that the interaction between E5 and the c subunit is important for cell transformation.

G. Structure and Function of Other V_0 Subunits

The 100 kDa Subunit

The 100 kDa subunit was first demonstrated to be a transmembrane glycoprotein from studies of the V-ATPase from clathrin-coated vesicles. Thus labeling of intact coated vesicles with membrane impermeant reagents resulted in some labeling of the 100 kDa subunit, with significantly greater labeling upon permeabilization of the vesicles, indicating some exposure of the 100 kDa polypeptide on the cytoplasmic surface as well as a significant lumenal domain (Arai et al., 1988). Further studies demonstrated that the 100 kDa subunit possesses a cytoplasmically exposed tryptic site which cleaves the protein into an 80 kDa and a 20 kDa fragment (Adachi et al., 1990a). It has been observed that the 100 kDa subunit displays increased sensitivity to protease digestion upon removal of the V_1 domain from the V_0 domain, suggesting that the V_1 domain may partly hinder access of the protease to the sensitive site on the cytoplasmic side of the membrane (Zhang, J. et al., 1992)). The 100 kDa subunit was also shown to possess covalently bound carbohydrate terminating in sialic acid, thus confirming the existence of a lumenally oriented domain (Adachi et al., 1990b).

The cDNA encoding the 100 kDa subunit was initially cloned from rat brain (Perin et al., 1991). The sequence revealed a bipartite structure in which the amino terminal half is hydrophilic while the carboxy terminal half is hydrophobic, containing six putative transmembrane helices. The 100 kDa protein has also been cloned from a mouse T cell library, where it was claimed to correspond to an immune-regulatory factor (Lee et al., 1990). In yeast, the 100 kDa subunit is encoded by two genes, vph1 (Manolson et al., 1992) and stv1 (Manolson et al., 1994). These two proteins display 54% identity with each other, while vph1 is 42% identical to the rat brain sequence. Disruption of vph1 alone gives a partial growth phenotype while disruption of stv1 gives a phenotype indistinguishable from the wild type (Manolson et al., 1994). Disruption of both genes, however, gives the typical vma phenotype and leads to the complete loss of vacuolar acidification. These results indicate that a 100 kDa subunit is essential for formation of a viable V-ATPase complex, but that the vph1 and stv1 gene products can at least partly substitute for each other. Overexpression of stv1 in the vph1 disrupted cells gives a wild-type phenotype. One possible function of these two isoforms of the 100 kDa subunit in yeast is in differential targeting of the V-ATPase complex (see pages 438-439).

Another possible function for the 100 kDa subunit has come from studies of the isolated, reconstituted V_0 subunits from the coated vesicle V-ATPase. It was found that the isolated, reconstituted 100 kDa subunit, like the intact V_0 domain, is able to protect the intact V-ATPase from inhibition by bafilomycin (Zhang, J. et al., 1994). These results suggest that the 100 kDa subunit possesses the binding site for bafilomycin on the V_0 domain. This binding site appears to be masked in reassembled V_0 complexes containing the full complement of V_0 subunits, although not in complexes containing only the 100 and 38 kDa subunits (Zhang, J. et al., 1994).

While two V-ATPase preparations have been reported to lack a 100 kDa subunit (Gluck and Caldwell, 1987; Ward and Sze, 1991), because of the extremely high sensitivity of this protein to protease digestion (Adachi et al., 1990a), it is possible that the 100 kDa subunits in these preparations have been proteolytically clipped, but that the fragments remain associated with the V-ATPase complex. Interestingly, a monoclonal antibody directed against the extracellular domain of the 100 kDa subunit has been shown to inhibit endosomal acidification (Sato and Toyama, 1994), supporting the view that the 100 kDa subunit is essential for function of the V-ATPases.

The 38 kDa Subunit

The 38 kDa subunit was initially cloned from bovine adrenal medulla (Wang et al., 1988) and subsequently from yeast (Bauerle et al., 1993), where it is encoded by the vma6 gene. Despite the fact that the 38 kDa subunit remains tightly associated with the V_0 domain upon removal of V_1 (Zhang, J. et al., 1992), the primary sequence reveals no putative transmembrane helices, suggesting that it remains associated with V_0 through protein-protein contacts with the other V_0 subunits. The accessibility of the 38 kDa subunit to protease digestion from the cytoplasmic side of the membrane indicates that it is exposed on the cytosolic surface (Adachi et al., 1990a), suggesting a possible role in attachment of the V_1 and V_0 domains. While the 38 kDa subunit does not bear sequence homology to any of the F_0 subunits, the F_0 b subunit, which possesses a cytoplasmically oriented loop which is critical in binding of F_1 to F_0 (Schneider and Altendorf, 1987), may represent a functional homolog to the 38 kDa subunit.

The 19 kDa Subunit (19 kDa)

In addition to the V_0 subunits listed above, a 19 kDa polypepide has been shown to copurify with the V_0 domain (Zhang, J. et al., 1992) and to co-immunoprecipitate with the intact V-ATPase from clathrin-coated vesicles (Arai et al., 1987b). This polypeptide is present in a stoichiometry of one copy per V-ATPase complex and was shown to be extremely hydrophobic in nature, both by its amino acid composition and by its extraction under some conditions with organic solvents (Arai et al., 1988). These properties are similar to those of the F-ATPase a subunit, which is also a highly hydrophobic polypeptide which is present in one copy per F_0 complex (Schneider and Altendorf, 1987; Fillingame, 1992). Mutagenesis studies have revealed that the F_0 a subunit is essential for proton translocation through the F_0 domain, particularly the C-terminal two transmembrane helices which contain multiple buried charged and polar residues which appear to participate in proton transport (Cain and Simoni, 1986). The 19 kDa subunit, while not absolutely essential for proton translocation through the V_0 domain, does nevertheless increase the rate of proton transport by the isolated c subunit by approximately 2-fold (Zhang, J. et al., 1994). Although the gene encoding this protein has not yet been isolated, one possible candidate in yeast is ppa1 (Apperson et al., 1990), which also has homology with the c subunit but contains five instead of four transmembrane helices, with the buried carboxyl group located in the third transmembrane helix.

The 45 kDa Subunit

A 45 kDa component of the chromaffin granule V-ATPase has been cloned and sequenced from adrenal medulla (Supek et al., 1994a). This protein contains a single transmembrane helix near the C-terminus together with multiple potential sites of N-linked glycosylation, suggesting a structure in which the bulk of the protein is lumenal but with a C-terminal anchor. It is synthesized as a precursor containing a signal sequence which presumably directs its translocation across the endoplasmic reticulum membrane. The 45 kD protein remains asociated with the V_0 sector upon dissociation of the V1 subunits. A crossreactive band was observed in V-ATPase preparations from kidney microsomes and synaptic vesicles (Supek et al., 1994a), and a polypeptide of similar molecular weight has also been observed in the purified coated vesicle V-ATPase (Forgac, unpublished observations), where its very diffuse appearance due to covalently bound carbohydrate has prevented its previous identification. The function of the 45 kDa polypeptide remains obscure.

V.   REGULATION OF VACUOLAR ACIDIFICATION

There is considerable evidence that different intracellular compartments are maintained at different pH values. Following endocytosis of fluorescently labeled ligands, the pH of the environment of the ligand decreases as the ligand moves from early to late endocytic compartments. Thus endocytic-coated vesicles (the earliest endocytic compartment) appear to have a near neutral pH *in vivo* (Anderson and Orci, 1988) whereas later endocytic structures (such as CURL) have a pH of ≤ 5.0-6.0 (Tycko and Maxfield, 1982) while lysosomes have a pH of 5.0 (Ohkuma and Poole, 1978). Studies employing the electron microscopic probe DAMP indicate that a similar pH gradient exists in comparing compartments along the secretory and intracellular targeting pathways (Anderson and Orci, 1988). Thus the endoplasmic reticulum and the cis and medial Golgi compartments appear to exist near neutral pH while the trans Golgi and Golgi derived vesicles are mildly acidic and later compartments en route to the lysosome as well as secretory vesicles themselves are more acidic still. As is clear from the discussion of the function of vacuolar acidification, the precise coordination of changes in vacuolar pH is essential for the correct functioning of a variety of membrane traffic processes, including receptor-mediated endocytosis and intracellular targeting. This section will focus on possible mechanisms that may be employed to regulate vacuolar acidification *in vivo*.

A.   Disulfide Bond Formation

We have demonstrated that the V-ATPase A subunit contains two cysteine residues (Cys254 and Cys532) which are capable of forming a disulfide bond that leads to reversible inactivation of the V-ATPase (Feng and Forgac, 1992b, Feng and Forgac, 1994). Cys254 is the cysteine residue responsible for the sensitivity of V-ATPase activity to sulfhydryl reagents (Feng and Forgac, 1992a) and is located in the glycine rich loop region of the protein. Cys254, Cys277and Cys532 are conserved as cysteine residues in all V-ATPase A subunit sequences, whereas other residues are substituted at these positions in the F-ATPase and *Archaebacterial* sequences (Puopolo et al., 1991). These results suggest that, while not essential for activity, these conserved cysteine residues play some important role in the V-ATPases.

It has been observed that a significant fraction (approximately 50%) of the V-ATPase in native clathrin-coated vesicles exists in the reversibly inactivated, disulfide bonded state (Feng and Forgac, 1992b). Because this fraction decreases with incubation of the vesicles at 4° C after isolation, 50% represents a lower limit to the fraction that is disulfide bonded *in vivo*. Moreover, because this disulfide bond becomes reduced with time even in the absence of reducing agents, reduction appears to occur through an internal thio-disulfide exchange involving some other reduced cysteine residue in the A subunit. These results suggest a model in which the V-ATPase exists *in vivo* in an equilibrium between two states, an active state in which Cys254 is reduced and an inactive state in which Cys254 and Cys532 are disulfide bonded (Figure 6). Because the cytoplasm is a highly reducing environment, this region of the A subunit must not sense the overall reducing potential of the cytoplasm in order to remain disulfide bonded. How this shielding of the catalytic site on the A subunit is accomplished and what triggers the thiodisulfide exchange which activates the V-ATPase *in vivo* are important but unanswered questions.

While there are no studies yet available which directly test this proposed mechanism for regulation, there are several relevant observations. First, in contrast to the

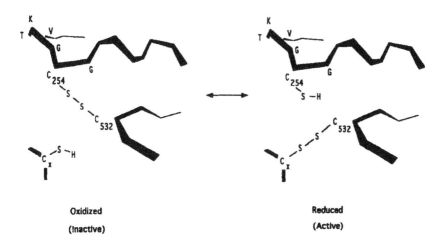

Oxidized (Inactive) — Reduced (Active)

Figure 6. Model for disulfide regulation of the V-ATPase. The V-ATPase was shown to exist in two states, one containing a disulfide bond between Cys254 and Cys532 in which the enzyme is reversibly inactive, and the other in which this disulfide bond is reduced through thio-disulfide exchange, resulting in activation (Feng and Forgac, 1992b; Feng and Forgac, 1994). Regulation is postulated to occur through control of this equilibrium. A significant fraction of the V-ATPase appears to exist in the disulfide bonded state in clathrin-coated vesicles *in vivo* (Feng and Forgac, 1992b).

V-ATPase in clathrin-coated vesicles, the V-ATPase in synaptic vesicles, isolated from the same source, was found to be fully reduced (Rodman et al., 1994). This is consistent with the expected constitutive activity of the V-ATPase in synaptic vesicles necessary for driving neurotransmitter uptake. Dschida and Bowman (1995) have observed that the *Neurospora* V-ATPase is also sensitive to oxidation and suggest that nitrate sensitivity of V-ATPase activity is due to its ability to act as an oxidant. In addition, they suggest that sulfite protects the enzyme from oxidative inactivation by acting as a reducing agent. This is in contrast to observations on the yeast V-ATPase which suggest that sulfite activates the enzyme by promoting release of inhibitory ADP (Kibak, et al. 1993). It is possible that both of these observations may be true if the oxidation state of the sulfhydryls at the catalytic site alters the affinity of the enzyme for ADP.

Mutations in the three highly conserved cysteine residues (corresponding to the bovine A subunit residues Cys254, Cys277, and Cys532) have been constructed in the yeast V-ATPase A subunit (Taiz et al., 1994). In each case the conserved cysteine was replaced with serine. Mutation of the residue corresponding to Cys254 gave a functional V-ATPase which was resistant to NEM, consistent with the previous identification of this cysteine as the residue responsible for the sensitivity of the V-ATPases to sulfhydryl reagents (Feng and Forgac, 1992a). Mutation of the other two cysteine residues to serine in each case gave an inactive V-ATPase. Because stability and assembly of the V-ATPase in the latter two cases was not checked, it is not possible to determine the cause of the observed loss of V-ATPase activity for these two mutants. In addition, no other phenotypic parameters were checked for the Cys254 mutant. Further analysis will therefore be required to determine the effect of mutations at the conserved cysteine residues on vacuolar acidification *in vivo*.

A novel isoform of the A subunit isolated from chicken osteoclasts was reported which may have relevance for oxidation-reduction as a mechanism for regulation (Hernando et al., 1995). This isoform, which is generated by alternative splicing of the transcript of the A subunit gene, possesses a unique 23 amino acid sequence in place of the glycine-rich loop region of the A subunit, thus generating a subunit lacking both the glycine rich loop and the conserved cysteine residue located in the GXGKTV sequence. While the catalytic properties of the enzyme containing this unique isoform have not yet been investigated, it is intriguing to speculate that because of the absence of the conserved cysteine residue, this enzyme would be unable to undergo disulfide bonding and would therefore exist in a constitutively active state.

Insight into how disulfide bond formation betweem Cys254 and Cys532 may lead to inactivation of the V-ATPase has come from photolabeling studies of the V-ATPase (Feng and Forgac, 1994) and the X-ray crystal structure of F_1 (Abrahams et al., 1994). It was found that disulfide bond formation of the V-ATPase A subunit did not interfere with nucleotide binding as monitored by covalent labeling using 2-azido[^{32}P]ATP (Feng and Forgac, 1994). Moreover, the crystal structure of

F1 indicates a marked asymmetry of the beta subunits, with a large movement of the amino terminal domain relative to the carboxyl terminal domain during the catalytic cycle (Abrahams et al., 1994). Thus, in the ATP and the ADP bound forms of the beta subunit, these two domains are quite close to each other, whereas in the beta subunit containing no nucleotide bound, these two domains are separated by approximately 20 Å. Since Cys254 is in the amino terminal domain of the A subunit while Cys532 is in the carboxyl terminal domain, disulfide bond formation between these two residues may lock the V-ATPase into the closed conformation, thus preventing the enzyme from completing its catalytic cycle.

B. Assembly and Dissociation of V_1 and V_0

A second possible mechanism for regulation of vacuolar acidification involves control of assembly of the V_1 and V_0 domains of the V-ATPase. Zhang, J. et al. (1992) demonstrated that clathrin-coated vesicles contain a significant population (approximately 50%) of free V_0 domains which are silent as proton channels. Because of the stability of the complex under the isolation conditions employed (that is, the absence of change in the distribution of assembled and dissociated V-ATPase with time after isolation), this free V_0 is unlikely to have been derived from dissociation of the enzyme during isolation. Immunoprecipitation of metabolically labeled V-ATPase from MDBK cells has also revealed the presence of a significant population of free V_1 domains in the cytosol (Myers and Forgac, 1993a).

Considerable information has been obtained on the assembly of the V-ATPase complex in yeast. It has been observed that for all but the 54 kD vma13 gene product, assembly of the complete V-ATPase on the vacuolar membrane requires the presence of all the V-ATPase subunits, but that the separate V_1 and V_0 domains can assemble independently (Kane et al., 1992; Doherty and Kane, 1993). Thus, in general, the absence of any of the V_1 subunits prevents assembly of the cytoplasmic V_1 domain but does not interfere with correct assembly and targeting of the V_0 domain. Similarly, disruption of the V_0 subunits prevents assembly and targeting of the V_0 domain but does not interfere with assembly of V_1 in the cytoplasm. There are several exceptions to this rule, however. Thus, deletion of the 54 kDa vma13 gene product (a V_1 subunit) does not block assembly of either V_1 or the intact V-ATPase complex, but does result in loss of V-ATPase activity (Ho et al., 1993b). Deletion of the 14 kDa subunit F (product of the vma7 gene), which is also a V_1 subunit, prevented not only assembly of V_1 onto vacuolar membranes but also resulted in the absence of the 17 kDa c subunit from vacuoles (Graham et al., 1994). The 100 kDa V_0 subunit, however, was present in the vacuoles from this strain. This is a very surprising result since it had previously been demonstrated that disruption of the vma3 gene (which encodes the c subunit) resulted in greatly diminished amounts of the 100 kDa subunit in vacuoles (Kane et al., 1992). This was explained as a lack of stability of the 100 kDa subunit in the absence of the other V_0 subunits. Nevertheless, it is clear that the V_1 and V_0 domains can assemble and exist independently in yeast.

Perhaps the strongest evidence that assembly of V_1 and V_0 plays a role in regulation of vacuolar acididification *in vivo* has also come from recent studies in yeast. Kane *(1995) has observed that growth of yeast in the absence of glucose results in a rapid dissociation of a significant fraction (70%) of the V-ATPase complex into its component V_1 and V_0 domains. This effect is observed within several minutes, is not affected by inhibitors of protein synthesis and is rapidly reversible upon resoration of glucose to the medium. Although this effect is observed after cell homogenization and isolation of the enzyme from vacuoles, experiments employing antibodies which only recognize the 100 kDa subunit in the free V_0 domain confirm that this rapid dissociation is also occurring in cells. Studies of the insect midgut V-ATPase (Sumner et al., 1995) have also suggested that dissociation of the V_1 domain accompanies loss of V-ATPase activity during moulting, but no information is available in this case to indicate that the dissociated V_1 and V_0 domains are reutilized at some point, as is clearly the case in yeast.

Although definitive information concerning how assembly of the V_1 and V_0 domains may be controlled *in vivo* is not currently available, it is intriguing that assembly of the coated vesicle V-ATPase *in vitro* depends upon the presence of the 50 kDa subunit of the AP-2 (Liu et al., 1994). This suggests the possibility that AP-2 present on the cytoplasmic surface of the clathrin-coated pit may act as a nucleation site for assembly of V_1 and V_0 domains destined for internalization.

C.   Targeting of V-ATPases

Although not demonstrated to serve a general role in regulation of vacuolar acidification, there are several well documented examples in which targeting of V-ATPases to a particular membrane serves to control both the number of V-ATPase molecules and the rate of proton transport across that membrane. Thus, in renal intercalated cells and related cells in the turtle bladder, the density of V-ATPases in the apical membrane is controlled by exocytic fusion of intracellular vesicles containing a high density of V-ATPases (Gluck et al., 1982; Brown et al., 1987). This insertion is rapidly reversible and occurs in response to an increased acid load. Control of transporter activity through such a rapid and reversible change in the number of transporters is similar to the mechanism by which glucose transporters respond to insulin stimulation of adipocytes (Cushman and Wardzala, 1980). The mechanisms controlling vesicle fusion with the plasma membrane remain to be determined but likely involve machinery similar to that implicated in the fusion of synaptic vesicles with the synaptic plasma membrane (Sollner et al., 1993).

One example in which V-ATPase density is controlled in an intracellular compartment is in phagosomes containing the intracellular parasite *Mycobacterium avium*. Following uptake of the parasite by macrophages, the phagosome containing the parasite is maintained at a pH of 6.5 by exclusion of V-ATPases from the phagosomal membrane (Sturgill-Koszycki et al., 1994). That controlling the density of V-ATPases cannot be the only mechanism controlling vacuolar pH is clear from the studies of Marquez-Sterling, et al. (1991) which demonstrate by immunocytochemistry the presence of V-ATPases in endocytic coated vesicles, despite the neutral pH of this compartment *in vivo*.

Another way in which targeting may play a role in regulation of vacuolar acidification is in the differential targeting of V-ATPases to particular intracellular membranes. This mechanism would necessitate structural differences in the V-ATPases which were targeted to different membranes. In plants, it has been demonstrated that two isoforms of the A subunit exist, one of which functions as part of the V-ATPase in the vacuole while the other is required for Golgi acidification (Gogarten et al., 1992). In yeast, differential targeting information appears to reside in the 100 kDa subunit. Vph-1 is localized exclusively to the vacuolar membrane (Manolson et al., 1992) while, at wild type levels, stv-1 is localized to some other intracellular membrane (Manolson et al., 1994). The nature of this second intracellular compartment remains to be determined. In this case, however, the two isoforms appear sufficiently homologous so that they are able to substitute for each other in the corresponding deletion strain.

In animal cells, the situation is less clear. Two isoforms of the B subunit have been identified (Bernasconi et al., 1990; Puopolo et al., 1992a), one of which is ubiquitously expressed while the other is expressed at highest levels in the kidney (Nelson, R. et al., 1992). Within the kidney, renal intercalated cells which target V-ATPases to the apical membrane express the kidney isoform at the apical membrane, although it has yet to be demonstrated that the targeting information actually resides in the B subunit.

How different isoforms of the V-ATPase might be differentially targeted within the cell is unclear, although associations with adaptor molecules that are themselves specifically targeted within the cell are suggestive. Thus, the AP-2 adaptor complex is specifically associated with clathrin-coated pits and coated vesicles at the plasma membrane (Pearse and Robinson, 1990), and may serve as a specific docking site for V-ATPases at the cell surface (Myers and Forgac, 1993b; Liu et al., 1994). A model for the role of the AP50 subunit of AP-2 in targeting and/or assembly of V-ATPases in clathrin-coated pits at the cell surface is shown in Figure 7. Similarly, the AP-1 adaptor complex may serve to sequester Golgi-associated V-ATPase molecules within clathrin-coated pits in the trans-Golgi. Interestingly, the B subunit has been localized to clathrin-rich endocytic-coated vesicles in certain renal epithelial cells in the apparent absence of other V-ATPase subunits (Sabolic et al., 1992). These results could be explained if association of the V-ATPase with AP-2 occurred through the B subunits, as is consistent with the assembly data available thus far (Myers and Forgac, 1993b). Further work will clearly be required to elucidate the mechanism by which V-ATPases are targeted to the appropriate intracellular membrane.

D. Activator and Inhibitor Proteins

Several low molecular proteins have been identified that are able to activate or inhibit the activity of purified V-ATPases. A heat-stable, trypsin-sensitive pro-

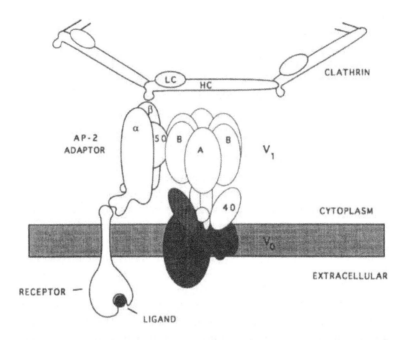

Figure 7. Model for the role of V-ATPase/AP-2 adaptor interaction in internalization and/or assembly of the V-ATPase in clathrin-coated pits at the cell surface. The AP-2 complex functions to bridge the cytoplasmic tails of internalized receptors with the heavy chain of clathrin (Pearse and Robinson, 1990). The V-ATPase interacts with AP-2 via AP50, which binds to either the A, B or 34 kDa subunits (Myers and Forgac, 1993b) and is required for activity and assembly of the V-ATPase complex *in vitro* (Liu et al., 1994). AP50 may either serve as a docking site for the V-ATPase during endocytic uptake of the V-ATPase at the cell surface or as a nucleation site for assembly of the V-ATPase in the endocytic coated pit. The Golgi V-ATPase is postulated to interact with the homologous AP47 protein which forms part of the Golgi-specific AP-1 adaptor complex. AP50 may transfer from AP-2 to the V-ATPase at some point following internalization.

tein which migrates by gel filtration with an apparent molecular weight of 35 kDa was partially purified from bovine kidney cytosol and shown to stimulate activity of the renal V-ATPase but only at pH values below 6.5 (Zhang, K. et al., 1992a). A 6 kDa protein which is also heat stable and trypsin-sensitive was isolated from bovine brain and shown to increase V-ATPase activity, again only at acidic pH (Xie et al., 1993). Unfortunately, addition of a very large excess of this protein was necessary in order to restore binding to the complex following its removal. A 6 kDa cytosolic protein which inhibits V-ATPase activity has also been isolated from bovine kidney (Zhang, K. et al., 1992b). The role that these activator and inhibitor proteins play in regulation of vacuolar acidification *in vivo* remains to be determined.

E. Other Mechanisms

Change in Coupling Efficiency

An additional mechanism which has been suggested for control of vacuolar acidification is a change in the coupling efficiency of proton transport and ATPase activity, otherwise referred to as the "slip" mechanism (Nelson, N., 1992). Several lines of evidence suggest that the V-ATPase is poised to exist in a partially coupled state in which the enzyme continues to hydrolyze ATP but without carrying out proton transport. First, it was found that DCCD inhibited both proton transport and V-ATPase activity in native clathrin-coated vesicles and in reconstituted vesicles containing the purified V-ATPase, but did not inhibit the purified enzyme in detergent solution (Arai et al., 1987a). This was the case despite the fact that the V_1 and V_0 domains were structurally assembled into a single complex (Arai et al., 1987b). To test whether DCCD failed to inhibit the solubilized enzyme because of a protective effect of the detergent, the detergent solubilized V-ATPase was first treated with DCCD and the DCCD-treated enzyme was then reconstituted. The reconstituted enzyme was completely inactive (Arai et al., 1987), indicating that DCCD reacted with the detergent solubilized protein but that effects on ATPase activity were not observed unless the enzyme was embedded in a membrane. A difference in the tightness of coupling between the V_1 and V_0 domains therefore exists in membranes and detergent micelles such that, although these domains are still physically assembled in detergent, ATP hydrolysis in the V_1 domain can occur independent of DCCD modification of V_0.

A second condition leading to partial uncoupling of ATPase activity and proton transport is a high concentration ATP (Arai et al., 1989). Measurement of ATPase activity of the coated vesicle V-ATPase as a function of ATP concentration revealed the presence of two kinetically distinguishable sites with K_m values of 80 and 800 μM. While both sites led to increased ATPase activity, only the higher-affinity site was associated with increased proton transport. Saturation of the 800 μM site resulted in decreased proton transport, indicating that at higher concentrations of ATP, ATPase activity was partially uncoupled from proton transport.

The final treatment which has been reported to partially uncouple the V-ATPase is mild proteolysis (Adachi et al., 1990a). Treatment of the purified, reconstituted V-ATPase with low concentrations of trypsin led to complete loss of proton transport with only a 50% loss of ATPase activity. This result was not due to increased proton permeability of the vesicles upon trypsin treatment. While none of the conditions described represent a likely physiological signal, the results do suggest that the coupling efficiency of the V-ATPase can be readily altered, raising the possibility that physiologically relevant signals may be able to influence vacuolar acidification by controlling this coupling.

Control of Chloride Conductance

Because the V-ATPase is an electrogenic proton pump (Forgac et al., 1983), continued proton transport requires a mechanism for dissipation of the membrane potential generated during ATP-dependent proton pumping. Dissipation of the membrane potential is accomplished through the action of a chloride channel which facilitates chloride uptake in parallel with proton transport (Glickman et al., 1983; Xie et al., 1983). This chloride channel was shown not to be a component of the V-ATPase itself since proton transport by the purified, reconstituted V-ATPase did not occur in the absence of potassium and valinomycin to dissipate the membrane potential, even in the presence of chloride (Arai et al., 1989).

Regulation of the chloride channels in both endosomes (Bae and Verkman, 1990) and coated vesicles (Mulberg et al., 1991) has been shown to be modulated by protein kinase A-dependent phosphorylation. Thus, dephosphorylation of the chloride channel results in decreased chloride conductance and decreased acidification, and these effects are reversed upon rephosphorylation with protein kinase A. Attempts to purify the coated vesicle chloride channel have not thus far provided information concerning its polypeptide composition (Xie et al., 1989). Using affinity chromatography, Redhead, et al. (1992) have identified a 64 kDa polypeptide which forms at least a component of the chloride channel from kidney microsomes based upon immunodepletion of a detergent solubilized preparation. Cloning and sequencing of this protein have revealed a hydropathy profile suggesting two putative transmembrane helices with consensus phosphorylation sites for protein kinase A and C (Landry et al., 1993). Expression of this clone in oocytes gave rise to a 64 kDa protein in microsomes rather than the plasma membrane, so no data on its chloride conductance properties were reported. Immunocytochemistry of renal cells, however, revealed staining at both the apical membrane and in intracellular vesicles (Redhead et al., 1992).

One interesting ramification of the dependence of vacuolar acidification on chloride conductance has emerged from studies of cystic fibrosis (CF). CFTR is known to be a gated chloride channel present in the apical membrane of epithelial cells, and defects in this channel lead to CF (Anderson et al., 1991). Barasch, et al. (1991) have observed that CF cells show defective acidification of endosomes and Golgi-derived vesicles, leading to alterations in glycosylation of secreted proteins, and have suggested that these results indicate that CFTR may play a role in providing the chloride conductance necessary for vacuolar acidification. Conflicting results, however, have been reported in transfected CHO cells (Lukacs et al., 1992), where no effect of CFTR on endosomal acidification was observed, suggesting the need for additional work to resolve the observed differences.

An additional mechanism for regulation of vacuolar acidification has been proposed that also depends upon the electrogenicity of the V-ATPases. It was observed that acidification of endosomes in CHO cells and A549 cells could be activated by treatments which inhibited Na^+-K^+-ATPase activity, including treatment with vanadate, the absence of Na^+ and K^+ and *in vivo* loading of endosomes with ouabain (Fuchs et al., 1989; Cain et al., 1989). These results were interpreted to indicate that Na^+-K^+-ATPase internalized along with the V-ATPase during endocytosis was able to establish an internally positive membrane potential in the endosome (by virtue of its stoichiometry of $3Na^+/K^+$) which opposed ATP-dependent proton influx. While such a mechanism does not eliminate the necessity of a compensating chloride conductance, it does impose another level of control on endosomal acidification. Further studies, however, have indicated that there are cell types in which this mechanism does not operate (Sipe et al., 1991).

VI.   CONCLUSIONS

The V-ATPases are emerging as one of the most exciting class of transport proteins. This stems not only from the diversity of intracellular environments in which they must operate but also from the range of cellular functions in which they participate. V-ATPases thus play multiple roles in endocytic and intracellular trafficking pathways, in secretory and degradative processes, and in cytoplasmic alkalinization and extracellular acidification. V-ATPases are also key participants in many pathobiological systems, including viral and toxin entry, osteoporosis, infection of cells by intracellular parasites and tumor metastasis. This diversity of functions in turn requires a complex mechanism for regulation, targeting and assembly of the V-ATPases. Unraveling the cellular mechanisms employed in controlling vacuolar acidification will represent a major advance in our understanding of cell biology.

ACKNOWLEDGMENTS

I wish to dedicate this article to Guido Guidotti, with whom I was lucky enough to do my graduate training. Guido's insight, humor and enthusiasm for science have remained a constant source of inspiration to me. The freedom he gives to members of his laboratory, his refusal to be listed as an author on papers to which he has not directly contributed experimental work, and his unwillingness to engage in the politics of science reflect a rare courage in today's research environment. His laboratory is a model for how science should be done. I also wish to thank Kathleen Forgac for her careful reading of this manuscript.

REFERENCES

Abrahams, J.P., Leslie, A.G., Lutter, R., & Walker, J.E. (1994). Structure at 2.8 Å resolution of F_1-ATPase from bovine heart mitoochondria. Nature 370, 621-628.

Aberer, W., Kostron, H., Huber, E., & Winkler, H. (1978). A characterization of the nucleotide uptake of chromaffin granules of bovine adrenal medullla. Biochem. J. 172, 353-360.

Adachi, I., Arai, H., Pimental, R., & Forgac, M. (1990a). Proteolysis and orientation on reconstitution of the coated vesicle proton pump. J. Biol. Chem. 265, 960-966.

Adachi, I., Puopolo, K., Marquez-Sterling, N., Arai, H., & Forgac, M. (1990b). Dissociation, crosslinking and glycosylation of the coated vesicle proton pump. J. Biol. Chem. 265, 967-973.

Anderson, M.P., Gregory, R. J., Thompson, S., Souza, D.W., Paul, S., Mulligan, R.C., Smith, A.E., & Welsh, M.J. (1991). Demonstration that CFTR is a chloride channel by alteration of its anion selectivity. Science 253, 202-205.

Anderson, R.G., & Orci, L. (1988). A view of acidic intracellular compartments J. Cell Biol. 106, 539-543.

Anraku, Y., Umemoto, N., Hirata, R., & Ohya, Y. (1992). Genetic and cell biological aspects of the yeast vacuolar H^+ATPase J. Bioenerg. Biomemb. 24, 395-406.

Apperson, M., Jensen, R.E., Suda, K., Witte, C., & Yaffe, M.P. (1990). A yeast protein, homologous to the proteolipid of the chromaffin granule proton-ATPase, is important for cell growth. Biochem. Biophys. Res. Commun. 168, 574-579. ·

Arai, H., Berne, M., & Forgac, M. (1987a). Inhibition of the coated vesicle proton pump and labeling of a 17, 000 dalton polypeptide by DCCD. J. Biol. Chem. 262, 11006-11011.

Arai, H., Berne, M., Terres, G., Terres, H., Puopolo, K., & Forgac, M. (1987b). Subunit composition and ATP site labeling of the coated vesicle (H^+)-ATPase. Biochemistry 26, 6632-6638.

Arai, H., Pink, S., & Forgac, M. (1989). Interaction of anions and ATP with the coated vesicle proton pump Biochemistry 28, 3075-3082.

Arai, H., Terres, G., Pink, S., & Forgac, M. (1988). Topography and subunit stoichiometry of the coated vesicle proton pump. J.Biol.Chem. 263, 8796-8802.

Bae, H.R., & Verkman, A.S. (1990). Protein kinase A regulates chloride conductance in endocytic vesicles from proximal tubule. Nature 348, 637-639.

Barasch, J., Kiss, B., Prince, A., Saiman, L., Gruenert, D., & AlAwqati, Q. (1991). Defective acidification of intracellular organelles in cystic fibrosis. Nature 352, 70-73.

Bauerle, C., Ho, M.N., Lindorfer, M.A., & Stevens, T.F. (1993). The S. cerevisiae vma6 gene encodes the 36-kDa subunit of the vacuolar H^+-ATPase membrane sector. J. Biol. Chem. 268, 12749-12757.

Beltran, C., Kopecky, J., Pan, Y.C., Nelson, H., & Nelson, N. (1992). Cloning and mutational analysis of the gene encoding subunit C of yeast vacuolar H^+-ATPase. J. Biol. Chem. 267, 774-779.

Bernasconi, P., Rausch, T., Struve, I., Morgan, L., & Taiz, L.(1990). An mRNA from human brain encodes an isoform of the B subunit of the vacuolar H^+-ATPase. J. Biol. Chem. 265, 17428-17431.

Berry, E.A., & Hinkle, P.C. (1983). Measurement of the electrochemical proton gradient in submitochondrial particles. J. Biol. Chem. 258, 1474-1486.

Birkenhager, R., Hoppert, M., Deckers-Hebestreit, G., Mayer, F., & Altendorf, K. (1995). The F_0 complex of the E. coli ATP synthase: Investigation by electron spectroscopic imaging and immunoelectron microscopy. Eur. J. Biochem. 230, 58-67.

Blair, H.C., Teitelbaum, S.L., Ghiselli, R., & Gluck, S. (1989). Osteoclastic bone resorption by a polarized vacuolar proton pump. Science 245, 855-857.

Bowman, B.J., Allen, R., Wechser, M.A., & Bowman, E.J. (1988). Isolation of the genes encoding the Neurospora vacuolar ATPase: analysis of vma-2 encoding the 57 kDa polypeptide and comparison to vma-1. J. Biol. Chem. 263, 14002-14007.

Bowman, B.J., Dschida, W.J., Harris, T., & Bowman, E.J. (1989). The vacuolar ATPase of Neurospora crassa contains an F_1-like structure. J. Biol. Chem. 264, 15606-15612.

Bowman, B.J., Vazquez-Laslop, N., & Bowman, E.J. (1992). The vacuolar ATPase of Neurospora crassa. J. Bioenerg. Biomemb. 24, 361-370.

Bowman, E.J. (1983). Comparison of the vacuolar membrane ATPase of Neurospora crassa with the mitochondrial and plasma membrane ATPases. J. Biol. Chem. 258, 15238-15244.

Bowman, E.J., Mandala, S., Taiz, L., & Bowman, B.J. (1986). Structural studies of the vacuolar membrane ATPase from *Neurospora crassa* and comparison with the tonoplast membrane ATPase from *Zea mays*. Proc. Natl. Acad. Sci. 83, 48-52.

Bowman, E.J., Siebers, A., & Altendorf, K. (1988a). Bafilomycins: A class of inhibitors of membrane ATPases from microorganisms, animal cells and plant cells. Proc. Natl. Acad. Sci. 85, 7972-7976.

Bowman, E.J., Tenney, K., & Bowman, B. (1988b). Isolation of the genes encoding the *Neurospora* vacuolar ATPase: analysis of vma-1 encoding the 66 kDa subunit reveals homolgy to other ATPases J. Biol. Chem. 263, 13994-14001.

Boyer, P.D. (1993). The binding change mechanism for ATP synthase Biochem. Biophys. Acta 1140, 215-250.

Brown, D., Gluck, S., & Hartwig, J. (1987). Structure of the novel membrane coating material in proton-secreting epithelial cells and identification as an H⁺ATPase. J. Cell Biol. 105, 1637-1648.

Brown, M.S., & Goldstein, J.L. (1986). A receptor-mediated pathway for cholesterol homeostasis. Science 232, 34-47.

Cain, B.D., & Simoni, R.D. (1986). Impaired proton conductivity resulting from mutations in the a subunit of F_1F_0ATPase in *E.coli*. J. Biol. Chem. 261, 10043-10050.

Cain, C.C., Sipe, D.M., & Murphy, R.F. (1989). Regulation of endocytic pH by the Na, K-ATPase in living cells. Proc. Natl. Acad. Sci. 86, 544-548.

Chappell, T.G., Welch, W.J., Schlossman, D.M., Palter, K.B., Schlesinger, M.J., & Rothman, J.E. (1986). Uncoating ATPase is a member of the 70 kDa family of stress proteins. Cell 45, 3-13.

Chatterjee, D., Chakraborty, M., Leit, M., Neff, L., Jamsa-Kellokumpu, S., Fuchs, R., & Baron, R. (1992). Sensitivity to vanadate and isoforms of subunits A and B distinguish the osteoclast proton pump from other vacuolar H⁺ATPases. Proc. Natl. Acad. Sci. 89, 6257-6261.

Clague, M.J., Urbe, S., Aniento, F., & Gruenberg, J. (1994). Vacuolar ATPase activity is required for endosomal carrier vesicle formation. J. Biol. Chem. 269, 21-24.

Creek, K.E., & Sly, W.S. (1984). The role of phosphomannosyl receptor in the transport of acid hydrolases to lysosomes. in Lysosomes in Biology and Pathology (J. Dingle et al., Eds.), pp.63-82. Elsevier, New York.

Crider, B.P., Xie, X.S., & Stone, D.K. (1994). Bafilomycin inhibits proton flow through the H⁺ channel of vacuolar proton pumps. J. Biol. Chem. 269, 17379-17381.

Cross, R.L, Cunningham, D., Miller, C.G., Xue, Z., Zhou, J.M., & Boyer, P.D. (1987). Adenine nucleotide binding sites on beef heart F_1 ATPase. Proc. Natl. Acad. Sci. 84, 5715-5719.

Cross, R.L., & Taiz, L. (1990). Gene duplication as a means for altering the H⁺/ATP ratios during the evolution of F_0F_1 ATPases and synthases. FEBS Lett. 259, 227-229.

Cushman, S.W., & Wardzala, L.J. (1980). Potential mechanism of insulin action on glucose transport in the isolated rat adipose cell. J. Biol. Chem. 255, 4758-4762.

Dautry-Varsat, A., Ciechanover, A., & Lodish, H.F. (1983). pH and the recycling of transferrin during receptor-mediated endocytosis. Proc. Natl. Acad. Sci. 80, 2258-2262.

Diaz, R., Mayorga, L., & Stahl, P. (1988). *In vitro* fusion of endosomes following receptor-mediated endocytosis. J. Biol. Chem. 263, 6093-6100.

Doherty, R.D., & Kane, P.M. (1993). Partial assembly of the yeast vacuolar H⁺-ATPase in mutants lacking one subunit of the enzyme. J. Biol. Chem. 268, 16845-16851.

Dong, J., Prence, E.M., & Sahagian, G.G. (1989). Mechanism for selective secretion of a lysosomal protease by transformed mouse fibroblasts. J. Biol. Chem. 264, 7377-7383.

Drose, S., Bindseil, K.U., Bowman, E.J., Siebers, A., Zeeck, A., & Altendorf, K. (1993). Inhibitory effect of modified bafilomycins and concanamycins on P and V-type ATPases. Biochemistry 32, 3902-3906.

Dschida, W.J., & Bowman, B.J. (1992). Structure of the vacuolar ATPase from *Neurospora crassa* as determined by electron microscopy. J. Biol. Chem. 267, 18783-18789.

Dschida, W.J., & Bowman, B.J. (1995). The vacuolar ATPase: Sulfite stabilization and the mechanism of nitrate activation. J. Biol. Chem. 270, 1557-1563.

Faust, J.R., Rodman, J.S., Daniel, P.F., & Bronson, R.T. (1994). Two related proteolipids and dolichol-linked oligosaccharides accumulate in mnd mice. J. Biol. Chem. 269, 10150-10155.

Feng, Y., & Forgac, M. (1992a). Cysteine 254 of the 73-kDa A subunit is responsible for inhibition of the coated vesicle (H^+)-ATPase upon modification by sulfhydryl reagents. J. Biol. Chem. 267 5817-5822.

Feng, Y., & Forgac, M. (1992b). A novel mechanism for regulation of vacuolar acidification. J. Biol. Chem. 267, 19769-19772.

Feng, Y., & Forgac, M. (1994). Inhibition of vacuolar H^+-ATPase by disulfide bond formation between Cysteine 254 and Cysteine 532 in subunit A. J. Biol. Chem. 269, 13224-13230.

Fillingame, R.H. (1992). H^+transport and coupling by the F_0 sector of the ATP synthase: Insights into the molecular mechanism of function. J. Bioenerg. Biomemb. 24, 485-491.

Forgac, M. (1988). Receptor-mediated endocytosis. In The Liver: Biology and Pathobiology (Arias, I., Jakoby, W., Popper, H., Schachter, D., and Schafritz, D. Eds.), pp.207-225. Raven Press, New York.

Forgac, M. (1989). Structure and function of the vacuolar class of ATP-driven proton pumps. Physiol. Rev. 69, 765-796.

Forgac, M. (1992). Structure and properties of the coated vesicle (H^+)-ATPase. J. Bioenerg. Biomemb. 24, 341-350.

Forgac, M., & Cantley, L. (1984). Characterization of the ATP-dependent proton pump of clathrin coated vesicles. J. Biol. Chem. 259, 8101-8105.

Forgac, M., Cantley, L., Wiedenmann, B., Altstiel, L., & Branton, D. (1983). Clathrin-coated vesicles contain an ATP-dependent proton pump. Proc. Natl. Acad. Sci. USA 80, 1300-1303.

Foury, F. (1990). The 32-kDa polypeptide is an essential subunit of the vacuolar ATPase in S.cerevisiae. J. Biol. Chem. 265, 18554-18560.

Fuchs, R., Schmid, S., & Mellman, I. (1989). A possible role for Na, K-ATPase in regulating ATP-dependent endosome acidification. Proc. Natl. Acad. Sci. 86, 539-543.

Futai, M., Park, M.Y., Iwamoto, A., Omote, H., & Maeda, M. (1994). Catalysis and energy coupling of H^+ATPase (ATP synthase): molecular biological approaches. Biochem. Biophys. Acta 1187, 165-170.

Geuze, H.J., Slot, J.W., Strous, G.J., Lodish, H.F., & Schwartz, A.L. (1983) Intracellular site of asialoglycoprotein receptor-ligand uncoupling:double-label immunoelectron microscopy during receptor-mediated endocytosis. Cell 32, 277-287.

Girvin, M.E., & Fillingame, R.H. (1995). Determination of local protein structure by spin label difference 2D NMR: The region neighboring Asp61 of subunit c of the F_1F_0 ATP synthase. Biochemistry 34, 1635-1645.

Glickman, J., Croen, K., Kelly, S., & Al-Awqati, Q. (1983). Golgi membranes contain an electrogenic proton pump in parallel to a chloride conductance. J. Cell Biol. 97, 1303-1308.

Gluck, S.L. (1992). The structure and biochemistry of the vacuolar H^+ATPase in proximal and distal urinary acidification. J. Bioenerg. Biomemb. 24, 351-360.

Gluck, S., & Caldwell, J. (1987). Immunoaffinity purification and characterization of vacuolar H^+ATPase from bovine kidney. J. Biol. Chem. 262, 15780-15789.

Gluck, S., Cannon, C., & Al-Awqati, Q. (1982). Exocytosis regulates urinary acidification in turtle bladder by rapid insertion of H^+ pumps into the luminal membrane. Proc. Natl. Acad. Sci. 79, 4327-4331.

Gogarten, J.P., Fichmann, J., Braun, Y., Morgan, L., Styles, P., Taiz, S.L., DeLapp, K., & Taiz, L. (1992). The use of antisense mRNA to inhibit the tonoplast H^+ATPase in carrot. Plant Cell 4, 851-864.

Goldstein, D.J., Finbow, M.E., Andresson, T., McLean, P., Smith, K., Bubb, V., & Schlegel, R. (1991). Bovine papillomavirus E5 oncoprotein binds to the 16 K component of vacuolar H^+-ATPases. Nature 352, 347-349.

Goldstein, D.J., Kulke, R., DiMaio, D., & Schlegel, R. (1992). A glutamine residue in the membrane-associating domain of the bovine papillomavirus type 1 E5 oncoprotein mediates its binding to a transmembrane component of the vacuolar H^+-ATPase. J. Virol. 66, 405-413.

Graf, R., Lepier, A., Harvey, W., & Wieczorek, H. (1994). A novel 14 kDa V-ATPase subunit in the tobacco hornworm midgut. J. Biol. Chem. 269, 3767-3774.

Graham, L.A., Hill, K.J., & Stevens, T.H. (1994). vma7 encodes a novel 14 kDa subunit of the *S. cerevisiae* vacuolar H^+-ATPase complex. J. Biol. Chem. 269, 25974-25977.

Haass, C., Capell, A., Citron, M., Teplow, D.B., & Selkoe, D.J. (1995). The V-ATPase inhibitor bafilomycin A1 differentially affects proteolytic processing of mutant and wild-type beta-amyloid precursor protein. J. Biol. Chem. 270, 6186-6192.

Hanada, H., Hasebe, M., Moriyama, Y., Maeda, M., & Futai, M. (1991). Molecular cloning of cDNA encoding the 16 kDa subunit of vacuolar H^+-ATPase from mouse cerebellum. Biochem. Biophys. Res. Commun. 176, 1062-1067.

Hanada, H., Moriyama, Y., Maeda, M., & Futai, M. (1990). Kinetic studies of chromaffin granule H^+-ATPase and effects of bafilomycin A1. Biochem. Biophys. Res. Commun. 170, 873-878.

Harford, J., Wolkoff, A.W., Aswell, G., & Klausner, R.D. (1983). Monensin inhibits intracellular dissociation of asialoglycoproteins from their receptor. J. Cell Biol. 96, 1824-1828.

Hasebe, M., Hanada, H., Moriyama, Y., Maeda, M., & Futai, M. (1992). Vacuolar type H^+-ATPase genes: presence of four genes including pseudogenes for the 16 kDa proteolipid subunit in the human genome. Biochem. Biophys. Res. Commun. 183, 856-863.

Hemken, P., Guo, X.L., Wang, Z.Q., Zhang, K., & Gluck, S. (1992). Immunologic evidence that V-ATPases with heterogeneous forms of 31 kDa subunit have different membrane distributions in mammalian kidney. J. Biol. Chem. 267, 9948-9957.

Hermolin, J., & Fillingame, R.H. (1989). H^+ATPase activity of *E. coli* F_1F_0 is blocked after reaction of DCCD with a single subunit c of F_0 complex. J. Biol. Chem. 264, 3896-3903.

Hernando, N., Bartkiewicz, M., Collin-Osdoby, P., Osdoby, P., & Baron, R. (1995). Alternative splicing generates a second isoform of the catalytic A subunit of the vacuolar H^+ATPase Proc. Natl. Acad. Sci. 92, 6087-6091.

Hirata, R., Ohsumi, Y., Nakano, A., Kawasaki, H., Suzuki, K., & Anraku, Y. (1990). Molecular structure of a gene, vma1, encoding the catalytic subunit of H^+ATPase from vacuolar membranes of *S.cerevisiae*. J. Biol. Chem. 265, 6726-6733.

Hirsch, S., Strauss, A., Masood, K., Lee, S., Sukhatme, V., & Gluck, S. (1988). Isolation and sequence of a cDNA clone encoding the 31-kDa subunit of bovine kidney vacuolar H^+-ATPase. Proc. Natl. Acad. Sci. 85, 3004-3008.

Ho, M.N., Hill, K.J., Lindorfer, M.A., & Stevens, T.H. (1993a). Isolation of vacuolar membrane H^+-ATPase-deficient yeast mutants; the vma5 and vma4 genes are essential for assembly and activity of the vacuolar H^+-ATPase. J. Biol. Chem. 268, 221-227.

Ho, M.N., Hirata, R., Umemoto, N., Ohya, Y., Takatsuki, A., Stevens, T.H., & Anraku, Y. (1993b). vma13 encodes a 54 kDa vacuolar H^+-ATPase subunit required for activity but not assembly of the enzyme complex in *S. cerevisiae*. J. Biol. Chem. 268, 18286-18292.

Holzenberg, A., Jones, P.C., Franklin, T., Pali, T., Heimburg, T., Marsh, D., Findlay, J., & Finbow, M.E. (1993). Evidence for a common structure for a class of membrane channels. Eur. J. Biochem. 213, 21-30.

Jiang, L.W., Maher, V.M., McCormick, J.J., & Schindler, M. (1990). Alkalinization of the lysosomes is correlated with ras transformation of murine and human fibroblasts. J. Biol. Chem. 265, 4775-4777.

Johnson, R.G., Beers, M.F., & Scarpa, A. (1982). H^+ATPase of chromaffin granules. J. Biol. Chem. 257, 10701-10707.

Johnson, R.G., Pfister, D., Carty, S.E., & Scarpa, A. (1979). Biological amine transport in chromaffin ghosts. J. Biol. Chem. 254, 10963-10972.

Jounouchi, M., Maeda, M., & Futai, M. (1993). The alpha subunit of ATP synthase: Lys175 and Thr176 are located in the domain required for stable subunit-subunit interaction. J. Biochem. 114, 171-176.

Kane, P.M. (1995). Disassembly and reassembly of the yeast vacuolar H^+-ATPase *in vivo*. J. Biol. Chem. 270, 17025-17032.

Kane, P.M., Kuehn, M.C., Howald-Stevenson, I., & Stevens, T. (1992). Assembly and targeting of peripheral and integral membrane subunits of the yeast vacuolar H^+-ATPase. J. Biol. Chem. 267, 447-454.

Kane, P.M., & Stevens, T.H. (1992). Subunit composition, biosynthesis and assembly of the yeast V-ATPase. J.Bioenerg.Biomemb. 24, 383-394.

Kane, P.M., Yamashiro, C.T., & Stevens, T.H. (1989). Biochemical characterization of the yeast vacuolar H^+-ATPase. J. Biol. Chem. 264, 19236-19244.

Kane, P.M., Yamashiro, C.T., Wolczyk, D.F., Neff, N., Goebl, M., & Stevens, T.H. (1990). Protein splicing converts the yeast TFP1 gene product to the 69 kDa subunit of the vacuolar H^+ATPase. Science 250, 651-657.

Kibak, H., Taiz, L., Starke, T., Bernasconi, P., & Gogarten, J.P. (1992). Evolution of structure and function of V-ATPases. J. Bioenerg. Biomemb. 24, 415-424.

Kibak, H., van Eckhout, D., Cutler, T., Taiz, S.L., & Taiz, L. (1993). Sulfite both stimulates and inhibits the yeast vacuolar H^+ATPase. J. Biol. Chem. 268, 23325-23333.

Klionsky, D.J., Nelson, H., & Nelson, N. (1992). Comartment acidification is required for efficient sorting of proteins to the vacuole in *S. cerevisiae*. J. Biol. Chem.267, 3416-3422.

Kornfeld, S. (1992). Structure and function of mannose-6-phosphate receptors. Ann. Rev. Biochem. 61, 307-330.

Lai, S., Watson, J.C., Hansen, J.N., & Sze, H. (1991). Molecular cloning and sequencing of cDNAs encoding the proteolipid subunit of the vacuolar H^+-ATPase from a higher plant. J. Biol. Chem. 266, 16078-16084.

Landry, D., Sullivan, S., Nicolaides, M., Redhead, C., Edelman, A., Field, M., Al-Awqati, Q., & Edwards, J. (1993). Molecular cloning and characterization of p64, a chloride channel protein from kidney microsomes. J. Biol. Chem. 268, 14948-14955.

Lee, C., Ghoshal, K., & Beaman, K.D. (1990). Cloning of a cDNA for a T cell produced molecule with a putative immune regulatory role. Molec. Immunol. 27, 1137-1144.

Leitch, B., & Finbow, M.E. (1990). The gap junction-like form of a vacuolar proton channel component appears not to be an artifact of isolation: an immunocytochemical study. Exp. Cell Res. 190, 218-226.

Liu, Q., Feng, Y., & Forgac, M. (1994). Activity and *In Vitro* reassembly of the coated vesicle (H^+)-ATPase requires the 50 kDa subunit of the clathrin assembly complex AP-2. J. Biol. Chem. 269, 31592-31597.

Lukacs, G.L., Chang, X.B., Kartner, N., Rotstein, O.D., Riordan, J.R., & Grinstein, S. (1992). CFTR is present and functional in endosomes. J. Biol. Chem. 267, 14568-14572.

Lukacs, G.L., Rotstein, O.D., & Grinstein, S. (1990). Phagosomal acidification is mediated by a vacuolar-type H^+ATPase in murine macrophages. J. Biol. Chem. 265, 21099-21107.

Maggio, M.B., Pagan, J., Parsonage, D., Hatch, L., & Senior, A.E. (1987). The defective proton-ATPase of uncA mutants of *E. coli* J. Biol. Chem. 262, 8981-8984.

Mandala, S., & Taiz, L. (1986). Characterization of the subunit structure of the maize tonoplast ATPase. J. Biol. Chem. 261, 12850-12855.

Mandel, M., Moriyama, Y., Hulmes, J.D., Pan, Y.C., Nelson, H., & Nelson, N. (1988). cDNA sequence encoding the 16-kDa proteolipid of chromaffin granules implies gene duplication in the evolution of H^+-ATPases. Proc. Natl. Acad. Sci. USA 85, 5521-5524.

Manolson, M.F., Ouellette, B.F., Filion, M., & Poole, R. (1988). cDNA sequence and homologies of the 57 kDa nucleotide binding subunit of the vacuolar ATPase from *Arabidopsis*. J. Biol. Chem. 263, 17987-17994.

Manolson, M.F., Proteau, D., Preston, R.A., Stenbit, A., Roberts, B.T., Hoyt, M.., Preuss, D., Mulholland, J., Botstein, D., & Jones, E.W. (1992). The vph1 gene encodes a 95 kDa integral membrane polypeptide required for *in vivo* assembly and activity of the yeast vacuolar H⁺-ATPase. J. Biol. Chem. 267, 14294-14303.

Manolson, M.F., Rea, P.A., & Poole, R.J. (1985). Identification of BzATP and DCCD-binding subunits of a higher plant H⁺-translocating ATPase. J. Biol. Chem. 260, 12273-12279.

Manolson, M.F., Wu, B., Proteau, D., Taillon, B.E:, Roberts, B.T., Hoyt, M.A., & Jones, E.W. (1994). stv1 gene encodes functional homologue of 95-kDa yeast vacuolar H⁺-ATPase subunit vph1p. J. Biol. Chem. 269, 14064-14074.

Marquez-Sterling, N., Herman, I.M., Pesecreta, T., Arai, H., Terres, G., & Forgac, M. (1991). Immunolocalization of the vacuolar-type (H⁺)-ATPase from clathrin-coated vesicles. (1991). Eur. J. Cell Biol. 56, 19-33.

Marsh, M., Bolzau, E., & Helenius, A. (1983). Penetration of Semliki forest virus from acidic prelysosomal vacules. Cell 32, 931-940.

Martinez-Zaguilan, R., Lynch, R., Martinez, G., & Gillies, R. (1993). Vacuolar-type H⁺-ATPases are functionally expressed in plasma membranes of human tumor cells. Am, J. Physiol. 265, C1015-C1029.

Miller, M.J., Oldenburg, M., & Fillingame, R.H. (1990). The essential carboxyl group in subunit c can be moved and H⁺-translocating function retained. Proc. Natl. Acad. Sci. USA 87, 4900-4904.

Moore, H.P., Gumbiner, B., & Kelly, R.B. (1983). Chloroquine diverts ACTH from a regulated to a constitutive secretory pathway in AtT-20 cells. Nature 302, 434-436.

Morano, K.A., & Klionsky, D.J. (1994). Differential effects of compartment deacidification on the targeting of membrane and soluble proteins to the vacuole in yeast. J. Cell Sci. 107, 2813-2824.

Moriyama, Y., Madea, M., & Futai, M. (1990). Energy coupling of L-glutamate transport and vacuolar H⁺ATPase in brain synaptic vesicles. J. Biochem. 108, 689-693.

Moriyama, Y & Nelson, N. (1987). Nucleotide binding sites and chemical modification of the chromaffin granule proton ATPase. J. Biol. Chem. 262, 14723-14729.

Moriyama, Y., & Nelson, N. (1989). Cold inactivation of vacuolar proton-ATPases. J. Biol. Chem. 264, 3577-3582.

Moriyama, Y., Tsai, H.L., & Futai, M. (1993). Energy-dependent accumulation of neuron blockers causes selective inhibition of neurotransmitter uptake by brain synaptic vesicles. Arch. Biochem. Biophys. 305, 278-281.

Mulberg, A.E., Tulk, B.M., & Forgac, M. (1991). Modulation of coated vesicle chloride channel activity and acidification by reversible protein kinase A-dependent phosphorylation. J. Biol. Chem. 266, 20590-20593.

Munn, A.L., & Riezman, H. (1994). Endocytosis is required for the growth of V-ATPase defective yeast. J. Cell Biol. 127, 373-386.

Myers, M., & Forgac, M. (1993a). Assembly of the peripheral domain of the bovine vacuolar H⁺-ATPase. J. Cell.Physiol. 156, 35-42.

Myers, M., & Forgac, M. (1993b). The coated vesicle vacuolar (H⁺)-ATPase associates with and is phosphorylated by the 50 kDa polypeptide of the clathrin assembly protein AP-2. J. Biol. Chem. 268, 9184-9186.

Nanda, A., Gukovsaya, A., Tseng, J., & Grinstein, S. (1992). Activation of vacuolar-type proton pumps by protein kinase C. J. Biol. Chem. 267, 22740-22746.

Nelson, H., Mandiyan, S., & Nelson, N. (1989). A conserved gene encoding the 57 kDa subunit of the yeast vacuolar H⁺ATPase. J. Biol. Chem. 264, 1775-1778.

Nelson, H., Mandiyan, S., & Nelson, N. (1994). The *S. cereviiae* vma7 gene encodes a 14 kDa subunit of the V-ATPase catalytic sector. J. Biol. Chem. 269, 24150-24155.

Nelson, H., Mandiyan, S., & Nelson, N. (1995). A bovine cDNA and a yeast gene (vma8) encoding the subunit D of the vacuolar H⁺-ATPase. Proc. Natl. Acad. Sci. USA 92, 497-501.

Nelson, H., Mandiyan, S., Noumi, T., Moriyama, Y., Miedel, M.C., & Nelson, N. (1990). Molecular cloning of cDNA encoding the C subunit of H⁺-ATPase from bovine chromaffin granules. J. Biol. Chem. 265, 20390-20393.

Nelson, H., & Nelson, N. (1990). Disruption of genes encoding subunits of yeast V-ATPase causes conditional lethality. Proc. Natl. Acad. Sci. USA 87, 3503-3507.

Nelson, N. (1992). Structural conservation and functional diversity of V-ATPases. J. Bioenerg. Biomemb. 24, 407-414.

Nelson, R.D., Guo, X.L., Masood, K., Brown, D., Kalkbrenner, M., & Gluck, S. (1992). Selectively amplified expression of an isoform of the V-ATPase 56 kDa subunit in renal intercalated cells. Proc. Natl. Acad. Sci. USA 89, 3541-3545.

Nolta, K.V., Padh, H., & Steck, T.L. (1991). Acidisomes in Dictyostelium: Initial biochemical characterization. J. Biol. Chem. 266, 18318-18323.

Noumi, T., Beltran, C., Nelson, H., & Nelson, N. (1991). Mutational analysis of yeast vacuolar H^+-ATPase. Proc. Natl. Acad. Sci. USA 88, 1938-1942.

Ohkuma, S., & Poole, B. (1978). Fluorescence probe measurement of the intralysosomal pH in living cells. Proc. Natl. Acad. Sci. USA 75, 3327-3331.

Palokangas, H., Metsikko, K., & Vaananen, K. (1994). Active vacuolar H^+ATPase is required for both endocytic and exocytic processes during viral infection of BHK-21 cells. J. Biol. Chem. 269, 17577-17585.

Parry, P.V., Turner, J.C., & Rea, P. (1989). High purity preparations of higher plant vacuolar H^+-ATPase reveal additional subunits. J. Biol. Chem. 264, 20025-20032.

Pearse, B.M., & Robinson, M.S. (1990). Clathrin, adaptors and sorting. Ann. Rev. Cell Biol. 6, 151-171.

Pedersen, P.L., & Amzel, L.M. (1993). ATP synthases: Structure, reaction center, mechanism and regulation of one of nature's most unique machines. J. Biol. Chem. 268, 9937-9940.

Pederson, P.L., & Carafoli, E. (1987). Ion motive ATPases. I.Ubiquity, properties and significance to cell function. Trends in Biochem. Sci. 12, 146-150.

Penefsky, H.S., & Cross, R.L. (1991). Structure and mechanism of F_0F_1-type ATP synthases and ATPases. Adv. Enzymol. 64, 173-214.

Peng, S.B., Stone, D.K., & Xie, X.S. (1993). Reconstitution of recombinant 40-kDa subunit of the clathrin-coated vesicle H^+-ATPase. J. Biol. Chem. 268, 23519-23523.

Peng, S.B., Zhang, Y., Tsai, S.J., & Stone, D.K. (1994). Reconstitution of recombinant 33-kDa subunit of the clathrin-coated vesicle H^+-ATPase. J. Biol. Chem. 269, 11356-11360.

Perin, M.S., Fried, V.A., Stone, D.K., Xie, X.S., & Sudhof, T.C. (1991). Structure of the 116 kDa polypeptide of the clathrin-coated vesicle/synaptic vesicle proton pump. J. Biol. Chem. 266, 3877-3881.

Perona, R., & Serrano, R. (1988). Increased pH and tumorigenicity of fibroblasts expressing a yeast proton pump. Nature 334, 438-440.

Pisoni, R.L., Theone, J.G., & Christensen, H.N. (1985). Detection and characterization of carrier-mediated cationic amino acid transport in lysosomes of normal and cystinotic human fibroblasts. J. Biol. Chem. 260, 4791-1798.

Puopolo, K., & Forgac, M. (1990). Functional reassembly of the coated vesicle proton pump. J. Biol. Chem. 265, 14836-14841.

Puopolo, K., Kumamoto, C., Adachi, I., & Forgac, M. (1991). A single gene encodes the catalytic "A" subunit of the bovine vacuolar H^+-ATPase. J. Biol. Chem. 266, 24564-24572.

Puopolo, K., Kumamoto, C., Adachi, I., Magner, R., & Forgac, M. (1992a). Differential expression of the "B" subunit of the vacuolar H^+-ATPase in bovine tissues. J. Biol. Chem. 267, 3696-3706.

Puopolo, K., Sczekan, M., Magner, R., & Forgac, M. (1992b). The 40 kDa subunit enhances but is not required for activity of the coated vesicle proton pump. J. Biol. Chem. 267, 5171-5176.

Randall, S.K & Sze, H. (1986). Properties of the partially purified tonoplast H^+-pumping ATPase from oat roots. J. Biol. Chem. 261, 1364-1371.

Randall, S.K., & Sze, H. (1987). Probing the catalytic subunit of the tonoplast H^+-ATPase from oat roots. J. Biol. Chem. 262, 7135-7141.

Redhead, C.R., Edelman, A.E., Brown, D., Landry, D.W., & Al-Awqati, Q. (1992). A ubiquitous 64 kDa protein is a component of a chloride channel of plasma and intracellular membranes. Proc. Natl. Acad. Sci. USA 89, 3716-3720.

Rhodes, C.J., Lucas, C.A., Mutkoski, R.L., Orci, L., & Halban, P.A. (1987). Stimulation by ATP of proinsulin conversion in isolated rat pancreatic islet secretory granules. J. Biol. Chem. 262, 10712-10717.

Rodman, J., Feng, Y., Myers, M., Zhang, J., Magner, R., & Forgac, M. (1994). Comparison of the coated vesicle and synaptic vesicle vacuolar (H⁺)-ATPases. Ann. N.Y. Acad. Sci. 733, 203-211.

Rothman, J.H., Yamashiro, C.T., Raymond, C.K., Kane, P.M., & Stevens, T.H. (1989). Acidification of the lysosome like vacuole and the vacuolar H⁺ATPase are deficient in two yeast mutants that fail to sort vacuolar proteins. J. Cell Biol. 109, 93-100.

Rudnick, G., & Clark, J. (1993). From synapse to vesicle: the reuptake and storage of biogenic amine neurotransmitters. Biochem. Biophys. Acta 1144, 249-263.

Sabolic, I., Wuarin, F., Shi, L., Verkman, A., Ausiello, D., Gluck, S., & Brown, D. (1992). Apical endosomes isolated from kidney collecting duct principal cells lack subunits of the proton pumping ATPase. J. Cell Biol. 119, 111-122.

Sahagian, G.G., & Novikoff, P.M. (1994). Lysosomes. In The Liver; Biology and Pathobiology (Arias, I., Boyer, J., Fausto, N., Jakoby, W., Schachter, D. and Shafritz, D. Eds.), pp.275-291. Raven Press, New York.

Sandvig, K., & Olsnes, S. (1980). Diptheria toxin entry into cells is facilitated by low pH. J. Cell Biol. 87, 828-832.

Sato, S.B., & Toyama, S. (1994). Interference with endosomal acidification by a monoclonal antibody directed toward the 100 kDa subunit of the vacuolar type proton pump. J. Cell Biol. 127, 39-53.

Schneider, E., & Altendorf, K. (1985). All three subunits are required for the reconstitution of an active proton channel (F0) of *E. coli* ATP synthase (F_1F_0). EMBO J. 4, 515-518.

Schneider, E., & Altendorf, K. (1987). Bacterial ATP synthetase (F_1F_0): Purification and reconstitution of F_0 complexes and biochemical and functional characterization of their subunits. Microbiol. Rev. 51, 477-497.

Schuldiner, S., & Rozengurt, E. (1982). Na⁺/H⁺ antiport in Swiss 3T3 cells: Mitogenic stimulation leads to cytoplasmic alkalinization. Proc. Natl. Acad. Sci. 79, 7778-7782.

Schwartz, A.L., Bolognesi, A., & Fridovich, S.E. (1984). Recycling of the asialoglycoprotein receptor and the effect of lysomotropic amines in hepatoma cells. J. Cell Biol. 98, 732-738.

Senior, A.E. (1990). The proton-translocating ATPase of *E. coli* Ann. Rev. Biophys. Biophys. Chem. 19, 7-41.

Sipe, D.M., Jesurum, A., & Murphy, R.F. (1991). Absence of Na, K-ATPase regulation of endosomal acidification in K562 cells. J. Biol. Chem. 266, 3469-3474.

Sollner, T., Whiteheart, S.W., Brunner, M., Erdjument-Bromage, H., Geromanos, S., Tempest, P., & Rothman, J.E. (1993). SNAP receptors implicated in vesicle targeting and fusion. Nature 362, 318-324.

Sturgill-Koszycki, S., Schlesinger, P.H., Chakraborty, P., Haddix, P.L., Collins, H.L., Fok, A.K., Allen, R.D., Gluck, S.L., Heuser, J., & Russell, D.G. (1994). Lack of acidification in Mycobacterium phagosomes produced by exclusion of the vesicular proton-ATPase. Nature 263, 678-681.

Sudhof, T.C., Fried, V.A., Stone, D.K., Johnston, P.A., & Xie, X.S. (1989). Human endomembrane H⁺ pump strongly resembles the ATP-synthase of Archaebacteria. Proc. Natl. Acad. Sci. USA 86, 6067-6071.

Sumner, J.P., Dow, J.A., Early, F.G., KleinU., Jager, D., & Wieczorek, H. (1995). Regulation of plasma membrane V-ATPase activity by dissociation of peripheral subunits. J. Biol. Chem. 270, 5649-5653.

Sun, S.Z., Xie, X.S., & Stone, D.K. (1987). Isolation and reconstitution of the DCCD sensitive proton pore of the clathrin-coated vesicle proton translocating complex. J. Biol. Chem. 262, 14790-14794.

Supek, F., Supekova, L., Mandiyan, S., Pan, Y.C., Nelson, H., & Nelson, N. (1994a). A novel accessory subunit for vacuolar H⁺-ATPase from chromaffin granules. J. Biol. Chem. 269, 24102-24106.

Supek, F., Supekova, L., & Nelson, N. (1994b). Features of vacuolar H^+-ATPase revealed by yeast suppressor mutants J.Biol.Chem. 269, 26479-26485.

Swallow, C.J., Grinstein, S., & Rotstein, O.D. (1990). A vacuolar type H^+ATPase regulates cytoplasmic pH in murine macrophages. J. Biol. Chem. 265, 7645-7654.

Sze, H., Ward, J.M., & Lai, S. (1992). Vacuolar H^+-translocating ATPases from plants: Structure, function and isoforms. J. Bioenerg. Biomemb. 24, 371-382.

Taiz, L., Nelson, H., Maggert, K., Morgan, L., Yatabe, B., Taiz, S.L., Rubinstein, B., & Nelson, N. (1994). Functional analysis of conserved cysteine residues in the catalytic subunit of the yeast vacuolar H^+ATPase. Biochem. Biophys. Acta 1194, 329-334.

Trowbridge, I.S.& Collawn, J.F. (1993). Signal-dependent membrane protein trafficking in the endocytic pathway. Ann. Rev. Cell Biol. 9, 129-161.

Tycko, B., & Maxfield, F.R. (1982). Rapid acidification of endocytic vesicles containing alpha2-macroglobulin. Cell 28, 643-651.

Uchida, E., Ohsumi, Y., & Anraku, Y. (1985). Purification and properties of H^+-translocating ATPase from vacuolar membranes of S. cerevisae. J. Biol. Chem. 260, 1090-1095.

Uchida, E., Ohsumi, Y., & Anraku, Y. (1988). Characterization and function of catalytic subunit A of H^+-translocating ATPase from vacuolar membranes of S. cerevisae. J. Biol. Chem. 263, 45-51.

Umemoto, N., Ohya, Y., & Anraku, Y. (1991). vma11, a novel gene that encodes a putative proteolipid, is indespensible for expression of yeast vacuolar membrane H^+-ATPase activity. J. Biol. Chem. 266, 24526-24532.

Umemoto, N., Yoshihisa, T., Hirata, R., & Anraku, Y. (1990). Roles of the vma3 gene product, subunit c of the vacuolar membrane H^+-ATPase on vacuolar acidification and protein transport. J. Biol. Chem. 265, 18447-18453.

van Dyke, R.W., Hornick, C.A., Belcher, J., Scharschmidt, B.F., & Havel, R.J. (1985). Identification and characterization of ATP-dependent proton transport by rat liver multivesicular bodies. J. Biol. Chem. 260, 11021-11026.

Walker, J.E., Fearnley, I.M., Gay, N.J., Gibson, B.W., Northrop, F.D., Powell, S.J., Runswick, M.J., Saraste, M., & Tybulewicz, V.L. (1985). Primary structure and subunit stoichiometry of F_1-ATPase from bovine mitochondria. J. Mol. Biol. 184, 677-701.

Wang, S.Y., Moriyama, Y., Mandel, M., Hulmes, J.D., Pan, Y.C., Danho, W., Nelson, H., & Nelson, N. (1988). Cloning of cDNA encoding a 32-kDa protein; An accessory polypeptide of the H^+-ATPase from chromaffin granules. J. Biol. Chem. 263, 17638-17642.

Wang, Z.Q., & Gluck, S. (1990). Isolation and properties of bovine kidney brush border vacuolar H^+-ATPase. J. Biol. Chem. 265, 21957-21965.

Ward, J.M., & Sze, H. (1991). Subunit composition and organization of the vacuolar H^+ATPase of plant roots. Plant Physiol. 99, 170-179.

Ward, J.M., Reinders, A., Hsu, H.T., & Sze, H. (1991). Dissociation and reassembly of the V-ATPase complex from oat roots. Plant Physiol. 99, 161-169.

Weber, J., Lee, R.S., Wilke-Mounts, S., Grell, E., & Senior, A.E. (1993). Combined application of site-directed mutagenesis, 2-azido-ATP labeling and lin-beenzo-ATP binding to study the noncatalytic sites of the E. coli F_1-ATPase. J. Biol. Chem. 268, 6241-6247.

Weber, J., Lee, R.S., Wilke-Mounts, S., Grell, E., & Senior, A.E. (1994). Tryptophan fluorescence provides a direct probe of nucleotide binding in the noncatalytic sites of E. coli F_1ATPase. J. Biol. Chem. 269, 11261-11268.

White, J.M. (1992). Membrane fusion Science 258, 917-924.

Wieczorek, H., Putzenlechner, M., Zeiske, W., & Klein, U. (1991). A vacuolar-type proton pump energizes K^+/H^+ antiport in an animal plasma membrane J. Biol. Chem. 266, 15340-15347.

Wilson, D.W., Lewis, M.J., & Pelham, H.R. (1993). pH-dependent binding of KDEL to its receptor in vitro. J. Biol. Chem. 268, 7465-7468.

Xie, X.S., Crider, B.P.& Stone, D.K. (1989). Isolation and reconstitution of the chloride transporter of clathrin-coated vesicles. J. Biol. Chem. 264, 18870-18873.

Xie, X, S., Crider, B.P., & Stone, D.K. (1993). Isolation of a protein activator of the clathrin-coated vesicle proton pump. J. Biol. Chem. 268, 25063-25067.

Xie, X.S., & Stone, D.K. (1988). Partial resolution and reconstitution of the subunits of the clathrin-coated vesicle proton ATPase responsible for Ca^{2+}-activated ATP hydrolysis J. Biol. Chem. 263, 9859-9867.

Xie, X.S., Stone, D.K., & Racker, E. (1983). Determinants of clathrin coated vesicle acidification. J. Biol. Chem. 258, 14834-14838.

Yilla, M., Tan, A., Ito, K., Miwa, K., & Ploegh, H.L. (1993). Involvement of the vacuolar H^+ATPases in the secretory pathway of HepG2 cells. J. Biol. Chem. 268, 19092-19100.

Yoshimori, T., Yamamoto, A., Moriyama, Y., Futai, M., & Tashiro, Y. (1991). Bafilomycin A1, a specific inhibitor of V-ATPase, inhibits acidification and protein degradation in lysosomes of cultured cells. J. Biol. Chem. 266, 17707-17712.

Zhang, J., Feng, Y., & Forgac, M. (1994). Proton conduction and bafilomycin binding by the V0 domain of the coated vesicle V-ATPase. J. Biol. Chem. 269, 23518-23523.

Zhang, J., Myers, M., & Forgac, M. (1992). Characterization of the V_0 domain of the coated vesicle (H+)-ATPase. J. Biol. Chem. 267, 9773-9778.

Zhang, J., Vasilyeva, E., Feng, Y., & Forgac, M. (1995). Inhibition and labeling of the coated vesicle V-ATPase by 2-azido-[^{32}P]ATP. J. Biol. Chem. 270, 15494-15500.

Zhang, K., Wang, Z.Q., & Gluck, S. (1992a). Identification and partial purification of a cytosolic activator of vacuolar H^+ATPases from mammalian kidney. J. Biol. Chem. 267, 9701-9705.

Zhang, K., Wang, Z.Q., & Gluck, S. (1992b). A cytosolic inhibitor of vacuolar H^+ATPases from mammalian kidney. J. Biol. Chem. 267, 14539-14542.

Zhang, Y., & Fillingame, R.H. (1994). Essential aspartate in subunit c of F_1F_0 ATP synthase. J. Biol. Chem. 269, 5473-5479.

Zhang, Y., & Fillingame, R.H. (1995). Changing the ion binding specificity of the *E. coli* ATP synthase by directed mutagenesis of subunit c. J. Biol. Chem. 270, 87-93.

Zimniak, L., Dittrich, P., Gogarten, J.P., Kibak, H. & Taiz, L. (1988). The cDNA sequence of the 69 kDa subunit of the carrot vacuolar H^+-ATPase: Homology to the beta-chain of F_0F_1-ATPases. J. Biol. Chem. 263, 9102-9112.

ANION TRANSPORT SYSTEMS

Parjit Kaur

I. INTRODUCTION

The phenomenon of membrane transport is integral to life. It has become clear in the last two decades that in spite of the low permeability of bilayers to charged or polar substances, certain highly specialized/dedicated transporters allow these

Advances in Molecular and Cell Biology
Volume 23B, pages 455-490.
Copyright © 1998 by JAI Press Inc.
All right of reproduction in any form reserved.
ISBN: 0-7623-0287-9

substances to cross the membrane. These transporters are protein molecules that are so constructed as to form hydrophilic channels in the highly hydrophobic environment. The genes directing the synthesis of many transporters have been cloned and the proteins isolated in a large number of cases. However, our knowledge about how these molecules actually function is still very rudimentary. It is not understood what determines the substrate specificity of these transporters, their polarity, or how the transport is coupled to consumption of energy. This chapter discusses the present state of knowledge of the molecular and biochemical properties of some representative anion transport systems found in prokaryotic and the eukaryotic cells. The major focus will be on primary active anion transporters while some discussion will involve anion-translocation by secondary porters, carriers, and channels.

The prokaryotic arsenite-translocating ATPase will be discussed as an example of an ATP-dependent export system, and the phosphate transporter as an example of the ATP-dependent import system. These two systems couple the transport of their substrates to ATP consumption via ATPases belonging to two different classes. The two eukaryotic systems discusssed in this chapter include the transporters for glutathione conjugates and methotrexate. Both import and export systems are known for these anions and will be discussed here. Table 1 gives a summary of the anion transport systems to be discussed in this chapter. It is however, by no means, a comprehensive list of all known anion-transport systems.

In nature different ion-transport problems have been tackled by employing different classes of transporters (channels, carriers, or pumps) and/or by coupling to different kinds of energy sources including light, membrane potential, or chemical energy in the form of ATP. When coupled to consumption of chemical energy, the energy-transducing component might belong to any of the major classes of transport ATPases (Pederson and Carafoli, 1987; Saier, 1994). It is not clear what could have led to selection of one kind of an ATPase over another. There also seems to be a need to redefine what constitutes a transporter vs. a channel. By definition, the three major classes of transporters can be differentiated on the basis of the uphill or downhill nature of the transport, and the substrate specificity exhibited by each class. However, the relationship between the channel-type or carrier-type transport mechanisms seems to be closer than was previously thought. Cystic fibrosis transmembrane regulator (CFTR) is an interesting example of an anion channel which has the properties of an active transporter (Higgins, 1992; Doige and Aimes, 1993). It is a chloride-selective ion channel, but belongs to the superfamily of traffic ATPases also called the ABC-type transporters. Several members of this superfamily have been shown to be ATP-dependent pumps. A brief description of this system and the mammalian P-glycoprotein, which is also implicated in chloride channel activity (Endicott and Ling, 1989; Hardy et al., 1995) is to be found in the last section of this chapter.

Table 1. Anion transport Systems

Transporter	Substrate	Energy Source (Type of ATPase)	Organism (Location)	Subunits Cyto.	Subunits Memb.	Periplasmic	Reference
Prokaryotic							
Exporters							
Arsenite transporter	AsO^{-12}, SbO^{-12}	ATP (A-type)	E. coli (plasmid)	ArsA	ArsB		Kaur and Rosen, 1992a
Asenite transporter	AsO^{-12}, SbO^{-12}	pmf	Staph. (plasmid)		ArsB		Silver et al., 1993
Asenite transporter	AsO^{-12}, SbO^{-12}	pmf	E. Coli		ArsB		Carlin et al., 1995
Chloride pump	Chloride	Light	Halobacterium		HR		Oesterhelt and Tittor, 1989
Importers							
Pit	Pi	pmf	E. Coli		Pit		Rosenberg et al., 1979
Pst	Pi	ATP (ABC-Type)	E. Coli	PstB	PstA PstC	PstS	Wanner, 1993
Ugp	G3P	ATP (ABC-Type)	E. Coli	UgpC	UgpE UgpA	UgpB	Wanner, 1993
Eukaryotic							
Exporters							
Organic-anion	GSH S-conjugates	ATP (unknown)	Rat liver, erythrocytes		?		Zimmniak et al., 1992
MOAT	GSH S-conjugates Cysteinyl leukotrienes GSSG	ATP (unknown)	Rat liver canalicular memb		MOAT		Meier, 1993; Pikula et al., 1994a

continued

457

Table 1. Continued

Transporter	Substrate	Energy Source (Type of ATPase)	Organism (Location)	Cyto.	Subunits Memb.	Periplasmic	Reference
Eukaryotic							
Exporters							
MRP	Cysteinyl leukotrienes	ATP (ABC-Type)	Human small cell lung cancer		MRP		Leier et al., 1994a
Methotrexate	MTX	ATP (unknown)	murine leukemia cells		?		Schlemmer and Sirotnak, 1992
Importers							
Organic anions	GSH S-conjugates GSH	Na+ Symport	Rat kidney		?		Lash and Jones, 1985
Methotrexate	MTX Reduced rolates	Anion exchange	Murine leukemia cells		RFC		Dixon et al., 1994
Channels							
CFTR	Chloride (Channel)	ATP (ABC-Type)	Human pancreatic epithelial cells		CFTR		Collins, 1992

458

II. PROKARYOTIC ANION TRANSPORT SYSTEMS

A. The Plasmid-Encoded Arsenite/Antimonite Transporter of *E. coli* (A-Type ATPase)

Among the anion transporting systems selected, the only plasmid-encoded transport system is the one conferring resistance to toxic anions, arsenite, arsenate, and antimonite in *E. coli* (Kaur and Rosen, 1992a) and *Staphylococci* (Novick and Roth, 1968; Silver et al., 1993). The chromosomal homolog of the arsenic resistance system has been found to exist in all strains of *E. coli* tested so far (Sofia et al., 1994, Carlin et al., 1995). The *E. coli* plasmid-encoded system is unique, however, in being a primary oxyanion pump. It utilizes ATP as the source of energy and is the only identified member of a family of ATP-dependent anion-translocating ATPases, now referred to as the A-type ATPases (Saier, 1994). This ATPase does not show homology to any of the known classes of transport ATPases including F_0F_1 (Senior, 1990), P-type (Pederson and Carafoli, 1987), or ABC-type (Fath and Kolter, 1993; Saier, 1994), even though it shows superficial similarity to the ABC-type of ATPases in its structure and overall assembly. Since the publication of the last major reviews on this system (Kaur and Rosen, 1992a; Rosen et al., 1992), considerable additional information has become available which allows us to conjecture on the functional and evolutionary aspects of this system.

The genes encoding for resistance to arsenic and antimony compounds, which were originally cloned from an *E. coli* R factor called R773, comprise an inducible operon (Hedges and Baumberg, 1973; Silver and Keach, 1982; Mobley et al., 1983). The *ars* operon consists of two regulatory genes, *arsR* and *arsD* in addition to three structural genes, *arsA*, *arsB*, and *arsC* (Chen et al., 1986b; San Francisco et al., 1990). The operon is induced by substrates of the pump: arsenite or antimonite (both oxyanions of the +3 oxidation state), and arsenate (a +5 oxidation state oxyanion). Even though uptake of arsenate is known to occur through the phosphate transport systems Pit and Pst (Bennet and Malami, 1970; Willsky and Malamy, 1981), the mechanism of uptake of arsenite into the cells is unknown. Mutations in the phosphate carrier Pit have been shown to result in arsenate resistance (Bennet and Malami, 1970). However, high-level resistance to both arsenite and arsenate is mediated by a specific efflux system encoded by the *ars* operon (Silver et al., 1981). A physical map of the *ars* operon and description of the gene products is given in Figure 1.

The product of the *arsA* gene is an ATPase which forms the energy-transducing component. It is a peripheral membrane protein and functions in association with the anion-conducting subunit ArsB, which is located in the inner membrane of *E. coli*. Together, the ArsA and the ArsB proteins form the basic anion pump. A model of the oxyanion pump depicting interaction between the ArsA and ArsB proteins is shown in Figure 2.

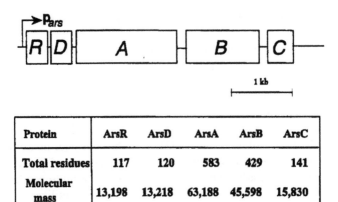

Protein	ArsR	ArsD	ArsA	ArsB	ArsC
Total residues	117	120	583	429	141
Molecular mass	13,198	13,218	63,188	45,598	15,830

Figure 1. Physical map of the *ars* operon. In the top line, the five genes of the operon are shown with the direction of transcription indicated by the arrow starting with the promoter, P_{ars}. In the bottom portion, the five gene products are listed with the number of amino acid residues, and molecular masses in daltons (Da).

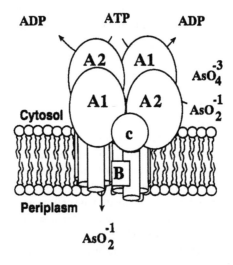

Figure 2. Model of the oxyanion pump. The complex of ArsA and ArsB proteins forms an oxyanion-translocating ATPase that catalyzes extrusion of oxyanions of the +III oxidation state. The ArsA protein is shown as a dimer with two catalytic sites formed as a result of interaction of A1 and A2 domain from subunits in trans. The ArsB protein is an inner membrane protein and serves both as a membrane anchor for the ArsA protein, and as the anion-conducting subunit of the pump. The ArsC protein is an arsenate reductase that couples the reduction of arsenate to extrusion of arsenite through the oxyanion translocating ATPase.

It has been clearly demonstrated that the extrusion of the anions by this pump is driven by the hydrolysis of ATP (Mobley and Rosen, 1982; Rosen and Borbolla, 1984). Uptake of arsenite into everted membrane vesicles prepared from cells containing the *arsA* and *arsB* genes, has been shown to be ATP-dependent. A non-hydrolyzable analog of ATP failed to act as a substitute. The arsenite uptake was not inhibited by addition of protonophores or ionophores (Dey et al., 1994) indicating that transport is independent of membrane potential. The ArsC protein has been shown to be a reductase (Oden et al., 1994) that catalyzes reduction of arsenate to arsenite. The ArsR and ArsD proteins (products of the two regulatory genes) act as repressors of the operon, negatively regulating it in the absence of the anions (Wu and Rosen, 1993a, 1993b). In the following sections, the salient features of the ArsA , ArsB, and the ArsC proteins will be discussed in detail.

Structure and Function of the ArsA Protein

The ArsA protein is a 63 kDa hydrophilic protein. Although normally membrane bound through protein-protein interactions with the ArsB protein, it can be purified from the cytosolic fraction of *E. coli* cells when overexpressed. It is the energy transducing component that couples the energy of ATP hydrolysis to the extrusion of oxyanions, arsenite or antimonite. Nucleotide sequence analysis of the *arsA* gene showed the presence of two putative nucleotide-binding domains (NBD; Chen et al., 1986b). Subsequently, the purified ArsA protein has been shown to be an anion-stimulated ATPase (Hsu and Rosen 1989). The ArsA protein belongs to a family of soluble ATPases which include proteins such as the MinD (involved in proper placement of the septum during cell division) (DeBoer et al., 1991), ParA (involved in partitioning of the plasmids to the daughter cells) (Davis et al., 1992), and NifH (involved in nitrogen fixation) (Mevarech et al., 1980). Even though these proteins do not share the ion transport function with the ArsA protein, these could have originated from the same ancestor as the ArsA protein. Dinitrogen reductase, the *nifH* gene product, couples the energy of ATP hydrolysis to the transfer of electrons to the Mo-Fe protein. The MinD and ParA proteins have been shown to be ATPases. However, the relationship of ATP hydrolysis to their biochemical function in the cell is poorly understood (De Boer et al., 1991; Davis et al., 1992).

ATP is the only nucleotide substrate for the ArsA protein. Inhibitors of other classes of transport ATPases such as DCCD and azide (inhibitors of F_0F_1-ATPase), nitrate (inhibitor of vacuolar proton pump), and vanadate (inhibitor of E1E2 ATPases) have no effect on the ATPase activity of the ArsA protein. This indicates that it does not belong to any of the known classes of transport ATPases (Hsu and Rosen, 1989). ATP binding to the ArsA protein has been demonstrated by photoaffinity labeling, where $[\alpha\text{-}^{32}P]$-ATP can be immobilized at its binding site by irradiation with ultraviolet light. It is believed that on photoactivation with UV light, ATP photoadduct formation occurs through the adenine ring to specific residues in the nucleotide-binding site of the protein (Kierdaszuk and Eriksson, 1988). The ArsA protein was shown to be photolabeled with ATP in a Mg^{2+} dependent manner (Rosen et al., 1988). Unlabeled ATP and ADP were the only effective inhibitors of labeling with ATP indicating that binding was specific for the nucleotide-binding site of the ArsA protein. Nucleotide photoaffinity labeling methods have been very useful for probing nucleotide-binding sites of proteins, including DNA Polymerase III (Biswas and Kornberg, 1984), RecA (Banks and Sedgwick, 1986), and β-tubulin (Linse and Mandelkow, 1988). This method proved valuable in differentiating between the two nucleotide-binding sites in the ArsA protein. As will be explained below, only one of the two nucleotide binding sites in the ArsA protein can be labeled by the UV photo-crosslinking method. This not only facilitates identification of the ATP-binding site, but also suggests a difference between the structure and/or affinity of the two sites for ATP. Both nucleotide-binding sites have been shown to bind ATP by a trinitrophenyl-ATP fluorescence assay (Karkaria and Rosen, 1991).

The ArsA protein has properties of an allosteric protein that goes through different conformational states on binding its substrates—the oxyanion and ATP (Hsu and Rosen, 1989). Different conformations of the protein have been detected by a variety of methods which include chemical cross-linking and trypsin protection experiments (Hsu and Rosen, 1989; Hsu et al., 1991). Simultaneous presence of the anion and ATP protects the ArsA protein from trypsin proteolysis, whereas each substrate by itself confers less protection (Hsu and Rosen 1989). These data indicate that not only are there separate binding sites for the two substrates, but that the two kinds of sites on the ArsA protein can interact. This is also obvious from the fact that the ArsA protein obligatorily requires binding of either arsenite or antimonite before it becomes competent to hydrolyze ATP. The stimulation of hydrolysis of ATP in the presence of antimonite is about 10-fold (V_{max}= 1.0 μmole/min/mg) as compared to in its absence. Arsenite is a poorer substrate and results in only a 5-fold stimulation (V_{max}= 0.3 μmole/min/mg) (Hsu and Rosen, 1989). Addition of arsenate does not result in stimulation of the ATPase activity of the ArsA protein. The reason for this will become clear below.

There is evidence suggesting that the ArsA protein functions as a dimer and that the conformational change from the monomeric to the dimeric state is brought about by binding of the anionic substrate. Chemical cross-linking experiments suggested that a small percentage of the ArsA protein can be chemically cross-linked into a dimeric form if the protein is preincubated with arsenite or antimonite (Hsu et al., 1991). This indicates a monomer-dimer equilibrium which shifts towards the dimeric state in the presence of the anion. Light scattering measurements with the ArsA protein in the presence or absence of the anionic substrate gave similar results (Hsu et al., 1991). Recent studies have suggested that the allosteric activation of the ArsA protein results from binding of the anion as a soft metal to the ArsA protein (Bhattacharjee et al., 1995). It has been proposed that the anion forms a tricordinate complex with three critical cysteine residues in the ArsA protein, and that the preferential binding of the anion to the dimer drives the reaction towards the dimeric state by mass action (Bhattacharjee et al., 1995). Metalloregulation of the arsenite-translocating ATPase at the transcriptional and enzymatic levels has been reviewed by Rosen et al. (1995).

The ArsA protein has two homologous halves, A1 and A2, corresponding to the N-(residues 1-280) and C-terminal (residues 321-583) domains of the protein. An alignment of the amino acid sequence of the two halves of the protein shows that the two domains are connected by a stretch of about 40 amino acid residues which have been hypothesized to form a flexible linker connecting the two domains (Kaur and Rosen, 1994a, 1994b). Each half carries one potential nucleotide-binding site which is readily identified by the presence of a consensus "glycine-rich" Walker A motif (Walker et al., 1982); once in the N-terminal, and once again in the C-terminal half of the protein (Figure 3). It has been proposed that the *arsA* gene arose by gene duplication and fusion of an ancestral gene that was only half the size of the present day *arsA* gene (Chen et al., 1986b). A comparison of the ArsA protein with one of its homologs, dinitrogen reductase, lends support to this idea. Dinitrogen reductase is half the size of the ArsA protein and it carries one nucleotide-binding domain. However, the functional protein is a dimer of two 32 kDa subunits (Mevarech et al., 1980) which shows a striking similarity to the ArsA protein. Gene fusion and duplication seems to have occurred in other transport ATPases as well. For exam-

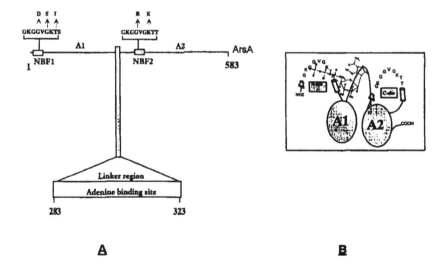

A **B**

Figure 3. A. Domain structure of the ArsA protein showing A1 and A2 halves with the two Walker A sequences (NBF1 and NBF2) and the linker region connecting the two halves. Mutations in certain residues in the two NFB's are shown by the arrows. Adenine binding site in the N terminal domain of ArsA lies within the linker region as identified by the light-activated reaction between ArsA or the subclones and ATP. B. A model showing the structure of the nucleotide binding pocket in the A1 domain of ArsA based on photolabeling data.

ple, the P-glycoprotein and CFTR, both members of the ABC superfamily (Gottesman and Pastan, 1993), carry two homologous halves—each containing six transmembrane α-helices and a nucleotide-binding domain (Chen et al., 1986a; Gottesman and Pastan, 1993).

The exact role of each domain of the ArsA protein and how they interact is not understood. However, the recent progress in successfully reconstituting the ATPase from peptide fragments is expected to facilititate studies on the contribution of each domain of the ArsA protein towards its function. In the following sections, the genetic and biochemical approaches employed to elucidate the domain structure of the ArsA protein and the role of the two nucleotide binding sites of ArsA in catalysis will be described.

Genetic Analysis.   The first approach was mutagenesis of the two domains of ArsA. Since each domain contains a conserved Walker A nucleotide-binding sequence (GXXGXGKT/S), these were targeted for mutagenesis. It is known that altering the residues in the conserved Walker A sequence of other ATP-binding proteins such as the F1-ATPase of *E. coli* (Parsonage et al., 1988; Rao et al., 1988), adenylate kinase (Reinstein et al., 1990), and the Ras protein p21 (Sigal et al., 1986) resulted in loss of function of the protein. Introduction of independent mutations in the glycine and lysine residues of the two Walker A sequences (GD18, GR18, GS20, KE21, TI22 in A1 and GR337, KE340 and KN340 in A2) (Figure 3) in ArsA indicated similarly that they were important for function of the protein (Karkaria et al., 1990; Kaur and Rosen, 1992b). Mutations in either nucleotide binding fold (NBF) led to partial or complete loss of resistance to arsenite in cells which contained a wild-type *arsB* gene. The purified mutant proteins showed loss of ATPase activity. Only one mutant, KE340, in the A2 NBF retained about 10% of the wild-type level of ATPase activity (Kaur and Rosen, 1992b), suggesting that this highly conserved lysine residue is important but not essential for binding the nucleotide. Since mutations in either domain resulted in loss of function, it was concluded that both the nucleotide-binding sites are required for function. ATP binding studies by the photoaffinity labeling method revealed that all the mutants in the A1 Walker A sequence had lost the ability to bind ATP while the mutants in the A2 domain could bind ATP as well as the wild-type protein (Karkaria et al., 1990; Kaur and Rosen, 1992b). These results indicated that the photoadduct formation with ATP occurs only at the first site. Two halves of the *arsA* gene were then subcloned to study function of each domain separately (Kaur and Rosen, 1993). Truncated proteins corresponding to the N-terminal or the C-terminal half were expressed in *E. coli* cells. The N-terminal 35 kDa (residues 1-323, which include the putative linker region) polypeptide was found to be independently capable of forming an adduct with ATP, whereas the C-terminal 46 kDa protein (residues 167-583) did not form an adduct with ATP under the same conditions (Kaur and Rosen, 1993). These results are in agreement with the results of the mutagenesis experiments where point mutants in the N-terminal NBF lost the ability to bind ATP while the C-terminal mutants did not. In addition, these experiments had another important implication—that the ArsA protein has two independent domains. The A1 domain is capable of binding ATP independently of the A2 domain. Neither the A1 nor A2 domain by itself had catalytic activity suggesting that an interaction between the two domains was required for full function of the protein.

If the ArsA protein is made up of two independent but interacting domains, point mutations in one domain (A1) should be complemented by mutations in the other domain (A2) as if they were products of two different genes. To test this hypothesis, the A1 and A2 mutant genes were introduced into compatible plasmids and the plasmids co-transformed into the same *E. coli* cell to look for complementation. Interestingly, point mutants in the N-terminal domain (namely GS20, TI22), when co-expressed with a point mutant KE340 in the C-terminal domain, resulted in almost full restoration of arsenite resistance in *E. coli* cells (Kaur and Rosen, 1993). The next step was to determine if subclones of the *arsA* gene corresponding to the N- and C-terminal halves, when co-expressed in the same cell, would confer arsenite resistance. Once again it was found that indeed co-expression of two subclones of the *arsA* gene conferred arsenite resistance. Furthermore, co-expression of an N-terminal subclone with a full-length point mutant (GS20 in A1), as well as co-expression of a C-terminal subclone with a point mutant (KE340 in A2), resulted in complementation of arsenite resistance (Kaur and Rosen, 1993).

Taken together, these results allow one to draw several conclusions: 1) Both the domains of the ArsA protein are required for function; 2) the two domains are independent, but they must interact; 3) the two domains need not be on the same polypeptide for an interaction between an A1 and an A2 domain to occur, and 4) the ArsA protein can oligomerize for domain interaction in trans. That the ArsA protein might be a functional dimer is also suggested from the chemical cross-linking and light scattering experiments mentioned earlier (Hsu et al., 1991).

Based on these studies, the current working hypothesis for the function of the Ars ATPase is the following: In a homodimer of the ArsA protein, there are two active sites—each site composed of an A1 and an A2 domain contributed by subunits in trans. According to the model presented in Figure 4, a combination of a point mutant in A1 domain with the N-terminal subclone would result in formation of only one active site, which is enough to confer arsenite resistance *in vivo*. The various possibilities depicted in this model correspond to the combinations of subunits tested by complementation experiments and found to confer arsenite resistance *in vivo*. Such an interaction between an A1 and A2 domain from subunits in trans can result from an antiparallel arrangement of the two subunits of the dimer. The antiparallel arrangement of subunits is commonly seen in dimeric proteins, and it has been suggested that such an arrangement confers an advantage as a single change during evolution might be enough to introduce an appropriate change in both sites (Klotz et al., 1975).

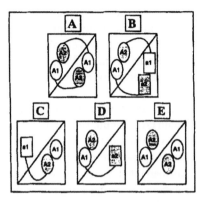

Figure 4. A model for the interaction of A1 and A2 nucleotide binding domains in an ArsA dimer. The ArsA protein is represented by two domains connected by a linker. The wild type domains are upper case in ellipses and the mutant domains are lower case in rectangles. In each dimer, there are two potential catalytic units each composed of an A1 and A2 domain contributed to by different polypeptides (A). The enzyme has activity with only one catalytic unit (B). From the genetic complementation data, the combinations of domains postulated to form active dimers are shown (B-D).

Biochemical Approaches. Biochemical approaches are complementary to genetic analysis and allow one to look at a process in isolation from other processes. Since the genetic experiments carried out with subclones of the *arsA* gene suggested that the A1 and A2 domains of the ArsA protein can interact to form a functional complex, identification of such a complex biochemically would provide evidence for such an interaction. A functional complex equivalent to the one formed under *in vivo* conditions should have ATPase activity. Using purified N- and C-terminal halves of the protein, conditions were tested under which mixing of the two halves would result in reconstitution of an active protein. It was found that in order to form an active complex, the two halves must be unfolded (by using a denaturant) and then allowed to fold together in the presence of each other. When two separately folded halves were mixed, no ATPase activity could be demonstrated indicating that an interaction was required during the folding process (Kaur and Rosen, 1994a). The reconstituted protein showed anion-stimulated ATPase activity, indicating that the activity was specific. The functional complex resulting from reconstitution of the two peptides could be identified as a distinct species on nondenaturing gels as the complex moved with a slower mobility compared to either of the two peptides by themselves (Kaur and Rosen, 1994a).

The *in vitro* reconstitution experiments were performed with polypeptides of various sizes corresponding to portions of the A1 and A2 halves of the ArsA protein. The location of the junction between the two polypeptides affected the ATPase activity of the reconstituted complex. Reconstitution was more efficient as the two polypeptides neared equal size, that is, as middle of the protein was approached. Best reconstituion was achieved with two nonoverlapping fragments which had their junction about 20 residues to the N-terminus of the start of the linker region (residues 281-321; Figure 3) (Kaur and Rosen, 1994a). This is in agreement with the hypothesis that the ArsA protein consists of two independent domains connected by a flexible linker. If the domains are separated and put back together, they fold back and form a functional ATPase. That they do fold back properly could also be shown by the ability of the reconstituted protein to form an adduct with ATP on UV photocrosslinking where either of the peptides alone did not form an adduct (Kaur and Rosen, 1994b).

Biochemical experiments described above have given interesting insights into the folding and structure of the nucleotide binding pocket in the A1 domain. An N-terminal peptide consisting of 323 residues (A1+linker) was independently capable of forming an adduct with ATP (Kaur and Rosen, 1993). The smaller N-terminal peptides which lacked the linker region were found not to be competent to form the adduct. However, when reconstituted *in vitro* with a C-terminal peptide that contained the linker region, the complex regained the ability to form a photo-crosslinked adduct with ATP. On denaturation of the complex the adduct was found to be on the C-terminal peptide. The peptide mapping experiments showed that the adduct in each case lay within the 40 amino acid linker region (Kaur and Rosen, 1994b). Since the C-terminal peptides did not form an adduct in the absence of the N-terminal peptide, the data suggest that an interaction of the phosphate groups of the ATP molecule, with residues in the Walker A sequence contained in the N-terminal peptide is essential for the adenine to become cross-linked near its binding site in the linker region. Under the *in vitro* experimental conditions, this could only result from an interaction of the ATP molecule with both peptides at once. These results suggest that the N-terminal nucleotide binding site in the ArsA protein is composed of regions of the protein far apart in the primary sequence. The P-loop region, or the Walker A sequence which would interact with the phosphate groups of the ATP molecule is localized at the N-terminus between residues 15-23, whereas a region 260 residues away from the P-loop is near the adenine binding site. Hence, in the tertiary structure, these two regions must be close together. An interaction between the two components of the first binding site generating the nucleotide binding pocket is depicted in Figure 3. Since the ArsA protein has been shown to undergo a conformational change upon binding of ATP (Hsu and Rosen, 1989), a location for the ATP-binding site in the linker region between the two domains might imply a regulatory role in transduction of conformational changes.

Even though a significant amount of information about the structure and function of the Ars ATPase has been obtained, not enough is known about its mechanism of action and the exact role of the two nucleotide-binding sites. It is not understood how the energy of ATP hydrolysis is coupled to the transport of the anions. A comparison with analogous systems suggests that the multiple nucleotide binding sites might be obligatorily required for the function of many ATPases. The most well characterized system with multiple nucleotide binding sites is the F_0F_1 proton translocating ATPase (Penefsky and Cross, 1991) with catalytic cooperativity between the three catalytic sites being well documented. Conservation of multiple nucleotide-binding sites is also seen in most members of the ABC superfamily (Doige and Ames, 1993; Gottesman and Pastan, 1993). The best-studied member of this family is the P-glycoprotein (Pgp) involved in multi-drug resistance in tumor cells. This protein is a functional analog of the ArsA protein. Like ArsA, it has two nucleotide binding sites and has been shown to be an ATPase that can catalyze extrusion of cytotoxic drugs from mammalian cells (Gottesman and Pastan, 1993).

Several exporters belonging to the ABC superfamily are structurally assembled in a fashion similar to the arsenical resistance pump which consists of a membrane-carrier protein that forms the channel, and a peripheral membrane protein that forms the energy-transducing component (Fath and Kolter, 1993). Because of the structural and functional similarity between ATPases, even though they belong to different classes, information from one system is expected to contribute toward an understanding of the mechanism of the other.

The ArsB Protein

The ArsB protein is an integral membrane protein. It has a predicted mass of 45.5 kDa (visualized as a 36 kDa protein on SDS-PAGE), and is a highly hydrophobic protein. Despite the fact that it is transcribed as part of a polycistronic message with other genes of the operon (*ars RDABC*), the expression of the ArsB protein is not proportional to the other gene products. Since ArsB is a membrane protein, overexpression of which is often toxic to the cell, certain regulatory mechanisms in the cell seem to selectively bring about the degradation of the message for the *arsB* gene preventing its overexpression (Owolabi and Rosen, 1990). Because of this reason, biochemical characterization of the ArsB protein has been limited.

Based on the analysis of a variety of gene fusions created in the *arsB* gene with the reporter genes such as *blaM*, *lacZ*, and *phoA*, it has been proposed that the ArsB protein consists of twelve membrane spanning regions (Wu et al., 1992) which is typical of several transport proteins belonging to the major facilitator superfamily (Marger and Saier, 1993; Saier, 1994). These studies also suggested the presence of 6 periplasmic and 5 cytoplasmic loop regions in the ArsB protein. A direct interaction of the ArsA protein with membranes containing the ArsB protein has been demonstrated (Tisa and Rosen, 1990a). It is proposed that the ArsA protein interacts with one of the cytoplasmic loops in the ArsB protein. However, the exact site of interaction between the two proteins is not yet known.

The ArsC Protein

The ArsC protein is a 16 kDa cytosolic protein. In addition to the ArsA and the ArsB proteins, the ArsC protein is required for conferring resistance to arsenate

(+5) and tellurite (+4). The ArsA and the ArsB proteins are sufficient to confer resistance to arsenite and antimonite—both oxyanions of the +3 oxidation state (Chen et al., 1985). It was originally proposed that the ArsC protein increased the substrate specificity of the pump. However, the actual function of the ArsC protein has become clear only recently. It has been demonstrated that the ArsC proteins encoded by both gram+ (discussed below) and gram- plasmids are reductases that catalyze reduction of arsenate to arsenite (Ji and Silver, 1992b; Oden et al., 1994). Arsenate is an inducer for the *ars* operon. However at the biochemical level, arsenate does not stimulate the ATPase activity of the ArsA protein, nor does it affect the monomer–dimer equilibrium of this ATPase (Hsu et al., 1991). This is in contrast to the action of the two substrates of the pump, arsenite or antimonite, which are known to be the allosteric modulators of this protein (Hsu and Rosen, 1989). The finding that the ArsC protein is a reductase made it possible to put pieces of the puzzle together. The actual inducers of the *ars* operon have now been shown to be oxyanions of +3 oxidation state: arsenite or antimonite (Wu and Rosen, 1993a). Arsenate acts as an inducer by first being converted to arsenite. Hence, arsenate is not a direct substrate for the oxyanion pump. After arsenate is reduced to arsenite by the ArsC protein, it is pumped out in an ATP-dependent manner using the ArsA and the ArsB proteins. The ArsC protein is also required for resistance to tellurite: the +4 oxidation state oxyanion of tellurium (Turner et al., 1992). It is postulated that the ArsC protein reduces tellurite to a more reduced anion which is then pumped out; however, the actual details are unknown.

The mechanism of action of the ArsC protein has been investigated. A requirement for reduced glutathione for resistance to arsenate and for reduction of arsenate to arsenite by the *arsC* gene product *in vivo* has been shown (Oden et al., 1994). Since the rate of reduction of arsenate to arsenite is almost similar in the wild-type *E. coli* cells no matter whether they carried the *arsC* gene or not, it has been proposed that the function of the ArsC protein is to couple the reduction of arsenate to the direct extrusion of arsenite through the ArsB protein resulting in resistance to about 10 mM arsenate. In the absence of the *ars* operon, an endogenous reduction system, which also utilizes glutathione as a reductant, confers a low-level resistance to about 2 mM arsenate (Oden et al., 1994). Mutations in the glutathione biosynthesis genes result in hypersensitivity to arsenate (Apontoweil and Berends, 1975; Oden et al., 1994).

In vitro studies carried out with the purified ArsC protein showed that the function of the protein requires glutaredoxin in addition to the reduced glutathione (Gladysheva et al., 1994). The reason for this difference in requirement under *in vivo* and *in vitro* conditions is unknown. Perhaps under *in vivo* conditions, another protein can substitute for thioredoxin. Since the reducing equivalents for the function of the ArsC protein are derived from the cysteine thiolates, the role of cysteine residues in the function of the ArsC protein was investigated. ArsC protein has two cysteine residues: Cys12 and Cys106. By site directed mutagenesis, it has been determined that Cys12 is the active cysteine site; mutation of this residue to a serine resulted in loss of arsenate resistance, and the purified mutant protein was shown to be catalytically inactive. The mutation of Cys106 to a serine had little effect on either resistance or the catalytic activity of the protein. Based on these studies, the ArsC protein has been proposed to be a single thiol reductase where the arsenylated thiolate anion of Cys12 would interact with glutaredoxin, resulting in reduction of arsenate (+5) to arsenite (+3) (Liu, J., Gladysheva, T.B., Leei, L., & Rosen, B.P, personal communication).

B.   Other Arsenite Transporters

Plasmid mediated resistance to arsenite and antimonite has also been found in *Staphylococcus aureus* harboring penicillinase plasmids (pI258, Ji and Silver, 1992a) and in a food-industry isolate of *Staphylococcus xylosus* (pSX267, Rosenstein *el al.*, 1992). Both of the *Staphylococcal ars* operons have been sequenced. The identity between the *E. coli* and the *Staphylococcal* genes is 30%, 58%, and 20% in the *arsR*, *arsB*, and the *arsC* genes, respectively. The most striking difference between the *E. coli* plasmid-encoded *ars* operon and the two *Staphylococcal* operons is the absence of the genes for the ArsD and the ArsA proteins. The *Staphylococcal* ArsR protein is a repressor protein that negatively regulates the expression of the operon (Ji and Silver, 1992a). The ArsC protein is a reductase that is required to confer resistance to arsenate (Ji and Silver, 1992b).

At the time the sequence of the two *Staphylococcal* operons became known, the absence of the *arsA* gene was difficult to understand, because the ArsA protein was known to be the energy-transducing component of the *E. coli* plasmid-encoded pump (Kaur and Rosen, 1992a). How then, do the *Staphylococcal ars* operons confer resistance to arsenite in the absence of the energy-transducing component? Two alternative hypotheses were proposed: one suggesting that the *Staphylococcal* ArsB functions in association with another chromosomally expressed ATPase, and the other that in the absence of the ArsA protein, the ArsB protein can function as a secondary carrier coupling the efflux of the anions to the electrochemical gradient (Rosen et al., 1992; Silver et al., 1993). The first hypothesis was supported by the observation that the expression of the *E. coli arsA* gene in trans with the *Staphylococcal arsB* gene resulted in an increase in the level of arsenite resistance as compared to that conferred by the *arsB* gene alone (Bröer et al., 1993; Dou et al., 1994), indicating that the *Staphylococcal* arsB protein had the capability to interact with an ArsA-like ATPase. The observation in support of the second hypothesis came from studies carried out with the mutant *arsA* genes which, in the presence of the wild-type *arsB* gene, conferred arsenite resistance to a level lower than that seen in the presence of the *arsB* gene alone (Kaur and Rosen, 1992b, 1993). It is possible that the mutant ArsA proteins block the channel function of the ArsB protein. This phenomenon has been observed with the F_0F_1 ATPase where mutant F_1 subunits prevent proton conduction by the F_0 subunit (Futai and Kanazawa, 1983).

The energetics of the *arsB* containing systems have now been investigated, and the available data seem to favor the second hypothesis. It has been shown that not only can the *Staphylococcal* ArsB protein carry out energy-dependent efflux in response to the membrane potential (Bröer et al., 1993), but that the *E. coli* plasmid-encoded ArsB protein can also function by itself as a secondary carrier in the absence of the ArsA protein (Dey and Rosen, 1995). The exclusion of arsenite by cells containing the *arsB* gene alone was found to be sensitive to the uncoupler CCCP indicating that it is dependent upon the membrane potential. In the presence of the ArsA protein, the ArsA-ArsB complex becomes an obligatory ATP-coupled pump and the transport is insensitive to the addition of uncouplers (Mobley and Rosen, 1982; Dey et al., 1994).

What could be the advantage conferred by the ATP-driven ArsA-ArsB active transport system vs. the ArsB secondary porter? The advantage becomes obvious if one compares the level of resistance to oxyanions conferred by the two systems. The *E. coli* cells carrying the *arsB* gene alone show resistance to about 3 mM arsenite, whereas the cells containing both the *arsA* and the *arsB* genes can withstand arsenite to about 10 mM or more. This is to be expected. An electrophoretic anion extrusion catalyzed by an ArsB protein would depend on the magnitude of the membrane potential, whereas an ATP-driven pump would be independent of that and would form a much more reliable mechanism of resistance (Rosen et al., 1992). Hence, the ArsA-ArsB type of system must have evolved under a selective pressure.

The fact that we have found examples in nature of both the "ArsB alone" and "ArsA-ArsB" systems, has fascinating implications for the evolution of protein complexes in general. It seems reasonable to assume that the system containing only the *arsB* gene led to the evolution of an *arsA-arsB* type of system. This could have resulted from recruitment of a soluble ATPase such as MinD, NifH, or another soluble ATPase after it had changed enough to allow its recognition and binding by the ArsB protein (Rosen et al., 1992). The interaction between the ArsA and ArsB proteins is specific, and the ancestral proteins would have evolved towards each other to develop affinity and specificity. What is interesting is the fact that the *Staphylococcal* ArsB protein (that has presumably never seen the ArsA protein) can recognize and interact with the *E. coli* ArsA protein (Bröer et al., 1993) indicating that the ArsB proteins have not changed or evolved very much at least in the domains required for interaction with the ATPase. It might imply unidirectional evolution where one protein adapts to the other.

Scientific pursuit is full of surprises. One such surprise came with the identification by Sofia and coworkers (Sofia et al., 1994) of homologs of the *E. coli* plasmid-encoded *arsRBC* genes on the chromosome of *E. coli*. Until recently, it was believed that arsenite transport and resistance is purely a plasmid-encoded function. The similarity between the chromosomal genes and the genes encoded on the *E. coli* R773 plasmid is 72%, 90%, and 94% in the *arsR*, *arsB*, and the *arsC* genes, respectively. These genes are arranged in an operon and once again, conspicuous by their absence, are the *arsD* and the *arsA* genes.

The chromosomal *ars* operon has been found to be responsible for the low-level intrinsic resistance to the arsenical and antimonial compounds that is seen in *E. coli* strains. Deletion of the *ars* genes from the *E. coli* chromosome resulted in hypersensitivity to arsenite, which corresponded to an increased accumulation of arsenite by these cells (Carlin et al., 1995). Presumably, in the absence of the *arsA* gene, the product of the *arsB* gene functions as a secondary carrier and couples the extrusion of arsenite to membrane potential as is seen in the case of the plasmid-encoded ArsB proteins (Bröer et al., 1993; Dey and Rosen,1995).

This finding adds yet another dimension to the evolution of the bacterial resistance operons and the transport systems. It lends credence to the fact that the bacterial genetic pool is mobile. The function of the chromosomal *ars* operon must have been to provide intrinsic low level protection against environmental toxic substances which later evolved into a much more efficient system capable of conferring a high level of resistance. This was achieved not only by moving the genetic markers to a plasmid with the double advantage of increased copy number and mobility, but also by acquisition of an additional component (the ATPase) to make it more efficient. This might have relevance to the origin of plasmid-encoded drug resistance systems in general. One wonders if the chromosomal ancestors of other plasmid-encoded systems, such as the cadmium- and mercury-resistance systems (Tisa and Rosen, 1990b), are yet to be discovered.

It is almost certain that the *E. coli* plasmid-encoded *ars* genes originated from the chromosomal operon by acquisition of the *arsD* and the *arsA* genes, but exactly how the *Staphylococcal ars* operon might have originated is not known. In terms of its structure, it is more closely related to the *ars* operon on the *E. coli* chromosome. It is not known at this time if the *Staphylococci* carry a chromosomal *ars* operon as well. Finally, why do we see the evolution of a repressor protein ArsD along with the ArsA protein? Perhaps it is a coincidence, or alternatively it has evolved with the explicit purpose of downregulating the expression of an ATPase. Some of these questions might find their answers as we stumble upon more surprises by way of discovering more transporters.

C.   The Phosphate Transport Systems of *E. coli (ABC-Type ATPases)*

Phosphorus is an essential cellular element as it is the structural component of vital cell macromolecules such as nucleic acids and phospholipids. *E. coli* has evolved an extensive regulatory network to be able to scavenge phosphate (Pi) under conditions of Pi starvation. This system called the *pho* regulon controls the expression of two major systems, the Pst system for transport of Pi, and the Ugp system for uptake of glycerol-3-phosphate (G3P). Both Pst and the Ugp transporters are derepressed on Pi starvation, and there is some preliminary evidence suggesting that both Pst and the Ugp systems utilize binding protein-dependent transporters in association with an ABC-type ATPase (Overduin et al., 1988). Another G3P transporting system in the cell is a pho-independent system called the

GlpT system (Ambudkar et al., 1986). Whereas the Ugp system is up-regulated by low Pi, GlpT is induced by the presence of G3P. The major function of GlpT appears to be the supply of a carbon source. It exchanges external G3P for internal Pi, thus preventing Pi imbalance in the cell (Ambudkar et al., 1986). The Pi exchange mechanism is also utilized by a sugar phosphate transporter Uhp (Sonna et al., 1988). Hence, UhpT and GlpT are in the category of the Pi-linked antiporters that mediate an exchange of Pi and the phosphorylated substrates while their main function is geared toward meeting the carbon requirement. For further information on these two systems and the Ugp system, the reader is referred to reviews by Maloney et al., (1990), Kadner et al., (1993), and Wanner, (1993).

The net Pi movement inside the *E. coli* cell is mediated by two different transporters: the high affinity transporter mentioned earlier, Pst; and another low-affinity transporter called Pit. Pit, the phosphate inorganic transport carrier, is a constitutively expressed transporter and is sufficient to meet Pi needs of the cell under conditions of Pi excess (Rosenberg, 1987). Under these conditions the diffusion of Pi through the OmpC and OmpF pores in the outer membrane is coupled to its transport by the Pit system. Pit has been shown to be a single-component carrier mediating a H^+/Pi symport (Rosenberg, 1987). This system will not be discussed further in this chapter.

The *pho* regulon constitutes the emergency system that deals with the situation created by low phosphate. It consists of a large number of genes scattered throughout the chromosome of *E. coli* encoding not only for the proteins of the high affinity transporters Pst and Ugp, but also a specific porin (PhoE) in the outer membrane (Tommassen and Lugtenberg, 1980) and an enzyme, alkaline phosphatase, in the periplasmic space (Torriani, 1990). A total of about 81 proteins have been identified as being phosphate regulated (Muda et al., 1992). However, not all of these belong to the *pho* regulon. The operons under the control of *pho* regulon are subject to regulation by two regulatory proteins: PhoR and PhoB (Makino et al., 1989; Rao and Torriani, 1990). These two regulatory proteins belong to the growing family of two-component signal transduction proteins that transmit the signal from the environment through a series of phosphorelay events to the site of transcription thus controlling the expression of the genes (Stock et al., 1989). PhoR is a histidine kinase, and an integral membrane protein that acts as a Pi sensor. PhoB is the response regulator that interacts with DNA to control the rate of transcription. Under conditions of low Pi, PhoR gets autophosphorylated which in turn phosphorylates PhoB. Phosphorylated PhoB is a transcriptional activator and binds efficiently to the "pho box' sequences that occur upstream of the *pho*-regulated promoters (Makino et al., 1988; Rao and Torriani, 1990). PhoR has an additional activity—that of being a phosphatase under conditions of Pi excess causing repression of the *pho* regulon (Yamada et al., 1990). It is believed that PhoR does not act as the Pi sensor directly; instead transmembrane signaling through the Pst complex located in the inner membrane is required for initially sensing the concentration of phosphate (Scholten and Tommassen, 1993; Torriani, A., 1994).

The genes for the high-affinity transporter Pst are arranged in an operon consisting of $pstS$, $pstA$, $pstB$, and $pstC$ genes and another gene of controversial function $phoU$ (Surin et al., 1985). The PhoU protein is thought to enhance the phosphatase function of PhoR when excess Pi is present causing repression of the pho regulon. A missense mutation of $phoU$ has no effect on the transport of Pi, but it results in the constitutive synthesis of alkaline phosphatase (Surin et al., 1985). The Pst transport complex consists of three proteins: two integral membrane proteins, PstA and PstC, and a peripheral membrane protein, PstB (Cox et al., 1988,1989; Silver and Walderhaug, 1992). The amino acid sequence of the PstB protein shows the presence of a consensus nucleotide-binding sequence called the Walker A motif (Walker et al., 1982), and belongs to the family of the binding protein-dependent ABC-type of transporters (Fath and Kolter, 1993). The consensus Walker A sequence is seen in most nucleotide-binding proteins (Walker et al., 1982). However, in members of the ABC superfamily of ATPases the homology extends further to a region of about 200 amino acids encompassing the Walker A motif. The biochemical analysis of the PstB protein, as well as the exact topology of the PstA and PstC proteins is awaited. Independent mutations introduced into the conserved Gly48 or Lys49 residues in the Walker A sequence in the PstB protein resulted in a loss of the Pi-transport activity through the Pst system (Cox et al., 1989). These results are an indication that the ATP-binding motif in the PstB protein is critical for the functioning of this transport complex, and that the PstB protein is most likely an ATPase. The PstA and PstC are hydrophobic proteins and have been proposed to form a membrane channel (Surin et al., 1987). The hydropathic analysis of the PstA protein suggests that this protein traverses the inner membrane of $E.$ $coli$ six times (Surin et al., 1987). The product of the gene $pstS$ is a Pi-binding protein (PBP) that resides in the periplasmic space and binds mono- and di-basic phosphates with high affinity (Luecke and Quiocho, 1990). Thus the similarity with the binding protein transport systems becomes evident. Other transporters that belong to the family of ABC transporters include the maltose and the histidine transport systems, and the mammalian multidrug resistance protein Pgp, for which a direct involvement of ATP hydrolysis concomitant with transport of their substrates has been demonstrated (Ames and Joshi, 1990; Davidson and Nikaido, 1991).

It seems reasonable to assume that the PstA, B, and C proteins together fulfill the minimum requirement of an ABC transporter. Analyses of the available molecular and biochemical information on several members of the ABC family provide a working hypothesis on the overall structure/function of an ABC transporter that has come to be widely accepted. An ABC transporter consists of a minimum of four domains: two membrane spanning (MSD) and two ABC nucleotide-binding domains (NBD) in various arrangements (Hyde et al., 1990). On one extreme is the mammalian P-glycoprotein, where all four domains are fused together in one large protein (Chen et al., 1986a). Each MSD of Pgp consists of six transmembrane α-helices giving a total of twelve transmembrane segments. Each group of six helices in Pgp is followed by a nucleotide-binding domain which occurs in the hydrophilic cytoplasmic loops. On the other extreme are some bacterial exporters such as the capsular polysaccharide exporter (KpsMT) of *E. coli* (Fath and Kolter, 1994) which consists of one ABC subunit and one membrane-bound subunit. The hemolysin transporter of *E. coli* HlyB consists of one ABC domain fused to MSD proposed to contain 6 α-helices (Fath and Kolter, 1994). In between is the maltose transport system which has two membrane subunits (MAlF, MalG) each proposed to contain six transmembrane α-helices (Boyd et al., 1993) and an ATP binding subunit (Mal K). It has been suggested that the bacterial transport complexes function as oligomers of the protein subunits in such a way that the total number of domains is four (Hyde et al., 1990). The evidence in support of this hypothesis comes from studies carried out with the maltose-transport complex that has been shown to consist of MalFGK$_2$ on reconstitution from the purified components (Davidson and Nikaido, 1991), and recent studies with the colicin export protein, CvaB, indicating that it can be identified as a dimer by SDS-PAGE (Dr. P.C. Tai, personal communication). The Pst system appears to resemble the maltose transporter in its assembly and structure.

The route for uptake of inorganic phosphate appears to be through the high-affinity polyanion porin PhoE, in the outer membrane to the periplasmic space of *E. coli*, where it is trapped by the PBP and presented to the Pst transport complex present in the inner membrane. If the analogy with the histidine and maltose transporters holds true, the PBP in the Pi-bound conformation will induce a specific change in the conformation of the membrane bound complex causing the PstB protein to become an ATPase. The hydrolysis of ATP by the PstB protein will couple the transport of Pi through the membrane channel consisting of the PstA and PstC proteins. These questions and several more need to be addressed biochemically before such a model can be viewed as feasible.

III. EUKARYOTIC ANION TRANSPORT SYSTEMS

A. Organic Anion Transporters (GSH S-Conjugates)

Glutathione has an important role to play in the protection of cells against oxidative damage and toxic substances (Ishikawa, 1992). For this reason, transport of glutathione conjugates has been the subject of intense investigation for quite some time. Liver is the major organ releasing GSH and its conjugates into bile and plasma. Several kinds of cells including the liver cells and erythrocytes have been shown to contain glutathione S-transferases—a group of enzymes that catalyze conjugation of GSH to toxic substances facilitating their removal from the cell. One example is the conjugation of glutathione to a xenobiotic 1-chloro-2,4-dinitrobenzene to form dinitrophenyl glutathione (DNP-SG; LaBelle et al., 1986). There is experimental evidence suggesting that such an dianionic species is ultimately transported out of the plasma membrane of these cells (LaBelle et al., 1986).

Biochemical analysis of the system responsible for transport of DNP-SG showed it to be an ATP-dependent efflux system in both the plasma membrane of the liver canalicular cells and erythrocytes. Using inside-out membrane vesicles prepared from human erythrocytes, the accumulation of DNP-SG was shown to be dependent upon the presence of magnesium and ATP. The inhibitors of Na^+-K^+-ATPase, or Ca^{2+}-ATPase had no effect on the transport (LaBelle, et al., 1986) indicating that the ion gradients are not essential. The transport was inhibited by vanadate which is a specific inhibitor of the P-type ATPases (Pederson and Carafoli, 1987). But vanadate has sometimes been shown to inhibit ATPases in general. Using DNP-SG affinity column, a protein of 38 kDa which showed ATPase activity stimulated by DNP-SG was purified (Sharma et al., 1990). The antibodies raised against the erythrocyte anion-ATPase also cross-reacted with the ATPase purified from the rat liver cells using the same basic methodology (Zimniak et al., 1992). The two ATPases showed the same characteristics indicating that they are probably the complementary systems. It seems reasonable to assume, as the authors have themselves speculated, that the 38 kDa protein may not be the whole transporter. It is possible that the purified protein represents only the energy transducing component, and that there might be another component of the transporter that would form the anion channel in the membrane.

It is now proper to discuss the multispecific organic anion transporter (MOAT) of liver canalicular membrane. MOAT has invoked much interest even though the existence of such a system had been inferred from several biochemical studies carried out with normal and mutant strains of rats for a long time. MOAT is one of the four major transporters in liver which include transport systems for drugs, phospholipids, bile acids, and the nonbile acid organic anions (Arias et al., 1993). MOAT is the ATP-dependent transporter of the nonbile acid organic anions in the canalicular membrane (Meier, 1993). It has been proposed that the activity of this multispecific transporter results in excretion of several different organic anions into bile which include GSH conjugates such as DNP-SG (Akerboom et al., 1991), cysteinyl leukotrienes (LTC_4) (Ishikawa, et al., 1990), and conjugated bilirubin. A genetic defect in the hepatobiliary excretion of organic anions in a TR-rat mutant strain results in severely reduced canalicular transport into bile of the above anionic species, in addition to transport of the oxidized and reduced glutathione (Oude-Elferink et al., 1989). This has been shown to be due to lack of an ATP-dependent transport activity in the membrane vesicles prepared from the mutant rats (Ishikawa et al., 1990). That a mutant rat with defects in transport of several organic anions has been isolated raises the question whether all of these anionic substances are transported across the canalicular membrane as a result of the activity of one multispecific transporter referred to as MOAT, or whether it involves a cluster of proteins with related function.

Purification of an anion-translocating ATPase from rat liver with characteristics and activity similar to those of the multispecific anion transporter (MOAT; Ishikawa et al., 1990; Akerboom et al., 1991) has been reported (Pikula et al., 1994a).

This protein is 90 kDA in size and shows ATP binding and DNP-SG-stimulated ATPase activity, which is inhibited by vanadate (Pikula et al., 1994a). In addition, the MOAT protein has been shown to be a glycoprotein with an oligosaccharide content of about 5% of the mass of the protein. The purified MOAT protein can be phosphorylated by protein kinase C (PKC) which results in stimulation of the ATPase activity of the protein. This might indicate a regulatory mechanism controlling the activity of MOAT *in vivo*.

The 90 kDa MOAT has also been reconstituted into proteoliposomes, and the ATP-dependent and vanadate-sensitive transport of DNP-SG into the vesicles demonstrated. Phosphorylation was shown to enhance the rate of uptake, but had no effect on the K_m for the substrate indicating that phosphorylation does not affect substrate binding. The transport was also shown to be independent of the membrane potential. Hence, this transporter is a primary active transport system (Pikula et al., 1994b). The question whether MOAT is indeed a multispecific transporter of the organic anions has not been addressed yet. However, with the availability of the purified protein, and a system where it can be reconstituted into proteoliposomes, it can be expected that such information would soon be forthcoming.

MOAT is presently believed to exist only in the liver canalicular membrane. However, transporters for the efflux of GSH S-conjugates do exist in other tissues including erythrocytes (LaBelle et al., 1986; Sharma et al., 1990; Heijn et al., 1992) and mastocytoma cells (Leier et al., 1994b). It is not known if the transport system(s) found in liver canalicular membrane referred to as MOAT is similar to those found in other tissues. An obvious question concerns the relation of the 38 kDa anion-stimulated ATPase purified from erythrocytes (discussed earlier in this section; Sharma et al., 1990; Zimniak et al., 1992) to MOAT. It is interesting to note that MOAT undergoes spontaneous proteolysis to yield 30 and 60 kDa products. The 30 kDa product is able to bind ATP as shown by photoaffinity studies carried out with an ATP analogue (Pikula et al., 1994b). Are the 30 kDa proteolysis product and the 38 kDa protein purified earlier the same proteins? This seems likely in view of the fact that antibodies to the 38 kDa protein cross-react with the 90 kDa MOAT (Pikula et al., 1994a, 1994b). It is possible that the 90 kDa protein is the complete transporter containing a membrane spanning domain connected to a rather large cytoplasmic domain with the ATP-binding site. The two domains are probably connected by a linker which is susceptible to proteolytic cleavage. However, this question cannot be resolved unless the gene for the MOAT transporter is cloned, and its nucleotide and/or amino acid sequence determined.

A cDNA encoding for a protein mediating LTC_4 transport has been cloned from a small cell lung cancer cell line (Cole et al., 1992; Leier et al., 1994a). This transporter referred to as multidrug resistance associated protein (MRP) belongs to the superfamily of ABC transporters, and has characteristics similar to the LTC_4 transporter of murine mast cells (Leier et al., 1994b) and the LTC_4 transporter of liver canalicular membrane (Ishikawa et al., 1990). Both MRP and the transporter from mast cells have been shown to be 190 kDa integral membrane phosphoproteins that mediate ATP-dependent transport of LTC_4 with a high affinity (Leier et al., 1994a, 1994b). MRP can also transport DNP-SG, but at a rate about ten times less than that for LTC_4. It is possible that MRP and the LTC_4 transporter of liver canalicular membrane might be the same proteins. If so, the multispecific organic anion transporter of liver would be predicted to consist of multiple loci, one of which encodes for the 90 kDa DNP-SG transporter (Pikula et al., 1994a). Additional proteins such as MRP might be responsible for mediating transport of LTC_4 and other organic anions.

The fate of glutathione and glutathione S-conjugates after they exit from the liver cells into plasma and bile has been the subject of considerable research over the past several years (Inoue et al., 1984; Akerboom et al., 1991). The GSH S-conjugates are translocated to either the small intesntine or kidney cells for final detoxification to cysteine-S conjugates (Jones et al., 1979; Inoue et al., 1984; Lash and Jones, 1985). Delivery of the GSH S-conjugates to the renal cells takes place by two mechanisms: degradation by γ-glutamyltransferase and dipeptidase activities, which are present predominantly on the brush-border membrane, followed by uptake of the constituent amino acids into the cells, or by an energy-dependent transport system in the basolateral plasma membrane of rat kidney proximal tubules (Lash and Jones, 1985). The transport process is coupled to the uptake of Na^+ ions and is electrogenic with a net transport of one positive charge ($2\,Na^+ : 1\,RSG^-$) into the cells. In fact, the properties of this transporter are the same as those of the Na^+/glutathione co-transporter (Lash and Jones, 1984). Mutual inhibition of transport between GSH and the GSH S-conjugates showed that the same system results in the transport of both compounds (Lash and Jones, 1985). Further molecular characterization of this transporter would be of great interest.

Another system for the transport of GSH exists in the brush border membrane of the renal cells (Inoue and Morino, 1985). This system is specific for the transport of glutathione and does not transport GSH derivatives. Transport by this system is electrogenic involving the transfer of negative charge across the membrane. Unlike the system in the basolateral membrane of the renal proximal tubule (Lash and Jones, 1985), this system is not coupled to the transport of sodium. Since the electrical potential of the lumen is positive relative to the intracellular milieu, the energetically favored direction is efflux. Hence the physiological function of this brush-border membrane system is for turnover of intracellular GSH as the transporter delivers GSH to the active site of γ-glutamyltransferase in the lumen. No sodium-dependent transport of GSH has been found in liver cells (Inoue et al., 1984), probably because liver is mostly involved in the release of GSH.

B.   Methotrexate Transporters

Methotrexate (MTX), an antimetabolite, is widely used for the treatment of cancers and also for the treatment of non-neoplastic diseases. Being a folate analogue, it targets the enzyme dihydrofolate reductase (DHFR) which results in decreased thymidylate and purine biosynthesis (Bertino, 1993). However, resistance to methotrexate in leukemia cells and sarcoma cells has been shown to develop in certain cases. The genetic basis for MTX resistance *in vitro* appears to be a selective gene amplification of the dihydrofolate reductase gene or point mutations resulting in amino acid changes in the DHFR enzyme (Bertino et al., 1992; Bertino, 1993: Banerjee et al., 1994). Both of the above mechanisms might result in an increased enzyme activity of DHFR or a decreased affinity of DHFR for MTX. In addition, other mechanisms seem to play a significant role in the development of resistance to MTX. These mechanisms include decreased uptake or an increased efflux of MTX from the cells, and lowered retention of the drug due to defective polyglutamylation in the target cells (Bertino, 1993). In order to be chemotherapeutically effective, the uptake of methotrexate into the cells and its retention is critical. Hence it is important to understand the biochemical mechanisms of transport.

The uptake of methotrexate into the cells takes place by two different pathways. It is believed that the MTX-resistant cells utilize mechanisms for the efflux of the drug that differ from those of MTX influx. One mechanism for the uptake of MTX in tumor cells involves the carrier used by reduced folate coenzymes (Goldman et al., 1968). The second pathway, employs an endocytic pathway for uptake of folates and MTX (Antony, 1992).

The mouse L1210 leukemia cells have been used as a model system for the study of the reduced folate transport system. The reduced folate carrier (RFC) is a high-affinity but a low-capacity carrier mediating the inward flow of MTX as a saturable function of the external concentration (Dembo et al., 1984). It is an energy-dependent uptake system, and shows greater preference for reduced folates as compared to the oxidized folates. When the effect of different concentrations of D-glucose on unidirectional influx of MTX in leukemia L1210 cells was studied, it was found that the K_m for the influx of MTX was independent of the D-glucose concentration. In contrast, a marked difference in dependence of efflux on D-glucose concentration was observed (Dembo et al., 1984). It was seen that the influx of glucose was maximal at low concentrations (2 mM) of glucose, and it began to decline at higher cencentrations (5-7 mM). On the other hand, the rate of efflux increased throughout the range of concentration of D-glucose tested (0-10 mM). It appears that the efflux of MTX through an ATP-dependent mechanism (see below) becomes much faster as compared to its influx through another carrier as the ATP levels in the cell are replenished at high glucose concentrations. Studies on energy requirements showed that L1210 cells can accumulate MTX to intracellular levels higher than would be predicted by passive influx of an anion. The energy source was found not to be a gradient of sodium ions as the transport was insensitive to ouabain (Henderson, 1986). Since glucose was shown to be an inhibitor of accumulation, due to an indirect effect of stimulation of efflux, it was concluded that the influx depends on an energy source that is not immediately affected by the energy levels of the cell. It is now believed that reduced folate/MTX uptake takes place by an anion-exchange mechanism where uptake of folates is balanced by exchange with an intracellular counter anion, such as phosphate or AMP, of an equal negative charge. Hence MTX transport depends on the energy present in the anion gradients, and is electroneutral (Henderson and Zevely, 1983).

Since the reduced folate transporter exists in the membrane at extremely low concentrations, a variant of the L1210 cells which exhibits an increase in the rate of influx for folates was used to demonstrate that the transport of MTX into these cells indeed takes place by the reduced folate carrier (Yang et al., 1988). Using plasma membranes prepared from this variant, the transporter for MTX could be covalently labeled with N-hydroxysuccinimide derivatives of MTX. Competition studies using other folate analogues showed that 5-formyl-tetrahydrofolate was an effective inhibitor of the affinity labeling, whereas folic acid had no effect. The affinity labeled protein corresponded to a protein of roughly 48 kDa in size (Yang et al., 1988).

Changes in the RFC activity have been correlated with acquisition of cellular resistance to MTX (Trippett et al., 1992). Using cell lines deficient in RFC activity, a cDNA clone that complements the defect has been isolated using mouse leukemia L1210 cDNA library (Dixon et al., 1994). This cDNA clone has been shown to complement RFC deficiency in a human breast cancer cell line. It codes for a protein of 58 kDa, and shows structural similarity to the human glucose transporter GLUT1. The hydropathic analysis suggests that it contains 12 transmembrane domains and belongs to the superfamily of transporters referred to as the major facilitator superfamily certain members of which carry out sugar transport in mammalian cells (Marger and Saier, 1993). More recently, a human cDNA clone that complements the RFC defect in chinese hamster ovary cells has also been isolated (Williams and Flintoff, 1995). The human protein has about 50% identity at the amino acid level with the mouse protein. Although the transfection of cells defective in RFC activity with the cloned cDNA results in restoration of MTX uptake, biochemical characterization of these proteins will reveal if the encoded protein is the RFC itself or an associated function.

There is controversy over the number of specific systems for the efflux of MTX by L1210 leukemia cells. Results from one group suggest the existence of a single ATP-dependent efflux route that accounts for 90% of the efflux of methotrexate whereas efflux by the reduced folate system accounts for the remainder of the total efflux (Sirotnak and Leary, 1991; Schlemmer and Sirotnak, 1992). By using inside out membrane vesicles from the L1210 leukemia cells, they showed that the accumulation of MTX into these vesicles was dependent on the presence of ATP. Both the nonhydrolyzable analog, ATPγS, and orthovanadate were potent inhibitors of ATP-dependent uptake suggesting a requirement for hydrolysis of ATP and the involvement of a plasma membrane ATPase. Efflux of MTX by the predominant ATP-dependent route was shown to be inhibited by organic anions such as probenecid (PBCD) and bromosulfophthalein (BSP). However, work of another group seems to suggest that a certain fraction of the PBCD-sensitive efflux is not sensitive to BSP. This indicates the existence of a second BSP-insensitive, PBCD-sensitive specific route for efflux of MTX (Henderson, 1992). The function of the BSP-sensitive transporter was also inhibited by cholate (Schlemmer and Sirotnak, 1992; Henderson and Hughes, 1993) suggesting that the efflux of cholate proceeds via the MTX efflux system, and that like the anion pumps for glutathione derivatives (Akerboom et al., 1991), these pumps mediating the efflux of MTX might also have a broad anion specificity.

IV. CFTR AND PGP: CHANNELS AND PUMPS

Until recently, the different transporters were classified as channels, carriers, or pumps, and it was believed that a transport protein would be either one or the other. The discovery of transport proteins with more than one kind of activity now raises several questions regarding their evolution, and the regulation of their activities. Are these proteins examples of transitional states during their evolution from one kind of a transporter to the other? Or does it mean that a transport protein can have more than one kind of activity which is subject to a regulatory switch?

Two very intriguing examples of such proteins are the multidrug resistance P-glycoprotein and the cystic fibrosis transmembrane regulator. Both of these transport proteins belong to the family of the ABC-type ATPases (Higgins, 1992), which includes members from the prokaryotic and the eukaryotic systems. Both the proteins are hypothesized to have evolved by a gene duplication and a fusion event as they contain two homologous halves, with each half containing one membrane spanning domain and one nucleotide binding domain (NBD; Chen et al., 1986a). The two halves are connected by a linker region of about 55 amino acids in Pgp, and by an R domain of about 241 amino acids in the CFTR protein (Orr et al., 1993; Collins, 1992). It has been shown that both Pgp and CFTR are phosphorylated by protein kinases in the linker region (Orr et al., 1993) and the R domain (Collins, 1992), respectively. Both the Pgp and the CFTR are glycosylated proteins of about 170 kDa in size.

The CFTR protein is required for normal flux of chloride ions across the respiratory epithelial cell membranes, and it has been identified as being nonfunctional in the cystic fibrosis patients. In 70% of CFTR patients, improper functioning of the CFTR protein results from a 3 bp deletion in exon 10 of the gene causing a deletion of a phenylalanine residue at position 508 in the 1480 amino acid long CFTR protein. This deletion was found to be responsible for lack of activation of a chloride conductance in response to the cAMP-activated protein kinase A (PKA). In spite of the strong similarity of the CFTR protein to an active transporter, there is little doubt that it can function as an anion-selective channel (Bear et al., 1992). The activity of this channel is under the dual control of PKA and ATP. The activity of cAMP-regulated PKA results in phosphorylation of several serine residues in the R domain that lies between the two homologous halves of the protein, and brings about activation of the chloride channel. The R domain seems to be inhibitory to channel function in its unphosphorylated state. Deletion of a substantial portion of the R domain results in a constitutively active chloride channel (Rich et al., 1991). It has been proposed that ATP binding to the phosphorylated form of CFTR and hydrolysis result in a conformational change causing the channel to acquire an open conformation allowing passive flow of chloride ions (Collins, 1992). Even though there is evidence that the two NBDs in CFTR interact with ATP, no direct evidence for the role of hydrolysis in opening the channel has been obtained. The indirect evidence in support of hydrolysis comes from studies demonstrating that hydrolyzable nucleotide triphosphates and divalent cations are required to open the chloride channel (Anderson et al., 1991). Hence, CFTR is an interesting example where gating of the anion channel seems to require binding and/or hydrolysis of ATP. Indeed in this context, the difference between channels and carriers seems to disappear.

The P-glycoprotein is overexpressed in cancer cells and was identified as a result of its ability to confer resistance on cancer cells to a variety of chemotherapeutic drugs (Chen et al., 1986a; Endicott and Ling, 1989). This activity of Pgp is ATP-dependent, and a direct role for hydrolysis of ATP in pumping out the drugs has been demonstrated using mouse Mdr3 Pgp expressed in yeast secretory vesicles (Reutz and Gros, 1994). In addition to the drug transport activity, it has been proposed that Pgp can either act as a volume-regulated chloride channel or it can regulate the activity of an endogenous chloride channel in the cell (Gill et al., 1992). Both the drug transport and the channel activities require ATP binding. However, the transport mode also requires hydrolysis of ATP. The active drug transport mode and the chloride channel activity of Pgp were shown to be mutually exclusive (Gill et al., 1992). In patch clamp experiments, the presence of both ATP and the drug prevented channel activation, whereas the presence of ATP alone under hypotonic conditions activated the channel. Certain mutations in the nucleotide-binding folds of Pgp which prevent ATP hydrolysis but not ATP binding, have also allowed separation of the two activities. Such mutant proteins seem to be locked in the conformation required for channel function (Gill et al., 1992). All of these observations imply a role for Pgp in chloride channel activity, but do not differentiate between the possibility of Pgp being the chloride channel itself or regulating the activity of another channel. This question can best be resolved by purifying Pgp to homogeneity and studying transport function in reconstituted proteoliposomes.

Recent studies carried out with HeLa cells argue against Pgp being the Cl⁻ channel itself. Instead in its phosphorylated form, Pgp might down-regulate the activity of an endogenous chloride channel (Hardy et al., 1995). HeLa cells normally do not express Pgp, but possess a chloride channel activity similar to the activity associated with the expression of Pgp. Transfection of HeLa cells with a plasmid containing the *MDR1* gene makes the channel activity sensitive to the action of a PKC activator, phorbol 12-myristate 13-acetate (TPA), indicating this might be due to phosphorylation of Pgp (Hardy et al., 1995). Mutagenesis of the serine residues in the Pgp linker region, which have been shown to be the sites for phosphorylation by PKC, results in insensitivity of the channel to the action of TPA. Hence phosphorylation of the linker region might be the crucial regulatory switch required to convert it from the transporter to the channel mode.

The physiological significance of regulation of the Cl^- channel activity by Pgp and the role of direct pumping activity of Pgp in conferring the MDR phenotype is not clear. Studies pertaining to the measurements of intracellular pH and plasma membrane electrical potential in MDR cell lines (Roepe et al., 1993,1994) seem to suggest that the MDR phenotype is quite complicated and might be the result of several different mechanisms. Roepe and coworkers have shown that the Pgp expressing cell lines have an increased intracellular pH and a lowered plasma membrane electrical potential. Based on these observations, they have proposed an alternative model, the altered partitioning hypothesis for the action of Pgp suggesting that the MDR phenotype is an indirect result of membrane depolarization, and that lowered drug accumulation may not be the result of drug pumping activity of Pgp. According to their hypothesis, the membrane depolarization could result from changes in chloride conductance (Gill et al., 1992; Hardy et al., 1995) or from the altered activity of an ion exchanger (Roepe et al., 1993) due to Pgp expression. That an increase in intracellular pH does indeed result in a lowered retention of a chemotherapeutic drug, has been shown in the case of daunomycin (Simon et al., 1994). It has also been demonstrated recently that depolarization of the plasma membrane in NIH3T3 cells that do not express Pgp, resulting from overexpression of CFTR, also gives the drug resistance phenotype (Dr. Paul Roepe, personal communication). These observations further complicate our understanding of the role of Pgp in MDR. Since Pgp is a known drug-stimulated ATPase (Ambudkar et al., 1992; Sharom et al., 1993), and since it can mediate ATP dependent transport of drugs into vesicles (Reutz and Gros, 1994), the role of Pgp as an active drug pump in the MDR phenotype can not be totally ignored. It is very likely that other mechanisms contribute significantly and caution in interpretation of the results is needed. Further experimentation would be required to clearly determine the contribution of each mechanism towards lowering drug retention in MDR cells.

V. CONCLUDING REMARKS

This chapter illustrates how much we already know about the structure and function of various transport proteins. However, the molecular details of how or whether a system can switch from being a channel activator to an active transporter, or what constitutes the specific recognition between a secondary porter and an ATPase that allows it to become an active transporter are far from known. One thing is clear: the relationships between the different ion transport systems in a cell and their regulation are more intricate than is obvious after an initial assessment.

ACKNOWLEDGMENTS

Thanks are due to Drs. Barry Rosen and P. C. Tai for a critical reading of the manuscript and suggestions; and to Dr. Paul Roepe for making available unpublished manuscripts from his laboratory for use in this article. I am indebted to Drs. Debu Banerjee and Amit Banerjee for sharing their knowledge and for useful discussions on methotrexate transport and drug resistance systems. Thanks are also due to Ashley Slappy for help in preparation of the manuscript in its final form.

REFERENCES

Akerboom, T.P.M., Narayanaswami, V., Kunst, M., & Sies, H. (1991). ATP-dependent S-(2,4-dinitrophenyl)glutathione transport in canalicular plasma membrane vesicles from rat liver. J. Biol. Chem. 266, 13147-13152.

Ambudkar, S.V., Larson, T.J., & Maloney, P.C. (1986). Reconstitution of sugar phosphate transport systems of Escherichia coli. J. Biol. Chem. 261, 9083-9086.

Ambudkar, S.V., Lelong, I. H., Zhang, J., Cardarelli, C.O., Gottesman, M.M., & Pastan, I. (1992). Partial purification and reconstitution of the human multidrug-resistance pump: Characterization of the drug-stimulatable ATP hydrolysis. Proc. Natl. Acad. Sci. USA 89, 8472-8476.

Ames, G. F.-L., & Joshi, A. (1990). Energy coupling in bacterial periplasmic permeases. J. Bacteriol. 172, 4133-4137.

Anderson, M.P., Berger, H.A., Rich, D.P., Gregory, R.J., Smith, A.E., & Welsh, M.J.(1991). Nucleoside triphosphates are required to open the CFTR chloride channel. Cell 67, 775-784.

Antony, A.C. (1992). The biological chemistry of folate receptors. Blood 79, 2807-2820.

Apontoweil, P., & Berends, W. (1975). Isolation and initial characterization of glutathione-deficient mutants of Escherichia coli K-12. Biochem. Biophys. Acta 399, 10-22.

Arias, I.M., Che, M., Gatmaitan, Z., Leveille, C., Nishida, T., & St Pierre, M. (1993). The biology of the bile canaliculus. Hepatology (Baltimore) 17, 318-329.

Banerjee, D., Schweitzer, B. I., Volkenandt, M., Li, M.X., Waltham, M., Mineishi, S., Zhao, S.C., Bertino, J.R. (1994). Transfection with a cDNA encoding a Ser31 or Ser34 mutant human dihydrofolate reductase into chinese hamster ovary and mouse marrow progenitor cells confers methotrexate resistance. Gene 139, 269-274.

Banks, G.R., & Sedgwick, S.G. (1986). Direct ATP photolabeling of Escherichia coli RecA proteins: Identification of regions required for ATP binding. Biochemistry 25, 5882-5889.

Bear., C.E., Li, C., Kartner, N., Bridges, R.J., Jensen, T.J., Ramjeesingh, M., & Riordan, J.R. (1992). Purification and functional reconstitution of the Cystic fibrosis transmembrane conductance regulator. Cell 68, 809-818.

Bennet, R.L., & Malamy, M.H. (1970). Arsenate resistant mutants of Escherichia coli and phosphate transport. Biochem Biophys. Res. Commun. 10, 496-503.

Bertino, J.R. (1993). Ode to methotrexate. J. Clin. Oncology 11, 5-14.

Bertino, J.R., Li, W.W., Lin, J., Trippett, T., Goker, E., Schweitzer, B., & Banerjee, D. (1992). Enzymes of the thymidylate cycle as targets for chemotherapeutic agents: mechanisms of resistance. Mount Sinai J. Med. 59 (5), 391-395.

Bhattacharjee, H., Li, J., Ksenzenko, M. Y., & Rosen, B.P. (1995). Role of the cysteinyl residues in metalloactivation of the oxyanion-translocating ArsA ATPase. J. Biol. Chem. 270, 11245-11250.

Biswas, S.B., & Kornberg, A. (1984). Nucleoside triphosphate binding to DNA polymerase III holoenzyme of Escherichia coli: a direct photoaffinity labeling study. J. Biol. Chem. 259, 7990-7993.

Boyd, D., Traxler, B., & Beckwith, J. (1993). Analysis of the topology of a membrane protein by using a minimum number of alkaline phosphatase fusions. J. Bacteriol. 175, 553-556.

Bröer, S., Ji, G., Bröer, A., & Silver, S. (1993). Arsenic efflux governed by the arsenic resistance determinant of Staphylococcus aureus plasmid pl258. J. Bacteriol. 175, 3480-3485.

Carlin, A., Shi, W., Dey, S., & Rosen, B.P. (1995). The *ars* operon of *Escherichia coli* confers arsenical and antimonial resistance. J. Bacteriol. 177, 981-986.

Chen, C.J., Chin, J.E., Ueda, K., Clark, D.P., Pastan, I, Gottesman, M.M., & Roninson, I.B. (1986a). Internal duplication and homology with bacterial transport proteins in the *MDR1* (P-glycoprotein) gene from multidrug-resistant human cells. Cell 47, 381-389.

Chen, C.M., Misra, T., Silver, S., & Rosen, B.P. (1986b). Nucleotide sequence of the structural genes for an anion pump: the plasmid-encoded arsenical resistance operon. J. Biol. Chem. 261, 15030-15038.

Chen, C.M., Mobley, H.L.T., & Rosen, B.P. (1985). Separate resistances to arsenate and arsenite (antimonate) encoded by the arsenical resistance operon of R factor R773. J. Bacteriol. 161, 758-763.

Cole, S.P.C., Bhardwaj, G., Gerlach, J.H., Mackie, J.E., Grant, C.E., Almquist, K.C., Stewart, A.J., Kurz, E.U., Ducan, A.M.V., & Deeley, R.G. (1992). Overexpression of a transporter gene in a multidrug-resistant human lung cancer cell line. Science 258, 1650-1654.

Collins, F.S. (1992). Cystic fibrosis: molecular biology and therapeutic implications. Science 256, 774-779.

Cox, G.B., Webb, D., Godovac-Zimmermann, J., & Rosenberg, H. (1988). Arg-220 of the PstA protein is required for phosphate transport through the phosphate-specific transport system of *Escherichia coli* but not for alkaline phosphatase repression. J. Bacteriol. 170, 2283-2286.

Cox, G.B., Webb, D., & Rosenberg, H. (1989). Specific amino acid residues in both the PstB and PstC proteins are required for phosphate transport by the *Escherichia coli* Pst system. J. Bacteriol. 171, 1531-1534.

Davidson, A.L., & Nikaido, H. (1991). Purification and characterization of the membrane-associated components of the maltose transport system from *Escherichia coli*. J. Biol. Chem. 266, 8946-8951.

Davis, M.A., Martin, K.A., & Austin, S.J. (1992). Biochemical activities of the ParA partition protein of the P1 plasmid. Mol. Microbiol. 6, 1141-1147.

De Boer, P.A.J., Crossley, R.E., Hand, A.R., Rothfield, L.I. (1991). The MinD protein is a membrane ATPase required for the correct placement of the *Escherichia coli* division site. EMBO J. 10, 4371-4380.

Dembo, M., Sirotnak, F.M., & Moccio, D.M. (1984). Effects of metabolic deprivation on methotrexate transport in L1210 leukemia cells: further evidence for separate influx and efflux systems with different energetic requirements. J. Memb. Biol. 78, 9-17.

Dey, S., Dou, D., & Rosen, B.P. (1994). ATP-dependent arsenite transport in everted membrane vesicles of *Escherichia coli*. J. Biol. Chem. 269, 25442-25446.

Dey, S., & Rosen, B.P. (1995). Dual mode of energy coupling by the oxyanion-translocating ArsB protein. J. Bacteriol. 177, 385-389.

Dixon, K.H., Lanpher, B.C., Chiu, J., Kelley, K., & Cowan, K.H. (1994). A novel cDNA restores reduced folate carrie activity and methotrexate sensitivity to transport deficient cells. J. Biol. Chem. 269, 17-20.

Doige, C.A., & Ames, G. F.-L. (1993). ATP-dependent transport systems in bacteria and humans: Relevance to cystic fibrosis and multidrug resistance. Ann. Rev. Microbiol. 47, 291-319.

Dou, D., Dey, S., & Rosen, B.P. (1994). A functional chimeric membrane subunit of an ion-transloating ATPase. Antonie van Leeuwenhoek 65, 359-368.

Endicott, J.A., & Ling, V. (1989). The biochemistry of P-glycoprotein-mediated multidrug resistance. Annu. Rev. Biochem. 58, 136-171.

Fath M., & Kolter, R. (1993). ABC transporters: bacterial exporters. Microbiol. Rev. 57, 995-1017.

Futai, M. & Kanazawa, H. (1983). Structure and function of proton-translocating ATPase (F_0F_1): biochemical and molecular biological approaches. Microbiol. Rev. 47,285-313.

Gill., D.R., Hyde, S.C., Higgins, C.F., Valverde, M.A., Mintenig, G.M., & Sepulveda, F.V. (1992). Separation of drug transport and chloride channel functions of the human multidrug resistance P-glycoprotein. Cell 71, 23-32.

Gladysheva, T.B., Oden, L.K., & Rosen, B.P. (1994). Properties of the arsenate reductase of plasmid R773. Biochemistry 33, 7288-7293.

Goldman, I.D., Lichenstein, W.S., & Oliveiro, V.T. (1968). Carrier mediated transport of the folic acid analog methotrexate in the l1210 leukemia cells. J. Biol. Chem. 243, 5007-5017.

Gottesman, M.M., & Pastan, I. (1993). Biochemistry of multidrug resistance by the multidrug transporter. Ann. Rev. Biochem. 62, 385-427.

Hardy, S.P., Goodfellow, H.R., Valverde, M.A., Gill, D.R., Sepulveda, F.V., & Higgins, C.F. (1995). Protein kinase C-mediated phosphorylation of the human multidrug resistance P-glycoprotein regulates cell volume-activated chloride channels. EMBO J. 14, 68-75.

Hedges, R.W., & Baumberg, S. (1973). Resistance to arsenic compounds conferred by a plasmid transmissible between strains of Escherichia coli. J. Bacteriol. 115, 459-460.

Heijn, M., Oude Elferink, R.P.J., & Jansen, P.L.M. (1992) ATP-dependent multispecific organic transport system in rat erythrocyte membrane vesicles. Am. J. Physiol. 262, C104-C 110.

Henderson, G.B. (1986). In: Folates and Pterins, Vol. 3 (Blakley, R.L., & Whitehead, V.M., Eds), pp. 207-250, John Wiley, New York.

Henderson, G.B. (1992). Separation of inhibitor specificity of a unidirectional efflux route for methotrexate in L1210 cells. Biochem. Biophys. Acta 1110, 137-143.

Henderson, G.B., & Hughes, T.R. (1993). Altered expression of unidirectional extrusion routes for methotrexate and cholate in an efflux variant of L1210 cells. Biochem. Biophys. Acta 1152, 91-98.

Henderson, G.B., & Zevely, E.M. (1983). Structural requirements for anion substrates of the methotrexate transport system in L1210 cells. Arch. Biochem. Biophys. 221, 438-446.

Higgins, C.F. (1992). ABC trasnporters: from microorganisms to man. Ann. Rev. Cell Biol. 8, 67-113.

Hsu, C.M., Kaur, P., Karkaria, C.E., Steiner, R.F., & Rosen, B.P. (1991). Substrate-induced dimerization of the ArsA protein, the catalytic component of an anion-translocating ATPase. J. Biol. Chem. 266, 2327-2332.

Hsu, C.M., & Rosen, B.P. (1989). Characterization of the catalytic subunit of an anion pump. J. Biol. Chem. 264, 17349-17354.

Hyde, S.C., Emsley, P., Hartshorn, M.J., Mimmack, M.M., Gileadi, U., Pearce, S.R., Gallager, M.P., Gill, D.R., Hubbard, R.E., & Higgins, C.F. (1990). Structural model of ATP-binding proteins associated with cystic fibrosis, multidrug resistance and bacterial transport. Nature 346, 362-365.

Inoue, M., Kinne, R., Tran, T., & Arias, M.I. (1984). Glutathione transport across hepatocyte plasma membranes; analysis using isolated rat liver sinusoidal membrane vesicles. Eur. J. Biochem. 138, 491-495.

Inoue, M., & Morino, Y. (1985). Direct evidence for the role of the membrane potential in glutathione transport by renal brush-border membranes. J. Biol. Chem. 260, 326-331.

Ishikawa, T. (1992). ATP-dependent glutathione S-conjugate export pump. Trend. Biochem. Sci. 17, 463-468.

Ishikawa, T., Muller, M., Klunemann, C., Schaub, T., & Keppler, D. (1990). ATP-dependent primary active transport of cysteinyl leukotrienes across liver canalicular membrane. Role of the ATP-dependent transport for glutathione S-conjugates. J. Biol. Chem. 265, 19279-19286.

Ji, G., & Silver, S. (1992a). Regulation and expression of the arsenic resistance operon from Staphylococcus aureus plasmid pI258. J. Bacteriol. 174, 3684-3694.

Ji, G., & Silver, S. (1992b). Reduction of arsenate to arsenite by the ArsC protein of the arsenic resistance operon of the Staphylococcus aureus plasmid pI258. Proc. Natl. Acad. Sci. USA 89, 9474-9478.

Jones, D.P., Moldeus, P., Stead, A.H., Ormstad, K., Jornvall, H., & Orrenius, S. (1979). Metabolism of glutathione and a glutathione conjugate by isolated kidney cells. J. Biol. Chem. 254, 2787-2792.

Kadner, R.J., Webber, C.A., & Island, M.D. (1993). The family of organo-phosphate transport proteins includes a transmembrane regulatory protein. J. Bioenerg. Biomemb. 25, 637-645.

Karkaria, C.E., Chen, C.M., & Rosen, B.P. (1990). Mutagenesis of a nucleotide binding site of an anion-translocating ATPase. J. Biol. Chem., 265, 7832-7836.

Karkaria, C.E., & Rosen, B.P. (1991). Trinitrophenyl-ATP binding to the wild type and mutant ArsA proteins. Arch. Biochem. Biophys. 288, 107-111.

Kaur, P., & Rosen, B.P. (1992a). Plasmid-encoded resistance to arsenic and antimony. Plasmid 27, 29-40.

Kaur, P., & Rosen, B.P. (1992b). Mutagenesis of the second putative nucleotide binding site of an anion-translocating ATPase. J. Biol. Chem. 267, 19272-19277.

Kaur, P., & Rosen, B.P. (1993). Complementation between nucleotide binding domains in an anion translocating ATPase. J. Bacteriol. 175, 351-357.

Kaur, P., & Rosen, B.P. (1994a). In vitro assembly of an anion-stimulated ATPase from peptide fragments. J. Biol. Chem., 269, 9698-9704.

Kaur, P., & Rosen, B.P. (1994b). Identification of the site of $[\alpha-^{32}P]$ ATP adduct formation in the ArsA protein. Biochemistry. 33, 6456-6461.

Kierdaszuk, B., & Eriksson, S. (1988). Direct photoaffinity labeling of ribonucleotide reductase from *Escherichia coli* using dTTP: characterization of the photoproducts. Biochemistry 27, 4952-4956.

Klotz, I.M., Darnall, D.W., & Langerman, N.R. (1975). In: The Proteins (Neurath, H., & Hill, R.L., Eds), pp. 293-411, Academic Press, New York.

LaBelle, E.F., Singh, S.V., Srivastva, S.K., & Awasthi, Y.C. (1986). Dintrophenyl glutathione efflux from human erythrocytes is primary active ATP-dependent transport. Biochem. J. 238, 443-449.

Lash, L. H., & Jones, D.P. (1984). Renal glutathione transport. Characteristics of the sodium-dependent system in the basal-lateral membrane. J. Biol. Chem. 259, 14508-14514.

Lash, L.H., & Jones, D.P. (1985). Uptake of the glutathione conjugate S-(1,2-dichlorovinyl)glutathione by renal basal-lateral membrane vesicles and isolated kidney cells. Mol. Pharmacol. 28, 278-282.

Leier, I., Jedlitschky, G., Buchholz, G., Cole, S.P.C., Deeley, R.G., & Keppler, D. (1994a). The MRP gene encodes an ATP-dependent export pump for leukotriene C_4 and structurally related conjugates. J. Biol. Chem. 269, 27807-27810.

Leier, I., Jedlitschky, G., Buchholz, G., & Keppler, D. (1994b). Characterization of the ATP-dependent leukotriene C_4 export carrier in mastocytoma cells. Eur. J. Biochem. 220, 599-606.

Linse, K., & Mandelkow, E.M. (1988). The GTP binding peptide of β-tubulin. Localization by direct photoaffinity labeling and comparison with nucleotide binding proteins. J. Biol. Chem. 263, 15205-15210.

Luecke, H., & Quiocho, F.A. (1990). High specificity of a phosphate transport protein determined by hyrogen bonds. Nature (London) 347, 402-406.

Makino, K.H., Shinagawa, H., Amemura, M., Kawamoto, T., Yamada, M., & Nakata, A. (1989). Signal transduction in the phosphate regulon of *Escherichia coli* involves phosphotransfer between PhoR and PhoB proteins. J. Mol. Biol. 210, 551-559.

Makino, K., Shinagawa, H., Amemura, M., Kimura, S., Nakata, A., & Ishihama, A. (1988). Regulation of phosphate regulon of *Escherichia coli*: activation of *pstS* transcription by PhoB protein in vitro. J. Mol. Biol. 203, 85-95.

Maloney, P.C., Ambudkar, S.V., Vellareddy, A., Sonna, L.A., & Varadhachary, A. (1990). Anion-exchange mechanisms in bacteria. Microbiol. Rev. 54, 1-17.

Marger, M.D., & Saier, M.H.,Jr. (1993). A major superfamily of transmembrane facilitators that catalyse uniport, symport and antiport. Trends Biochem. Sci. 18, 13-20

Meier, P. (1993). In: Hepatic Transport and Bile Secretion: Physiology and Pathophysiology (Tavoloni, N., & Berk, P.D., eds), pp. 587-596, Raven Press, New York.

Mevarech, M., Rice, D., & Haselkorn, R. (1980). Nucleotide sequence of a cyanobacterial *nifH* gene coding for nitrogenase reductase. Proc. Natl. Acad. USA 77, 6476-6480.

Mobley, H.L.T., Chen, C.M., Silver, S., & Rosen, B.P. (1983). Cloning and expression of R-factor mediated arsenate resistance in *Escherichia coli*. Mol. Gen. Genet. 191, 421-426.

Mobley, H.L.T., & Rosen, B.P.(1982). Energetics of plasmid-mediated arsenate resistance in *Escherichia coli*. Proc. Natl. Acad. Sci. USA. 79, 6119-6122.

Muda, M., Rao, N.N., & Torriani, A. (1992). Role of PhoU in phosphate transport and alkaline phosphatase regulation. J. Bacteriol. 174, 8057-8064.

Novick, R.P., & Roth, C. (1968). Plasmid-linked resistance to inorganic salts in *Staphylococcus aureus*. J. Bacteriol. 95, 1335-1342.

Oden, K.L., Gladysheva, T.B., & Rosen, B.P. (1994). Arsenate reduction mediated by the plasmid-encoded ArsC protein is coupled to glutathione. Mol. Microbiol. 12, 301-306.

Orr, G.A., Han, E.K.H., Browne, P.C., Nieves, E., O'Connor, B.M., Yang, C.P.H., & Horwitz, S.B. (1993). Identification of the major phosphorylation domain of murine mdr1b P-glycoprotein. Analysis of the protein kinase A and protein kinase C phosphorylation sites. J. Biol. Chem. 268, 25054-25062.

Oesterhelt, D., & Tittor, J. (1989). Two pumps, one principle: light driven ion transport in halobacteria. Trends Biochem. Sci. 14, 57-61.

Oude-Elferink, R.P.J., Ottenhof, R., Liefting, W., De Haan, J., & Jansen, P.L.M. (1989). Hepatobiliary transport of glutathione and glutathione conjugates in rats with hyperbilirubinemia. J. Clin. Invest. 84, 476-483.

Overduin, P., Boos, W., 7 Tommassen, J. (1988). Nucleotide sequence of the *ugp* genes of *Escherichia coli* K-12: homology with the maltose system. Mol. Microbiol. 2, 767-775.

Owolabi, J.B., & Rosen, B.P. (1990). Differential mRNA stability controls relative gene expression within the plasmid-encoded arsenical resistance operon. J. Bacteriol. 172, 2367-2371.

Parsonage, D., Al-Shawi, M.K., & Senior, A.E. (1988). Directed mutations of the conserved lysine 155 in the catalytic nucleotide binding domain of β-subunit of F1-ATPase from *Escherichia coli*. J. Biol. Chem. 263, 4740-4744.

Pederson, P.L., & Carafoli, E. (1987). Ion motive ATPases. I. Ubiquity, properties, and significance to cell function. Trends Biochem. Sci. 12, 146-150.

Penefsky, H.S., & Cross, R.L. (1991). Structure and mechanism of F_0F_1-type ATP synthases and ATPases. Adv. Enzymol. Relat. Areas Mol. Biol. 64, 173-214.

Pikula, S., Hayden, J.B., Awasthi, S., Awasthi, Y.C., & Zimniak, P. (1994a). Organic anion-transporting ATPase of rat liver. I. Purification, photoaffinity labeling, and regulation by phosphorylation. J. Biol. Chem. 269, 27566-27573.

Pikula, S., Hayden, J.B., Awasthi, S., Awasthi, Y.C., & Zimniak, P. (1994b). Organic anion-transporting ATPase of rat liver. II. Functional reconstitution of active transport and regulation by phosphorylation. J. Biol. Chem. 269, 27574-27579.

Rao, R., Pagan, J., & Senior, A.E. (1988). Directed mutagenesis of the strongly conserved lysine 175 in the proposed nucleotide binding domain of α-subunit from *Escherichia coli* F1-ATPase. J. Biol. Chem. 263, 15957-15963.

Rao, N.N., & Torriani, A. (1990). Molecular aspects of phosphate transport in *Escherichia coli*. Mol. Microbiol. 4, 1083-1090.

Reinstein, J., Schlichting, I., & Wittinghofer, A. (1990). Structurally and catalytically important residues in the phosphate binding loop of adenylate kinase of *Escherichia coli*. Biochemistry 29, 7451-7459.

Rich, D.P., Gregory, R.J., Anderson, M.P., Manavalan, P., Smith, A.E., & Welsh, M.J. (1991). Effect of deleting the R domain on CFTR-generated chloride channels. Science 253, 205-207.

Roepe, P.D., Wei, L.Y., Cruz, J., & Carlson, D. (1993). Lower electrical membrane potential and altered pHi homeostasis in multidrug-resistant (MDR) cells: Further characterization of a series of MDR cell lines expressing different levels of P-glycoprotein. Biochemistry 32, 11042-11056.

Roepe, P.D., Weisburg, J.H., Luz, J.G., Hoffman, M.M., & Wei, L.Y. (1994). Novel Cl$^-$-depenedent intracellular pH regulation in murine MDR 1 transfectants and potential implications. Biochemistry 33, 11008-11015.

Rosen, B.P., Bhattacharjee, H., & Shi, W. (1995). Mechanisms of metalloregulation of an anion-translocating ATPase. J. Bioenerg. Biomemb. 27, 85-91.

Rosen, B.P., & Borbolla, M.G. (1984). A plasmid-encoded arsenite pump produces resistance in *Escherichia coli*. Biochem. Biophys. Res. Commuun. 124, 760-765.

Rosen, B.P., Dey, S., Dou, D., Ji, G., Kaur, P., Ksenzenko, M., Silver, S., & Wu, J. (1992). Evolution of an ion-translocating ATPase. Ann. New York Acad. Sci. 671, 257-272.

Rosen, B.P., Weigel, W., Karkaria, C. & Gangola, P. (1988). Molecular characterization of an anion pump. The *arsA* gene product is an arsenite(antimonate)-stimulated ATPase. J. Biol. Chem. 263, 3067-3070.

Rosenberg, H. (1987). In: Ion Transport in Prokaryotes (Rosen, B.P., & Silver, S., Eds.), pp. 205-248, Academic Press, New York.

Rosenstein, R., Peschel, P., Weiland, B., & Gotz, F. (1992). Expression and regulation of the antimonite, arsenite and arsenate resistance operon in *Staphylococcus xylosus* plasmid pSX267. J. Bacteriol. 174, 3676-3683.

Reutz, S., & Gros, P. (1994). Functional expression of P-glycoproteins in secretory vesicles. J. Biol. Chem. 269, 12277-12284.

Saier, M.H., Jr. (1994). Computer aided analysis of transport protein sequences: gleaning evidence concerning function, structure, biogenesis, and evolution. Microbiol. Rev. 58, 71-93.

San Francisco, M.J.D., Hope, C.L., Owolabi, J.B., Tisa, L.S., & Rosen, B.P. (1990). Identification of the metalloreglatory element of the plasmid-encoded arsenical resistance operon. Nucleic Acids Res. 18, 619-624.

Schlemmer, S.R., & Sirotnak, F.M. (1992). Energy-dependent efflux of methotrexate in L1210 leukemia cells. Evidence for the role of an ATPase obtained with inside-out plasma membrane vesicles. J. Biol. Chem. 267, 14746-14752.

Scholten, M., & Tommassen, J. (1993). Topology of the PhoR protein of *E. coli* and functional analysis of internal deletion mutants. Mol. Microbiol. 8, 269-275.

Senior, A. (1990). The proton-translocating ATPase of *Escherichia coli*. Ann. Rev. Biophys. Chem. 10, 7-41.

Sharma, R., Gupta, S., Singh, S.V., Medh, R.D., Ahmad, H., LaBelle, E.F., & Awasthi, Y.C. (1990). Purification and characterization of dinitrophenylglutathione ATPase of human erythrocytes and its expression in other tissues. Biochem. Biophys. Res. Commun. 171, 155-161.

Sharom, F.J., Yu, X., & Doige, C.A. (1993). Functional reconstitution of drug transport and ATPase activity in proteoliposomes containing partially purified P-glycoprotein. J. Biol. Chem. 268, 24197-24202.

Sigal, I.S., Gibbs, J.B., D'Alonzo, J.S., Temeles, G.C., Wolanski, B.S., Socher, S.H., & Scolnick, E.M. (1986). Mutant ras-encoded proteins with altered nucleotide binding exert dominant biological effects. Proc. Natl. Acad. Sci. USA 83, 952-956.

Silver, S., Budd, K., Leahy, K.M., Shaw, W.V., Hammond, D., Novick, R.P., Willsky, G.R., Malamy, M.H., & Rosenberg, H. (1981). Induible plasmid-determined resistance to arsenate, arsenite and antimony (III) in *Escherichia coli* and *Staphylococcus aureus*. J. Bacteriol. 146, 983-996.

Silver, S., Ji, G., Broer, S., Dey, S., Dou, D., & Rosen, B.P. (1993). Orphan enzyme or patriarch of a new tribe: the arsenic resistance ATPase of bacterial plasmids. Mol. Microbiol. 8, 637-642.

Silver, S., & Keach, D. (1982). Energy-dependent arsenate efflux: the mechanism of plasmid-mediated resistance. Proc. Natl. Acad. Sci. USA 79, 6114-6118.

Silver, S., & Walderhaug, M. (1992). Gene regulation of plasmid- and chromosome-determined inorganic ion transport in bacteria. Microbiol. Rev. 56, 195-228.

Simon, S., Roy, D., & Schindler, M. (1994). Intracellular pH and the control of multidrug resistance. Proc. Natl. Acad. Sci. USA 91, 1128-1132.

Sirotnak, F.M., & O'Leary, D.F. (1991). The issues of transport multiplicity and energetics pertaining to methotrexate efflux in L1210 cells addressed by an analysis of cis and trans effects of inhibitors. Cancer. Res. 51, 1412-1417.

Sofia, H.J., Burland, V., Daniels, G., Plunkett, III, & Blattner, F.R. (1994). Analysis of the *Escherichia coli* genome. V. DNA sequence of the region from 76.0 to 81.5 minutes. Nucleic Acids Res. 22, 2576-2586.

Sonna, L.A., Ambudkar, S.V., & Maloney, P.C. (1988). The mechanism of glucose 6-phosphate transport by *Escherichia coli*. J. Biol. Chem. 263, 6625-6630.

Stock, J.B., Ninfa, A.J., & Stock, A.M. (1989). Protein phosphorylation and regulation of adaptive responses in bacteria. Microbiol. Rev. 53, 450-490.

Surin, B.P., Cox, G.B., & Rosenberg, H. (1987). In: Phosphate Metabolism and Cellular Regulation in Microorganisms. (Torriani, A., Rothman, F.G., Silver, S., Wright, A., & Yagil, E., Eds.), pp. 145-149, American Society for Microbiology, Washington, D.C.

Surin, B.P., Rosenberg H., & Cox, G.B. (1985). Phosphate-specific transport system of *Es-cherichia coli*: nucleotide sequence and gene-polypeptide relationships. J. Bacteriol. 161, 189-198.

Tisa, L.S., & Rosen, B.P. (1990a). Molecular characterization of an anion pump: The ArsB protein is the membrane anchor for the ArsA protein. J. Biol. Chem. 265, 190-194.

Tisa, L.S., & Rosen, B.P. (1990b). Plasmid-Encoded Transport Mechanisms, J. Bioenerg. Biomemb. 22, 493-507.

Tommassen, J., & Lugtenberg, B. (1980). Outer membrane protein e of *Escherichia coli* K-12 is co-regulated with alkaline phosphatase. J. Bacteriol. 143, 151-157.

Torriani, A. (1990). From cell membranes to nucleotides: the phosphate regulon in *Es-cherichia coli*. Bioessays 12, 371-376.

Torriani, A. (1994). In: Phosphate in Microorganisms. Cellular and molecular biology (Torriani, A., Yagil, E., & Silver, S., eds), ASM Press, Washington DC, pp. 1-4.

Trippett, T., Schlemmer, S.R., Elisseyeff, Y., Goker, E., Wachter, M., Steinherz, P., Tan, C., Berman, E., Wright, J.E., Rosowky, A., Schweitzer, B., & Bertino, J.R. (1992). Defective transport as a mechanism of acquired resistance to methotrexate in patients with acute lymphocytic leukemia. Blood 80, 1158-1162.

Turner, R.J., Hou, Y., Weiner, J.H., & Taylor, D.E. (1992). The arsenical ATPase efflux pump mediates tellurite resistance. J. Bacteriol. 174, 3092-3094.

Walker, J.E., Saraste, M., Runswick, M.J., & Gay, N.J. (1982). Distantly related sequences in the α- and β5-subunits of the ATP synthase, myosin kinases and other ATP-requiring enzymes and a common nucleotide binding fold. EMBO J. 1, 945-951.

Wanner, B.L. (1993). Gene regulation by phosphate in enteric bacteria. J. Cell. Biochem. 51, 47-54.

Williams, F.M.R., & Flintoff, W. F. (1995). Isolation of a human cDNA that complements a mutant hamster cell defective in methotrexate uptake. J. Biol. Chem. 270, 2987-2992.

Wu, J.H., & Rosen, B.P. (1993a). Metalloregulated expression of the *ars* operon. J. Biol. Chem. 268, 52-58.

Wu, J.H., & Rosen, B.P. (1993b). The *arsD* gene encodes a second *trans*-acting regulatory protein of the plasmid-encoded arsenical resistance operon. Mol. Microbiol. 8, 615-623.

Wu, J.H., Tisa, L.S., & Rosen, B.P. (1992). Membrane topology of the ArsB protein, the membrane component of an anion-translocating ATPase. J. Biol. Chem. 267, 12570-12576.

Yamada, M., Makino, K., Shinagawa, H., & Nakata, A. (1990). Regulation of the phosphate regulon of *Escherichia coli*: Properties of *phoR* deletion mutants and subcellular localization of PhoR protein. Mol. Gen. Genet. 220, 366-372.

Yang, C.H., Sirotnak, F.M., & Mines, L.S. (1988). Further studies on a novel class of genetic variants of the L1210 cell with increased folate analogue transport inward. J. Biol. Chem. 263, 9703-9709.

Zimniak, P., Ziller, S.A., Panfil, I., Radominska, A., Wolters, H., Kuipers, F., Sharma, R., Saxena, M., Moslen, M.T., Vore, M., Vonk, R., Awasthi, Y.C., & Lester, R. (1992). Identification of an anion-transport ATPase that catalyzes glutathione conjugate-dependent ATP hydrolysis in canalicular plasma membranes from normal rats and rats with conjugated hyperbilirubinemia (GY mutant). Arch. Biochem. Biophys. 292, 534-538.

INDEX

Printed and bound by CPI Group (UK) Ltd, Croydon, CR0 4YY

03/10/2024

01040433-0013